BETWEEN MAKING AND KNOWING

Tools in the History of Materials Research

A World Scientific Encyclopedia of the Development and History of Material Science

---

Published

Between Making and Knowing: Tools in the History of Materials Research
edited by Joseph D Martin and Cyrus C M Mody

A World Scientific Encyclopedia of the Development and History of Material Science
Editor-in-Chief: Arne Hessenbruch

BETWEEN MAKING AND KNOWING

Tools in the History of Materials Research

Editors

Joseph D Martin
Durham University, UK

Cyrus C M Mody
Maastricht University, The Netherlands

World Scientific

NEW JERSEY • LONDON • SINGAPORE • BEIJING • SHANGHAI • HONG KONG • TAIPEI • CHENNAI • TOKYO

Published by

World Scientific Publishing Co. Pte. Ltd.
5 Toh Tuck Link, Singapore 596224
USA office: 27 Warren Street, Suite 401-402, Hackensack, NJ 07601
UK office: 57 Shelton Street, Covent Garden, London WC2H 9HE

British Library Cataloguing-in-Publication Data
A catalogue record for this book is available from the British Library.

A World Scientific Encyclopedia of the Development and History of Material Science
BETWEEN MAKING AND KNOWING
Tools in the History of Materials Research

ISBN 978-981-120-762-4 (hardcover)
ISBN 978-981-120-763-1 (ebook for institutions)
ISBN 978-981-120-764-8 (ebook for individuals)

For any available supplementary material, please visit
https://www.worldscientific.com/worldscibooks/10.1142/11482#t=suppl

Desk Editor: Cheryl Heng

Typeset by Stallion Press
Email: enquiries@stallionpress.com

Dedicated to the memory of Ann Johnson

ACKNOWLEDGEMENTS

For both of us, this volume was a somewhat unexpected undertaking prompted by a request from the series editor, Arne Hessenbruch. Both of us, though, had been thinking about materials research for a long time, and especially the role of instruments and other tools in quilting the materials community together. So we were delighted by Arne's proposal for a multi-volume series on the history of that community and we gladly accepted responsibility for the volume on tools.

One of the attractive aspects of this assignment was that it provided an opportunity to showcase, in one place, the work of colleagues and friends who have inspired us over the years. Equally exciting, though, was that the volume gave us an excuse to meet a new group of scholars. This group includes some junior scholars who represent the forefront of the history of science and technology, as well as some practicing scientists and engineers looking after the historical memory of their own fields. We are honored that this volume can form a bridge among disciplines and between junior and senior scholars, and we thank all of our contributors for making this possible.

We would also like to thank some of the people who helped us find those contributors: Carin Berkowitz, Hyungsub Choi, Kate de Luna, Yoshiyuki Kikuchi, Klaus Hentschel, Dave Brock, Patrick McCray, Christine Nawa, Jutta Schickore, Hasok Chang, Graham Machin, Matthias Dörries, Elisabeth Vaupel, and Charlotte Bigg. Ernst Homburg and Bernadette Bensaude-Vincent offered valuable advice at the outset, and some of the groundwork for this undertaking was laid by an earlier volume that one of us (Mody) co-edited with Pierre Teissier and Brigitte van Tiggelen. We are also indebted to Davis Baird, Hans Glimell, Christian Joas, Terry Shinn, and others for conversations over the years as well as introductions to some of the authors in this volume. All academic projects are social in nature, but this one — with its forty-two contributors — is even more of a community effort than most.

There is, however, one individual we should thank who unfortunately is not among our authors but whose ideas and support inform this volume throughout: Ann Johnson. Ann was an ever-present supporter and innovative practitioner of the history of applied science in general and histories of materials research communities and the tools they use and develop in particular. She boosted the careers of many of the people in this volume including one of us (Mody) by providing introductions, organizing and co-organizing conferences and panels, diverting resources from her many grants, and generally through her interest, conversation, and care. Unfortunately, Ann passed away just as we were starting this project. We could've used her help! But, in her absence, we offer this volume as a tribute to her role in bringing together a network of scholars interested in the history of tools in materials research, and a token of the affection members of that network have for her.

NOTES ON CONTRIBUTORS

Joanna Behrman is a PhD candidate at Johns Hopkins University in the Department of History of Science and Technology. Her research focuses on gender and education in the history of science.

Mark D. Bowles is Professor of History at American Public University System. He is the author of fifteen books on the history of science and technology, including *Science in Flux: NASA's Nuclear Program at Plum Brook Station, 1955–2005* (NASA History Office, 2006).

Julia R. S. Bursten is a historian and philosopher of science based in the Department of Philosophy at the University of Kentucky. Her research characterizes methods for making knowledge in the chemical and physical sciences. She is particularly interested in knowledge structures in sciences that draw upon multiple theoretical frameworks, such as materials science and nanoscience.

Justin Carone is a graduate student in Sociology at the University of California, Davis. Their research draws on resources from science and technology studies, the history and philosophy of science, and cultural sociology to understand the role that measures of human capacity play in the distribution of material and symbolic resources. Their current project focuses on the rhetoric of science as an emotion management strategy and as a way to legitimate inequalities in and beyond online gaming communities.

Robert P. Crease is a professor in the Department of Philosophy of Stony Brook University in Stony Brook, New York. He has written, translated, or edited over a dozen books on history and philosophy of science. He writes a monthly column, "Critical Point," for *Physics World*. His books include *Making Physics: A Biography of Brookhaven National Laboratory, 1946–1972* (Chicago, 1999), and most recently, *The Workshop and the World: What Ten Thinkers Can Teach Us about Science and Authority* (W. W. Norton & Co., 2019).

Andrew Ede is the Director of the Science, Technology and Society Program at the University of Alberta, Canada. He teaches history science and technology in the Department of History and Classics and does research on the physical sciences in the early twentieth century.

Matthew N. Eisler studies the relationship between public policy and science, technology, and industrial engineering. He is a Strathclyde Chancellor's Fellow and Lecturer at the University of Strathclyde.

Cyrille Foasso is Curator of the Scientific Instruments Collection at the Musée des arts et Métiers in Paris. After earning a Diploma in materials engineering with a specialty in semiconductor devices, he wrote a PhD thesis on the history of nuclear power safety in France. He is the author of *Atomes sous surveillance. Une histoire de la sûreté nucléaire en France* (PIE Peter Lang, 2012).

Gerald Gallwas is a physical chemist by training and a thirty-year employee of Beckman Instruments (Beckman Coulter). His major interest is the development of medical devices, namely clinical laboratory analyzers. He served twenty-two years on the Arnold and Mabel Beckman Board of Directors, from which he retired in 2016.

Apostolos Gerontas has studied food chemistry at the Aristotle University of Thessaloniki and history of science at the University of Stuttgart and the Norwegian University of Science and Technology (NTNU) in Trondheim. He has twice been a fellow of the Science History Institute in Philadelphia, PA. Currently, he is lecturing on history and philosophy of science at the Coburg University of Applied Sciences and on history of science at the PhD program of the Faculty of Natural Sciences at NTNU.

Hermione Giffard is a historian of technology in Utrecht. Her research focuses on twentieth-century invention with a special interest in military technology and production history. She is the author of *Making Jet Engines in World War Two: Britain, Germany, and the United States* (Chicago, 2016).

Andreas T. Haka is a lecturer at the University of Stuttgart at the Section for the History of Science and Technology. He studied wood and fiber materials engineering at the Technical University of Dresden where he earned a Master's degree in modern and recent history, history of technology, and sociology. Haka undertook his doctoral thesis, "Social Networks in Engineering at German Research Facilities in Academia and Beyond 1920–1970," at the University of Stuttgart. His current project is "Constructed Stability: Technical Development of Materials and Design Applications of Hybrid Materials in the 20th Century."

Olof Hallonsten is a sociologist of science and organization at the Lund University School of Economics and Management. He is specialized in the history, politics, and organization of contemporary Big Science, and institutional analysis of the governance of publicly funded science. He is the author of *Big Science Transformed: Science, Politics and Organization in Europe and the United States* (Palgrave Macmillan, 2016).

Dan Holbrook is Professor of History at Marshall University. His research revolves around the topic of contamination, broadly defined, especially including the control thereof in manufacturing and other areas of human interaction. He is the author of "Controlling Contamination: The Origins of Clean Room Technology" *History and Technology* 25:3.

Catherine M. Jackson is Associate Professor of the History of Science at the University of Oxford. Originally trained as a synthetic organic chemist, she retrained as a historian and now researches the history of modern chemistry. Her publications to date use a focus on chemical practice, pedagogy, and material culture to elucidate the work done by chemists including Justus Liebig, August Hofmann, and Emil Fischer, opening the path toward a new account of the origins and development of organic synthesis.

Thomas Kaiserfeld is a historian of ideas and sciences at the Department of Arts and Cultural Studies, Lund University. His research focuses primarily on institutional aspects of science and technology since the eighteenth century. He is the author of *Beyond Innovation: Technology, Institution and Change as Categories for Social Analysis* (Basingstoke, 2015).

Debra Kolah received her Master of Library and Information Science in 1995 from the University of Texas. She is currently the Head of User Experience and the Physics, Math, Astronomy Librarian at Rice University. She has served as served on the American Physical Society's Publication Oversight Committee, Library Advisory Board of the Institute of Physics Publishing (IOPP), American Institute of Physics Ebook Program, and is Past Chair of the Physics-Astronomy-Mathematics Division of the Special Libraries Association.

Sharon Tsai-hsuan Ku is an Assistant Professor at the Department of Engineering & Society, University of Virginia. Her research focuses on geopolitics of standardization and innovation, particularly in the US and China. She is also an active member in the international standard community, including ASTM International, ANSI/ISO and IEEE-SA. As a standard reviewer and educator, she uses standardization as a platform to stimulate engineering students' critical thinking and value-sensitive design toward real-world problem solving.

Christophe Lécuyer is professor of the history of science and technology at the Sorbonne Université. His research explores the history of computing, microelectronics, advanced materials, and scientific instrumentation. He is the author of *Making Silicon Valley: Innovation and the Growth of High Tech, 1930–1970* (MIT, 2006) and the co-author (with David C. Brock) of *Makers of the Microchip: A Documentary History of Fairchild Semiconductor* (MIT, 2010).

Victoria Lee is Assistant Professor in the Department of History at Ohio University, where she teaches and writes about modern science and technology. She is currently writing a book about Japanese society's engagement with microbes in science, industry, and environmental management. She has published in *Historical Studies in the Natural Sciences*, *New Perspectives on the History of Life Sciences and Agriculture* (eds. Phillips and Kingsland), and *Osiris*, and was a postdoctoral fellow at the Max Planck Institute for the History of Science.

Joseph D. Martin is an assistant professor in the Department of History, Durham University, and the co-editor-in-chief of *Physics in Perspective*. His research focuses on the history of solid state and condensed matter physics and materials science in the twentieth century. He is the author of *Solid State Insurrection: How the Science of Substance Made American Physics Matter* (Pittsburgh, 2018).

Philippe Martin is an associate researcher of science and technology at the Centre François Viète, Department of Epistemology, History of Science and Technology, University of Nantes. His research focuses on the history of the fertilizer industry. The subject of his PhD was "History of fertilizer industry in the Loire estuary in the contemporary era."

Cyrus C. M. Mody is an historian of recent science and technology at Maastricht University, focusing on the applied physical sciences in the United States since 1965. He studies the commercialization of academic research, countercultural science and technology, and the *longue durée* of responsible research and innovation. His books include *Instrumental Community: Probe Microscopy and the Path to Nanotechnology* (MIT, 2011) and *The Long Arm of Moore's Law: Microelectronics and American Science* (MIT, 2016).

Max Moeller is the Curator of Published Collections at Hagley Museum and Library in Wilmington, DE. Previously, he worked as the Director of Research Services at the Historical Society of Pennsylvania in Philadelphia, PA. He holds a Master's degree in history from Villanova University and a Master's degree in library science from Clarion University.

Hubertus Nederbragt is a retired biomedical scientist and lecturer in Pathobiology from the Faculty of Veterinary Medicine, and is an external member of the Descartes Centre of Utrecht University, the Netherlands. His current interest is the history and philosophy of animal diseases in the nineteenth century. He is now editing the text of a hand-written lecture notebook of General Pathology of Alexander Numan (1780–1852).

Keith A. Nier was a documentary editor with the Thomas A. Edison Papers project for many years. As an independent historian, he is focused on the development of mass spectrometry and on the roles of instrumentation in general as foundational for the nature of science, crucial in objectivity, and leading to the unification of science.

Loïc Petitgirard is an historian of recent science, technology and innovation (assistant professor at the *Conservatoire national des arts et métiers* in Paris, France). He investigates the evolution of scientific instruments, the developments of metrology, and the history of computer science. He also studies the history of scientific institutions in France and is currently editor-in-chief of the *Cahiers d'histoire du Cnam*.

John C. Powers is an Associate Professor of History at Virginia Commonwealth University. His recent work focuses on the history of chemistry in the eighteenth century, specifically the impact of thermometers and new theories of heat on the practices of the chemical arts and the development of chemistry as a scientific discipline.

Simón Reif-Acherman is a Titular Professor at the School of Chemical Engineering, Universidad de Valle, in Cali, Colombia. Besides the teaching courses on material and energy balances in process industries, he studies several subjects in the history of science and technology. He has published more than twenty papers in international scientific journals on the history of theories and scientific instruments, and biographical profiles, among other subjects.

Pedro Ruiz-Castell is Professor of History of Science and member of the Institut Interuniversitari López Piñero d'Estudis Històrics i Socials sobre Ciència, Tecnologia, Medicina i Medi Ambient at the University of Valencia (Spain). His research focuses on the history of physics and allied sciences and scientific instruments.

Robin Wolfe Scheffler is the Leo Marx Career Development Associate Professor in History and Culture of Science and Technology in the Program in Science, Technology, and Society at the Massachusetts Institute of Technology. His work, such as his first book, *A Contagious Cause* (Chicago, 2019), examines the political, cultural, and technical dimensions of molecular medicine and biotechnology.

Brit Shields is a historian and sociologist of science, mathematics, and technology and a Senior Lecturer at the University of Pennsylvania in the School of Engineering and Applied Science. Her scholarship focuses on the cultural history of scientific communities, institutions and ideas. She holds a PhD in the History and Sociology of Science from the University of Pennsylvania, an MA in Humanities and Social Thought: Science Studies from New York University, and a BS in Mass Communications from Miami University.

Indianara Silva is a tenured professor of physics teaching at State University of Feira de Santana, and of history of science at Graduate Program in History, Philosophy and Science Teaching at Federal University of Bahia, Brazil. Her current research interests include history of modern physics, gender and science, and physics teaching.

Amy E. Slaton is a professor of history at Drexel University and co-editor of the journal *History+Technology.* She has written on historical understandings of race, disability, and other identities in engineering education and labor.

Pierre Teissier is associate professor in history of science and technology at the University of Nantes (France). His research interests are linked to history of materials sciences, especially solid state chemistry, and history of alternative energies in the modern period.

Anthony S. Travis is based at the Sidney M. Edelstein Center for the History and Philosophy of Science, Technology and Medicine, The Hebrew University of Jerusalem, Edmond J. Safra Campus, Jerusalem 9190410, Israel. Travis has published widely on the history of chemistry and chemical technology in the nineteenth and twentieth centuries. Other interests include civil engineering and railway history.

Brian Tyrrell is a Visiting Assistant Professor of History and Environmental Studies at Reed College. He completed his PhD in history of science at the University of California, Santa Barbara, where, from 2013 to 2016, he was a graduate fellow at the UCSB Center for Nanotechnology in Society (National Science Foundation Grant No. SES 051184).

Jorijn van Duijn has worked at Maastricht University and Rijksmuseum Boerhaave, the Dutch National Museum of Natural Science & Medicines in Leiden. His PhD research, *Fortunes of High Tech*, explored dynamics of high-tech innovation through the history of semiconductor equipment manufacturer ASM International from 1958 until 2008.

Thomas Vogt is a chemist at the Department of Chemistry and Biochemistry at the University of South Carolina who works on structure-property correlations and an adjunct faculty in the Department of Philosophy with an interest in the history and philosophy of science. He is a fellow of the IAS at Durham University, the American Association for the Advancement of Science, the American Physical Society and the American Society for Neutron Scattering.

Phillip C. Wankat is retired from the School of Chemical Engineering, Purdue University. He has dabbled in the history of separation methods and of chemical engineering at Purdue. He is the coauthor with Cristina D. Farmus of *A Pictorial History of the School of Chemical Engineering at Purdue University, 1911–2011* (Purdue, 2011).

Catherine Westfall is Professor Emerita of History, Philosophy, and Sociology of Science at Michigan State University. Her research has focused in the history of the US national laboratories. She is currently investigating a new era in the history of large instruments, the New Big Science, which is dominated by materials science.

Benjamin Wilson is an Assistant Professor of the History of Science at Harvard University. His research focuses on the history of modern physics and the intellectual history of the nuclear age. His first book, in preparation for Harvard University Press, offers a new social and intellectual history of nuclear arms control and strategy.

CONTENTS

Acknowledgements vii

Notes on Contributors ix

Introduction: Tools in Materials Research 1
Joseph D. Martin and Cyrus C. M. Mody

Part I: Always Already Tools **13**

Always Already Tools: Introduction 15
Hermione Giffard

1.1 Glassware 21
Catherine M. Jackson

1.2 Thermometers 33
John C. Powers

1.3 Simple Heating 43
Cyrus C. M. Mody

1.4 Scales and Balances 53
Joseph D. Martin

1.5 Light Microscopy 63
Hubertus Nederbragt

1.6 Manuals, Handbooks, and Recipes 73
Joanna Behrman

1.7 Distillation 83
Phillip C. Wankat

1.8 Mixing Equipment 97
Philippe Martin

Part II: Invisible and Infrastructural Tools 109

Invisible and Infrastructural Tools: Introduction 111
Cyrus C. M. Mody

2.1 Buildings 121
Brit Shields

2.2 Safety Equipment 129
Amy E. Slaton

2.3 Vacuum Chambers, Pumps, Gauges, and Systems 141
Cyrus C. M. Mody

2.4 Contamination Control 151
Dan Holbrook

2.5 Trade Catalogs 159
Max Moeller

2.6 Bibliographic Tools 169
Debra Kolah

2.7 International Standards 181
Sharon Tsaai-Hsuan Ku

2.8 Computer Simulations 195
Julia R. S. Bursten

2.9 Information Display and Recording 207
Cyrille Foasso

2.10 Mechanical Testing 219
Justin Carone

2.11 Climate Test Chambers and Exposure Testing 227
Andreas T. Haka

Part III: Tools for Making Materials 239

Tools for Making Materials: Introduction 241
Cyrus C. M. Mody

3.1 Modern Heating Technologies 257
Pierre Teissier

3.2 Refrigerators and Vacuum Flasks 273
Simón Reif-Acherman

3.3 High-Pressure Apparatus 287
Anthony S. Travis

3.4 Bioorganic Synthesis 299
Victoria Lee

3.5 Chromatography 315
Apostolos Gerontas

3.6 Lithography 327
Joseph D. Martin and Cyrus C. M. Mody

3.7 Ion Implantation 341
Christophe Lécuyer

3.8 Power Sources 349
Matthew N. Eisler

3.9 Molecular Beam Epitaxy 357
Brian Tyrrell

3.10 Vapor Deposition 367
Jorijn van Duijn

3.11 Nuclear Research and Test Reactors 377
Mark D. Bowles

3.12 Centrifuges 391
Andrew Ede

Part IV: Tools for Characterizing Materials 401

Tools for Characterizing Materials: Introduction 403
Joseph D. Martin

4.1 X-Ray Crystallography of Biomaterials 425
Robin Wolfe Scheffler

4.2 Raman Spectroscopy 435
Indianara Silva

4.3 Infrared Spectroscopy 443
Loïc Petitgirard

4.4 Ultraviolet Spectroscopy 457
Gerald Gallwas

4.5 Mössbauer Spectroscopy 475
Catherine Westfall

4.6 Lasers 483
Benjamin Wilson

4.7 Probe and Other Microscopies 495
Cyrus C. M. Mody

4.8 Scanning Electron Microscopy 503
Pedro Ruiz-Castell

4.9 Aberration-Corrected Electron Microscopy 513
Thomas Vogt

4.10 Mass Spectrometry 527
Keith A. Nier

4.11 Thermography 539
Andreas T. Haka

4.12 Neutron Sources 553
Olof Hallonsten and Thomas Kaiserfeld

4.13 Nuclear Magnetic Resonance Spectroscopy 561
Joseph D. Martin

4.14 Synchrotron Light Sources 571
Robert P. Crease

Index 579

INTRODUCTION: TOOLS IN MATERIALS RESEARCH

JOSEPH D. MARTIN AND CYRUS C. M. MODY

The science of materials has contributed to changes in our civilization as pervasive as they are profound. The ways we travel, communicate, wage war, build buildings, dress, heal, play sports, read, listen to music, use energy, and care for the young, the old, and the vulnerable have all been shaped and reshaped by our knowledge and mastery of metals, semiconductors, organic and biocompatible materials, gels, plastics, polymers, plasmas, and other substances. But our large-scale historical understanding of materials research is still surprisingly flimsy. We might say of materials research, as common as it is, what Clifford Geertz said of common sense: "it lies so artlessly before our eyes it is almost impossible to see."[4 p92]

The WSPC Encyclopedia of the Development and History of Materials Science aims to make materials — which we might otherwise overlook for their familiarity — more visible by charting their history. This volume focuses on the tools and practices that have guided materials research. Where did they come from and how were they enrolled in the cause of understanding, manipulating, and fabricating the stuff of the modern world? Laboratories dedicated to studying materials proliferated in the mid- to late twentieth century. They were sponsored by government and military organizations, assembled within universities, and established by industrial firms.

And they succeeded in reforming our understanding of matter and changing the material profile of our technological world because a diverse assortment of tools was successfully coordinated within them.

Imagine walking into one of these labs — at Cornell University, or the Centre National de la Recherche Scientifique, or General Electric — in the 1970s or 1980s and looking around. You are surrounded by a wide assortment of tools. Some — glass flasks and beakers, thermometers, microscopes — have been shaped by centuries of development and modification. Others, particularly those taking advantage of various scattering and diffraction phenomena, are recent developments. Still others are so unassuming that you might not register them as tools at all, from the trade catalogues that researchers use to browse new prefabricated materials and equipment, all the way up to the building itself, which was designed to instrumentalize the interactions of the researchers within it. This volume tells their stories.

What Is Materials Research?

A volume of this nature must wrestle with the question of what counts as materials research. Using materials constitutes one of humanity's oldest technological enterprises. If we wanted to be truly expansive, we could point out that successful flint knapping requires a remarkable sensitivity to the material properties of some types of stone (if not a theoretical knowledge of its structure). And so, from one standpoint, materials research can refer to traditions that are many millennia old and span several radical shifts in craft and scientific understanding of the stuff that makes up our technologies.

From another standpoint, however, materials research is a much more recent enterprise. Fields that trade principally in material properties, such as metallurgy or inorganic chemistry, can boast centuries-old research traditions. Only after World War II, however, did institutions and communities emerge that were dedicated to systematically applying physical and chemical understanding of solids to the characterization, manipulation, and synthesis of materials in service of human ends.[1] When we speak of "materials science" or "materials research" today, we usually mean this tradition of much more recent vintage.

Some tools are better captured by the more expansive definition of materials research, others by the narrower, contemporary perspective. Rather than imposing a particular view of materials research on this volume as a whole, we have left the decision about how to understand the scope of materials research largely up to individual contributors — different understandings of the field are better suited to telling the stories of different tools. This volume, therefore, is not just a catalogue of the histories of individual tools; it is a study of how those tools have helped to define the boundaries of the field. We believe this approach reflects the reality that the precise parameters of materials research are difficult to establish, especially in historical inquiry. Rather than presenting a narrow vision of what materials research is, we hope the following essays will inspire you to ask yourself what we can gain by considering these tools, many of which have long histories and diverse applications, as part of the story of materials research.

Themes and Patterns

One of the joys of assembling a multi-author work such as this one is watching how diverse perspectives combine to illustrate larger themes and patterns that might not have been evident to one or two authors on their own. In reading the contributions that follow, we have noticed a few such themes and patterns that bear emphasizing — although we also encourage readers to be attentive to those that we do not highlight here.

First, we note the rising prominence of the black box. The entries in Part 1 frequently reference experimenters who were also toolmakers. Chemical analysts and synthesists were often also glassblowers, as Catherine Jackson discusses. Critical advances in thermometry were made by the people interested in studying the phenomena of heat, as we learn from John Powers. These early tools were usually bespoke, designed for particular and often narrow tasks. As we move into the middle twentieth century, however, we begin to see more and more tools coming in standard form, designed to be used in standardized ways and to free experimenters from the tasks of instrument design.

This is not to say, however, that the black-boxing of instruments reduced the importance of tacit experimental sense. We only need to read Cyrille Foasso's entry on information display and recording to see how even standardized instruments require the same subtle sensibility that might have informed an expert chemist-*cum*-glassblower in the mid-1800s. Following the theme of black-boxing instruments therefore indicates both a discontinuity and a continuity. It shows a growing division of labor between tool makers and tool users, but it also represent a continuity of the sorts of skills and sensibilities required to approach the science of materials.

Second, we discern an increasing internationalism in the practice of materials research. This is of little of surprise, insofar as it mirrors the extent to which most of science and technology became more international over the span of the twentieth century. But it is notable that materials research followed a similar trajectory given the extent to which it is and has been bound up in national and commercial interests. Tools for materials research were emerging from international collaborations and moving across borders even as they were being directed toward classified military research and being subjected to intellectual property regimes, and their stories have much to tell us about the ways in which those regimes of secrecy and control shaped the internationalization of science and technology.

The third theme we discern is an arc that describes the types of materials at which tools tended to be directed. Throughout the twentieth century, this arc moved from regular crystals and molecules to amorphous solids to squishy biomaterials. A technique like X-ray diffraction, as Robin Scheffler notes in Part 4, was seen as ideal early in the twentieth century for studying the regular crystal structure of solids, and molecules with repeating structures like protein and DNA. Later in the century, tools like nuclear magnetic resonance spectroscopy provided the different variety of data that was necessary to tackle the somewhat different challenge presented by amorphous materials and organic matter. The story of tools in materials research, that is, was not just a story of new instruments and techniques opening up new vistas. The materials themselves pushed back. Evolving notions about what constituted a useful material influenced what sorts of tools gained prominence, and what developmental trajectories those tools would take.

Finally, we note that we have imposed a somewhat artificial distinction on this volume between tools for making materials (Part 3) and tools for characterizing materials (Part 4). Many of the tools discussed in these final two section straddled that boundary, or had distinct applications for manufacturing materials and exploring their properties. The artificiality of that distinction indicates the ways in which materials research challenges some of the distinctions that, for better or worse, have shaped much of our thinking about science and technology in the twentieth century — in particular between basic and applied research. It is de rigueur for today's historians to observe that this distinction is artificial and does not appear so sharply in scientists' practice as we see it expressed in their rhetoric. What we see in the contributions to this volume, however, is the value of further attention to those fields where the distinction was not only moot, but was not even a rhetorical advantage.

How to Use This Book

The short essays that follow each introduce a few essential historical features of the instruments, techniques, and practices that make up materials research. These tools have rich and intricate histories, which cannot be done justice in just a few pages. These essays, do, however, show us why the stories of these tools matter. You can use them to learn how particular tools fit into some of the larger stories of the history of science and technology and to find guidance for more in-depth study. We hope that this volume can be comprehensive in one, limited sense. It will not tell you everything that's worth knowing about thermometers, or lasers, or mass spectroscopes. But by discussing how all of these tools, and many others, contributed to some of the most important developments in the history of science and technology, this volume provides the most inclusive historical compendium to date on the techniques people have used to pursue their fascination with materials and their uses.

We have sorted the tools of materials research into four categories. The first, "Always Already Tools," describes apparatus that all materials researchers must master as part of basic laboratory procedure. Scales and balances, glassware, thermometers, microscopes, and other such tools have long histories and are likely to be familiar to secondary school chemistry students around the world. They are also relevant to every stage of the research process and every type of materials research. Metallurgists fabricating new alloys and biomedical scientist designing novel implants will need precision weighing devices just as much as the basic physical or chemical researcher interested in a substance's essential properties does. These humble tools thereby have tremendous influence over the practice of materials research.

Part 2, "Invisible and Infrastructural Tools," confronts those tools that are necessary before research can get off the ground. These include the buildings and trade catalogues mentioned above, as well as safety equipment, clean rooms, information display and recording devices, and standards, to name a few more. These tools and procedures, although they rarely arise when we think about how we research materials, are far from ancillary. They are critical both to the process of investigation and to the social relations that make it possible.

"Tools for Making Materials" is the subject of Part 3. It is easy to take for granted that the things we use to navigate the world have properties suited for the tasks to which we turn them. But the precision and consistency of those properties more often than not depend on the mastery of some delicate laboratory procedures. Tools like lithographs, centrifuges, distillation columns, and specialized heating and cooling devices have developed to ensure that we can shape the properties of materials to meet specific demands, and that those materials will perform predictably and reliably in our technologies.

Finally, Part 4 addresses "Tools for Characterizing Materials." The better understanding that emerged through the twentieth century of various types of radiation, the micro structure of matter, and how these two interact, paved the way for a variety of new, non-destructive techniques for investigating the structure and properties of materials. Investigations of the properties of materials have advanced our basic understanding of matter. They also allow us to manipulate it more easily. The story of these tools is the story of how abstract knowledge and practical know-how can proceed in lockstep.

Each of these sections begins with an introduction that provides general background for the essays that follow. Whereas the shorter essays focus on particular tools, these longer introductory essays offer readers a wider-angle view of the types of tools investigated in each part. They discuss existing historical work in more depth and comment on the big historical questions that the histories of these tools raise. These essays provide essential context for the shorter vignettes they introduce and prompt readers to understand the often-detailed history of particular tools within the larger sweep of the histories of science and technology.

Situating the Volume

Readers will note that the each of the volume's four sections is longer than the previous section. This is *not* a reflection of the editor's valuation of these four topics. We see all four categories of tools as equally important for understanding the historical evolution of materials research. Rather, the greater weight accorded to synthesis and characterization tools reflects the current breadth of scholarly interest in those topics. Significantly more studies thus far have scrutinized high-tech instruments and techniques used in materials science, condensed matter physics, colloid and polymer chemistry, and related fields than have examined ancient or infrastructural tools used in those same disciplines. Of course, ancient alchemical glassware and other older experimental equipment do have literature devoted to them[7] — but few studies connect those techniques to modern materials research. Similarly, quite a bit has been written about some kinds of infrastructural tools — for example, laboratory buildings[3;5] — but, again, not specifically linked to the history of materials research. One aim of this volume, therefore, is to bring greater attention to the role of these kinds of tools in materials research and to encourage our colleagues to incorporate their histories into general narratives of the field's development.

The reasons this has not yet happened already are many. One is that materials research, as such, is only about a half-century old. It is perhaps more surprising that there are any historians of materials research at all than that some corners of that history have not attracted scholars. The more developed topics — histories of individual instruments or techniques — are good targets for PhD dissertations, so many of the extant studies grew out of thesis topics.[11; 12] A modern instrument or technique usually comes with an easily identifiable published literature and often with a dedicated community that is willing to be interviewed or even generates its own first draft of its history. By contrast, studies that span the centuries- or even millennia-long development of ancient tools that live on in materials research require a depth of expertise that PhD students would be challenged to develop. Similarly, infrastructural tools are associated with historical sources that are difficult to come by and therefore risky for a PhD student to investigate.

We could also point to a number of what one of us (Martin) has called "prestige asymmetries," which explain why histories of instruments and techniques for materials synthesis and characterization are more prevalent than histories of older and infrastructural tools.[8] High-tech instrument-making is strongly associated with disciplines — especially physics — that have long dominated the history of science. Histories of large instruments designed for the physical sciences, such as telescopes and particle accelerators,[6; 9] were common up to the beginning of the present century. Histories of smaller-scale condensed matter physics instruments were less abundant; but because many of these instruments were linked to a Nobel Prize in Physics, most historians of physics recognized the histories of these instruments as integral to their field. Ancient and infrastructural tools, by contrast, rarely merited Nobel Prizes in anything. The histories of such tools would be harder to justify, at least in the eyes of other historians, and would therefore be riskier, especially for PhD students and early career scholars.

But why would there be fewer Nobel Prizes for ancient and infrastructural tools? The question perhaps answers itself, but it also reveals important dynamics, both in history and in science, that this volume attempts to contest. One dynamic is the asymmetric prestige accorded to technological innovation as opposed to maintenance, adaptation, conservation, and simple enduring use, at least in the global north.[2] New technologies, and especially new research technologies, are seen as contributing more to economic and social life than old ones, even if all indications are that the old technologies dominate both the laboratory and ordinary life. Hermione Giffard's introduction to Part 1 elaborates on this point.

Another related dynamic is that new characterization and synthesis tools often are — or are often seen as — the creations of celebrated geniuses who just happen to be middle-class men, whereas ancient and infrastructural tools have a long association with women, the working class, and other subalterns. One aim of this volume is therefore to highlight the ubiquity and centrality of collective effort in the development of all the tools used in materials research, including characterization and synthesis tools. Nobel Prizes might go to groups of no more than three, but historians' narratives should reflect a more distributed process of design, use, and refinement. A second aim of this volume is to connect the history of materials research to labor history, and in particular to the labor history of groups whose members rarely win Nobel Prizes. Entries that could be grouped under this theme include (but are not limited to) those by Justin Carone (mechanical testing), Amy Slaton (safety), Julia Bursten (computing and simulation), Debra Kolah (bibliographical tools), Joanna Behrman (recipes and manuals), and Catherine Jackson (glassware).

It has not escaped the editors' notice that this theme should also apply reflexively to this volume itself. We believe that socially responsible materials research must be inclusive and therefore requires a diverse group of practitioners.[13] We also believe that the entries in this volume demonstrate that there has been greater diversity in the history of materials research than many of its practitioners (and historians!) acknowledge, and that that diversity has given rise to epistemically more robust knowledge of materials. The same should be true of the histories of materials research, and we have therefore tried to recruit a diverse group of authors. Readers will find a few well-established authors, such as Catherine Westfall and Christophe Lécuyer, who have been writing on this topic for much of their careers. Their voices are joined by scholars, such as Brit Shields and Jorijn van Duijn, who have entered the historical profession more recently.

We also sought essays by practicing materials researchers, such as Philip Wankat and Gerald Gallwas, as well as practitioners-turned-historians such as Catherine Jackson and dual practitioner-philosophers like Thomas Vogt. As mentioned above, practitioners usually (and rightly) write the first histories of the tools their communities have developed. Those participant histories offer compelling firsthand accounts of developments. But immediacy is not the only reason why practitioners must be part of writing the history of materials research. Another is that good history of science must grapple with arcane technical detail. Researchers are not human beings in one corner and technical experts in another; they are always both together, and crafting a historical narrative requires continual melding of the two. It is no coincidence that a substantial number of our authors, including one of us (Mody), have at least one degree in a field related to materials research. That is not a prerequisite, of course. Two people cited throughout this volume, Ann Johnson and David Brock, developed their deep understanding of materials science after becoming historians, as did many of the authors included in the volume. Again, we aimed for a diverse group of authors because we believed that would lead to a more comprehensive and robust understanding.

One place where we, as editors, failed in that aim is in making the history of tools in materials research a global story. Only two of our essays, by Victoria Lee and Indianara Lima Silva, devote much attention to places outside of Europe and North America. That is partly a reflection of inadequate effort and imagination on our part, as well as a reflection of the time required to write even a short historical essay with global, or even transnational reach. It is also a reflection of a real gap in the literature. If prestige asymmetries have led to greater emphasis on the histories of some instruments more than others, they have also led to greater emphasis on the history of instrument development in some places more than others. We do not, however, believe that the geographic narrowness of this volume in any way reflects the reality of instrument development. As Amit Prasad argues, any topic of research is always already global, that is, sustained by networks with participants from all over the world.[10] It is therefore never possible to say that the development of a particular tool "started in" one place or another. Over time, global economic and political power asymmetries mean that some nodes in those networks come to be seen as more central than others, and some nodes therefore drop out of historical memory. In Prasad's case study of magnetic resonance imaging (MRI), the ironic result was that many Indian MRI researchers believed that the technique was invented in the United States and United Kingdom and were unaware of its deep history in India itself! Much of the same could no doubt be shown for many of the tools represented here, and our hope his that further research will correct this volume's shortcomings in that respect.

In closing, we should note that examining materials research through tools is just one way to access its history. This book is part of a series that explores some other approaches in parallel. We encourage you to consider the history of materials research through the lenses of the institutions that supported it, the markets that drove its development, and the materials themselves. Together, we hope that these volumes can map out an agenda for further historical work into a field that has made countless contributions to our scientific understanding of the world, and whose technological developments pervade our lives, but whose history we have only begun to systematically explore.

References

1. Bensaude-Vincent B. The construction of a discipline: materials science in the United States. Historical Studies in the Physical and Biological Sciences. 2001;31(2):223–48.

2. Edgerton D. The shock of the old: technology and global history since 1900. Oxford: Oxford University Press; 2007.
3. Galison P, Thompson E, editors. The architecture of science. Cambridge, MA: MIT Press; 1999.
4. Geertz C. Local knowledge: further essays in interpretive anthropology. New York: Basic; 1983.
5. Gieryn TF. What buildings do. Theory and Society. 2002;31:35–74.
6. Hoddeson L, Kolb AW, Westfall C. Fermilab: physics, the frontier and mega-science. Chicago: University of Chicago Press; 2008.
7. Holmes FL, Levere TH, editors. Instruments and experimentation in the history of chemistry. Cambridge, MA: MIT Press; 2000.
8. Martin JD. Prestige asymmetry in American physics: aspirations, applications, and the purloined letter effect. Science in Context. 2017;30(4):475–506.
9. McCray WP. Giant telescopes: astronomical ambition and the promise of technology. Cambridge, MA: Harvard University Press; 2006.
10. Prasad A. Imperial technoscience: transnational histories of MRI in the United States, Britain, and India. Cambridge, MA: MIT Press; 2014.
11. Rasmussen N. Picture control: the electron microscope and the transformation of biology in America, 1940–1960. Stanford: Stanford University Press; 1997.
12. Reinhardt C. Shifting and rearranging: physical methods and the transformation of modern chemistry. Sagamore Beach, MA: Science History Publications; 2006.
13. Stilgoe J, Owen R, Macnaghten P. Developing a framework for responsible innovation. Research Policy. 2013;42(9):1568–80.

PART I

ALWAYS ALREADY TOOLS

ALWAYS ALREADY TOOLS: INTRODUCTION

HERMIONE GIFFARD

In 2016, articles appeared that mentioned a six-thousand-year-old amulet shaped like a wheel that had been discovered on an archeological dig in Pakistan in 1980.[10] The amulet was not a beautiful example of metalworking. It was made of pure copper — and hence contained large crystals — rather than the alloys later favored for such work. Nor was it symbolically significant. It did not herald, say, the importance of wheels to ancient agriculture.[2] Rather, the amulet was brought to the public's attention because the technique used to make it — lost-wax casting — is supposed to be indicative of a highly innovative society. More importantly still: NASA still uses a process based on lost-wax casting — investment casting — to make equipment that has flown to the International Space Station and to Mars.

No matter how important, things like processes and tools, the focus of this book, are often overlooked by historians. This section covers the oldest of these: handbooks and recipes, glassware, microscopes, thermometers, mixing and distillation. Such tools and methods are a good example of "low-profile technologies," technologies that are ubiquitous yet have attracted the attention of neither users nor historians. There are a number of reasons that a tool might have a low profile: its ubiquity might lead it to go unnoticed or its use unrecorded; its oldness might make it less novel; who it was generally used by, if those users were not elites in the West, can make it less visible.

Writing a history based on how technologies were used shifts attention from the new to the old. Whereas invention and innovation are concentrated in relatively few places, use takes place around the world and throughout time.[4 pxiv] Dismissing the assumption that only new technology matters opens up entirely new vistas of history, vistas that include many more people and places than the histories that we have.[8] Recording these histories, as this volume begins, is therefore of crucial importance.

Tools are often overlooked because of our preference for stories of innovation and novelty. The history of technology tends to focus on invention, on new machines.[4] In the example above, we can see that even though a method that has been around for six thousand years is not new, the story of lost-wax casting can be pitched as a story of innovation. Yet if we were to study the technique throughout history — all six thousand years of it — we might see a very different picture. Lost-wax casting could have played a role in the Peloponnesian Wars as well as in World War I; it might have been used by individuals in ancient Samaria as well as in modern day Italy to effect political change or to increase exports. Studying tools lends itself to use-based narratives, rather than innovation-centric narratives.

The study of raw materials demonstrates the value of studying things with low profiles. Raw materials are beginning to draw more historical attention because scholars have realized the potential that following raw materials give them to create valuable transnational histories that reveal the ways in which the world was connected in the past. A history of sugar, for example, goes from the Caribbean to the eastern Mediterranean, and the story of tea from China to Britain.[6; 7] Having recognized that raw materials and materials in general were important carriers of knowledge,[15] scholars have attached this also to recognition of their economic importance. Yet the tools required for working and creating these materials remain under-researched and under-appreciated — hence the current volume. The few exceptions are when a particular tool (like the thermometer) was connected to a specific task or discipline (like metrology).[12]

Writing a history of tools that spans many centuries will bring attention to new areas of activity. Early tools were not primarily used to create or spread knowledge, natural scientific or otherwise, but to make or process things that moved around the world, like tea, sugar, porcelain, and textiles.[1; 9; 14] As historians of science have observed, the world of arts and crafts, of artisans, overlapped and was in some cases indistinguishable from that of natural scientists in the early modern period. Materials and tools served to create this intersection of artisanal labor, natural history, and experimental history. Indeed, the history of tools reveals part of the overlap of science with the everyday practices of commerce and consumption.[11] The knowledge created by using tools was passed on in recipes and handbooks explaining how to make new materials; and, in this way, different parts of the world became connected to a larger world of the circulation of knowledge. Indeed, some argue that recipes, as a chapter in this section recounts, removed skill from practitioners of chemistry.

Objects and methods are particularly valuable to historians because, as anthropology teaches us, you can only understand an artifact by understanding the society that made it — that is, its values. The framework into which a tool is set gives it meaning, and, importantly, it has no meaning apart from this. Histories of tools in the first century BCE and in the twentieth century are radically different because the societies that used them — although the tool may be the same — are radically different. The tools that define our civilization — a crucial part of human culture — despite being pervasive, risk being consigned to historical oblivion. In part for this reason, there should be a stronger link between anthropology, with its interest in the meaning of old artifacts, and the history of technology.

A history based on use opens for us the route to investigating places and times in which tools were created and used without understanding how or why they worked; this also changed over time. As this section's chapter on optical microscopy reveals, the theoretical underpinnings to the use of the microscope (and what you can see with it) was a much later innovation than the tool. Concrete things often remained in use while theories changed around them. Thus the use of the microscope for biological research as we understand it is not properly understood as the work of the seventeenth century (although Antonie van Leeuwenhoek provided an important precedent), but that of the nineteenth century, when the scientific theory of biology behind the use of microscopes was developed into what we recognize today. New methods of glasswork, despite being dismissed by historians, also ushered in revolutions in chemistry by leading to the creation of new methods and new ways of thinking, as the chapter on glasswork argues. It is important that we understand the materiality of the world and its connection to theory because material things influence people's thoughts and actions.

The history of tools before organized natural philosophy allows the historian a glimpse into the evolution of human knowledge. Reversing a typical hierarchy, the history of tools is valuable because it gives us access to a world in which technology proceeded in advance of science. In later times, the history of tools allows us to explore the question of how certain ways of thinking — those associated with science — influenced action in other parts of users' lives — how chemical recipes influenced food recipes, for example, which didn't have units in the nineteenth century.[5] A given tool can only be fully understood if we take the intellectual framework that supported it into account.

Some tools that have existed in something like their present form for many centuries still clutter the lab bench of many a practicing materials scientist today. They are a standing rebuke to accounts of technology that assume that visibility is related to importance. Can we say that something so ubiquitous has not shaped the modern world? And even more importantly, as the chapters in this section relate, those tools have shaped patterns of thought in the modern world. Recognizing this, the history of science has placed instruments at the center of natural knowledge production in the early modern era.[3] Because these tools were in use before natural philosophy arose, they formed in a very real way the bedrock of natural philosophers' knowledge. The following chapters show just how new tools led investigators to ask new questions.

This volume is a contribution to the realization that tools and techniques are important and that we need to know more about them if we wish to understand human civilization. This section begins the story of the larger volume by looking at the simplest tools and methods in use for analyzing materials: handbooks/recipes, glassware, mixing, microscopes, thermometers, and distillation. Although often overshadowed by later discoveries and inventions, the widespread use of these tools is testimony to their importance.

Many of the contributions to this section cover tools or methods in their more recent forms, yet all of the tools covered in this section also have histories going back in time, which demonstrates how knowledge of different sorts — most importantly, craft knowledge rather than that later codified as materials science, was transmitted through and with tools. The simplest tools that have been developed for investigating materials are the tools (and methods) that in some form were very often the first way that individuals interrogated and shaped the natural world. These were tools of alchemists as they toiled to change lead into gold, for example, but also of traders, and natural philosophers whose goals in shaping materials were different.

Because tools offer us an entry point into the histories of practices that are otherwise unrecorded, they are particularly useful. The difficulties of writing the full history of a tool by no means make it a useless endeavor. The history of tools offers us many insights: into a world before the development of science; into a world in which commerce was more important than anything else; into a world in which great men are virtually absent.

References

1. Brewer J, Trentmann F. Consuming cultures, global perspectives: historical trajectories, transnational exchanges. London: Berg; 2006.
2. Bulliet RW. The wheel: inventions and reinventions. New York: Columbia University Press; 2016.
3. Daston L, Park K. The Cambridge history of science: volume 3, Early Modern science. Cambridge: Cambridge University Press; 2006.

4. Edgerton D. Shock of the old: technology and global history since 1900. Oxford: Oxford University Press; 2007.
5. Eighmey RK. Abraham Lincoln in the kitchen. New York: Penguin Random House; 2014.
6. Freese B. Coal: a human history. London: Arrow; 2016.
7. Fromer JE. A necessary luxury: tea in Victorian England. Athens: Ohio University Press; 2008.
8. Giffard H. Making jet engines in World War II: Britain, Germany, and the United States. Chicago: University of Chicago Press; 2016.
9. Lemire B. Fashion's favourite: the cotton trade and the consumer in Britain, 1660–1800. Oxford: Oxford University Press; 1991.
10. Kaplan S. This 6,000-year-old amulet is the oldest example of a technology still used by NASA. Washington Post. 2016 Nov 15.
11. Klein U, Spary EC. Materials and expertise in early modern Europe: between market and laboratory. Chicago: University of Chicago Press; 2010.
12. Middelton WEK. A history of the thermometer and its use in metrology. Baltimore: Johns Hopkins University Press; 1966.
13. Mintz S. Sweetness and power: the place of sugar in the modern world. New York: Viking Penguin; 1985.
14. Riello G, Parthasarathi P. The spinning world: a global history of cotton textiles, 1200–1850. Oxford: Oxford University Press; 2011.
15. Smith P. Knowledge in motion: following itineraries of matter in the early modern world. In Rodgers DT, Rama B, Reimitz H, editors. Cultures in motion. Princeton, NJ: Princeton University Press; 2014. 109–33.

CHAPTER 1.1

GLASSWARE

CATHERINE M. JACKSON

The Age of Glassware

From the mid-nineteenth century onwards, every laboratory was filled with glassware and every trainee scientist learned to work with glass. Not any more. The age of glassware is passing, progressively displaced by the black boxes, new materials, and digital displays of twenty-first century laboratory instrumentation.[23] To anyone under the age of fifty, the contents of this chapter are in danger of appearing like a strange tale from a strange land. We have already lost much knowledge about glass — all the more reason to write this brief history of glassware in modern science.

What Is — and Was — Special about Glass?

Known since antiquity, glass as material spans both art and science. Made predominantly from sand and formed into objects as decorative as they were functional, glass was prized for its transparency, chemical inertness, and forgiving working properties. Drop a glass flask and it will shatter. But if you heat glass with sufficient skill and care, you can make it take just about any form you choose.

Glass is an almost infinitely variable material. Small changes in the recipe can alter its properties, making it heat resistant, shatter proof, or colored. Glass manages light. In perfectly flat float-glass form, it is so close to invisible that doors and windows are routinely etched at eye-level to prevent accidents. Silvered on one side, it becomes a mirror. Formed into lenses and prisms, it produces a stunning range of visual effects. When people speak of scientific glass, they almost always mean optical instruments like telescopes, microscopes, and spectroscopes, in which glass renders other dimensions visible to the human eye.[1] This chapter is not about such instruments. Nor is it about thermometers, barometers, and other precision instruments used to study heat, temperature, and pressure.[4] This chapter is about chemical glassware made by bending and blowing glass tubing in the flame of a glassblower's torch. Moving into glass was a fundamental change in the material basis of chemical practice. Its consequences for chemistry were so significant that I have coined the term "glassware revolution" to describe it. The glassware revolution not only changed chemistry, but also physiology and physics. The glassware revolution ushered in an age of glassware that lasted for 150 years.[15]

The *Glassware Revolution*: Time and Place

The glassware revolution began when two famous chemists, Jöns Jacob Berzelius in Sweden and Michael Faraday in London, independently tackled the cutting-edge chemical problem of organic analysis. Both sought to discover by analysis how much of the elements carbon, hydrogen, and oxygen particular organic compounds contained. It is important that neither man was in Paris, city of the great Antoine Lavoisier and still the center of the chemical universe in the early 1800s. Both Faraday and Berzelius were isolated from the elite chemical community — but they still managed to do brilliant analytical work. How? They were extremely good glassblowers and they used this skill to develop a new kind of apparatus.

Fortunately for our historical understanding, they also wrote textbooks, which explain how they used glass and why they thought this was so important. Being able to make his own apparatus was essential for Berzelius because — unlike in Paris — there were very few instrument makers in Stockholm.[2] Faraday had a democratizing goal: he wanted people to be able to do chemistry regardless of their wealth and position. Each chemist valued the fact that making his own apparatus put the chemist in closer touch with his experimental design. And each of their books described basic glassblowing equipment and provided instructions for doing what Faraday called "tube chemistry."[7]

These developments were widely emulated. Between about 1825 and 1835, numerous glassblowing manuals were published across northern Europe. Written by glassblowers and intended for chemists, especially chemistry students, these manuals provide important historical evidence for the glassware revolution. Their authors were well aware of the limitations of text as a medium for transmitting manipulative skill. Publishing a translation of the Parisian instrument maker Ferdinand Danger's *The Art of Glassblowing* in 1831, the British chemist John Griffin explained that: "Unquestionably the best method of learning to work glass is to obtain personal instructions from one who is conversant with the art but such instructions are not easily obtained."[6 piii] That same year in Germany, the glassblower Friedrich Körner published a glassblowing manual to remind his students of what they had seen in class. By 1831, Körner was already teaching glassblowing to chemistry students at the University in Jena.[18 p2–4]

At the heart of this change was the German chemist Justus Liebig's invention of a new piece of analytical apparatus, a small triangular piece of hollow glassware that became known as the *Kaliapparat* (Figure 1.1.1). Liebig developed the *Kaliapparat* in late 1830 during experiments to determine the chemical composition of a series of medicinal alkaloids including morphine and quinine. This new device measured the carbon content of organic compounds reliably and much more simply than was possible with previous methods. Liebig's accomplishment is best known because he integrated analysis using the *Kaliapparat* within an innovative approach to teaching organic chemistry. Liebig's Giessen laboratory became the first site for large-scale practical training in research in the sciences.[12; 20; 22]

Liebig's innovation changed much more than laboratory training. It represented a completely new approach to organic chemistry and a radical solution to the problem of how to provide manpower

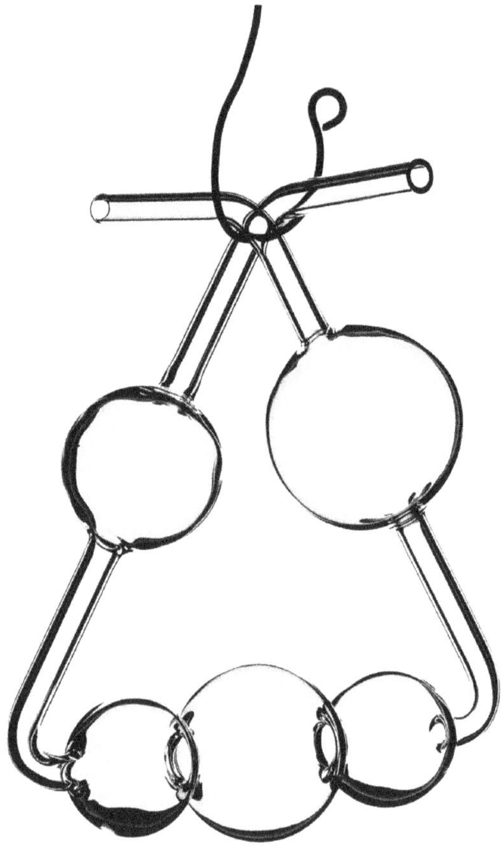

Fig. 1.1.1. Reconstruction of Liebig's *Kaliapparat* in borosilicate glass, by Tracy Drier at the University of Wisconsin–Madison, 2016. Liebig's *Kaliapparat* measured carbon by trapping carbon dioxide gas produced when an organic substance burned in potassium hydroxide (potash) solution. This was how the *Kaliapparat* got its English name: potash apparatus or potash bulbs. Weighing the *Kaliapparat* before and after combustion gave the mass of carbon dioxide, and hence the mass of carbon in the sample. *Credit*: Photograph courtesy of Ilia Guzei.

for that extremely labor-intensive endeavor.[14] It also functioned as an important driver of the glassware revolution. Chemists elsewhere were quick to adopt Liebig's *Kaliapparat* and they soon began to teach its use. In learning how to make and use the *Kaliapparat*, chemists learned that working in hollow glassware could solve problems that much larger, more expensive apparatus could not. This was how glassblowing became an essential component of the chemist's skillset. Chemists everywhere began to learn glassblowing, and from the 1830s onwards they worked almost exclusively in flame-worked glass apparatus.

The *Glassware Revolution*: Cause and Effect

Like Berzelius before him, Liebig turned to home-blown hollow glassware in response to the problems of geographical isolation and limited resources. The Parisian chemists who taught Liebig the state of the art of organic analysis could call upon some of the best instrument makers in the world. When Liebig arrived in Giessen, a small market town in rural Germany, he had no laboratory and no instruments, and his income derived entirely from student fees. These were the circumstances that drove Liebig to innovate in glass.

Having made that important move, Liebig soon learned that a skilled glassblower could improve on his ideas. Within a few years, his assistant Karl Ettling had perfected the design and construction of Liebig's *Kaliapparat*.[3 p50] In 1833, Liebig published instructions for how to make the *Kaliapparat* in his journal (*Annals of Chemistry and Pharmacy*) alongside an introduction to glassblowing and by 1837 similar instructions were published in his textbook of analysis (Figure 1.1.2).[19; 21] From the mid-1830s, chemists across Europe were analyzing organic substances using the *Kaliapparat*, and they were learning how to blow glass in order to do so.

The success of Liebig's *Kaliapparat* has usually been taken to indicate that it made fixing the composition and formula of organic substances a matter of routine. In fact, this was not the case. Chemists continued to experience great difficulty in determining reliable formulae for organic compounds for many decades. By the early 1840s, using the *Kaliapparat* taught Liebig that it was impossible to make progress in organic chemistry by purely analytical means. In response, he set his best student, August Hofmann, to develop a new and powerful investigative method: synthesis. Far from solving the problem of organic analysis, Liebig's *Kaliapparat* drove the introduction of synthetic organic chemistry.[13]

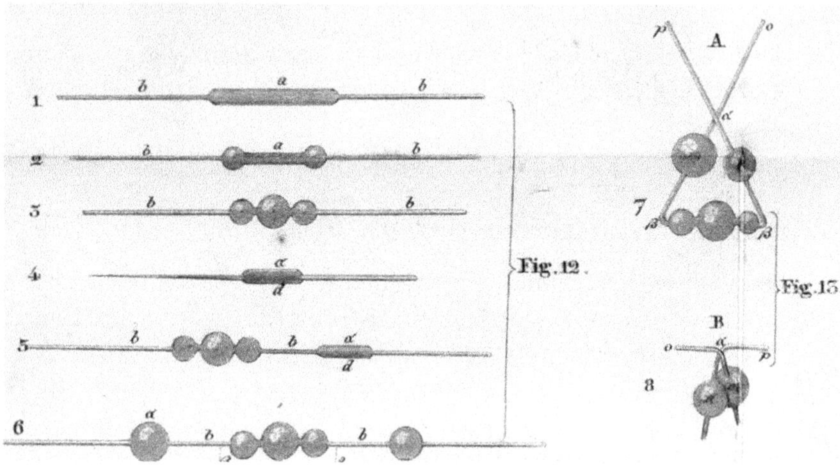

Fig. 1.1.2. Liebig's 1837 textbook of organic analysis included step by step instructions for how to make a *Kaliapparat* by glassblowing.[21 Tafel I] *Credit:* Image courtesy of the Special Collections Research Center, University of Chicago Library.

Organic synthesis soon began producing a large number of new substances and it hugely expanded the domain of organic chemistry in both the academic and industrial spheres. Doing organic synthesis had the potential for significant commercial rewards and this attracted increased student numbers to the field from the mid-century onwards. Synthesis proved even more labor-intensive than analysis, driving a wide range of developments in everything from practical methods to the laboratory itself. It also became the powerhouse of German industry.[16; 25]

The vast majority of organic chemistry's new practical methods were implemented in glass. Nineteenth-century organic chemists did reactions in sealed glass tubes and they used distillation and recrystallization to purify what those reactions produced. These combinations of flammable organic substances in glass with extremes of temperature and pressure were potentially explosive, especially in the hands of students.[11]

These were powerful factors in the development of the nineteenth-century chemical laboratory. Though it has usually been portrayed as merely a change of scale driven by competition between German states pre-1871, understanding how synthetic organic chemists worked makes clear that the rise of the institutional chemical laboratory was in fact a necessary tool in the development of the discipline. By 1900 chemical laboratories were certainly larger than they had ever been, but chemists' use of glass apparatus to control and manipulate organic nature meant that these laboratories were also different in character. Chemistry had changed dramatically during the preceding century and these changes prompted fundamental alterations in how laboratories were built, equipped, and operated.[10]

Chemists' increasing reliance on glassware had immediate consequences for suppliers of chemical apparatus. Early nineteenth-century firms supplied items such as flasks and retorts, all produced by working glass gathered from a glassmaking furnace. By mid-century, these were supplemented by a growing diversity of apparatus made from glass tubing, including Liebig's *Kaliapparat*. Would-be analysts in the 1830s had to make their own *Kaliapparat*, or know someone who could. From about 1850, you could buy a *Kaliapparat* from suppliers across Europe and North America.[5; 8]

The glassware revolution resulted in the emergence of a new category of professional scientific glassblowers. Just as Ettling had helped Liebig to transform his original innovation into a standard item, so others begin to work with skilled glassblowers to realize their experimental visions in glass. Sometimes these collaborations produced entirely new pieces of apparatus. More often they introduced variations on existing pieces of glassware.

One example will clarify why such innovations were important. Liebig's *Kaliapparat* quite soon transcended its original context of invention, organic analysis, instead becoming associated with its gas-absorbing function. When Julius Plücker examined the magnetic behavior of carbon monoxide gas around 1850, he had to be sure the sample was absolutely free from carbon dioxide. Plücker's preparation of carbon monoxide also produced carbon dioxide, which had to be removed. Plücker achieved this by working with the Bonn glassblower Heinrich Geissler. Geissler developed Liebig's original design, producing a new variation of the *Kaliapparat* with significantly improved gas absorption. After passing through two of Geissler's modified devices, Plücker's carbon monoxide was no longer contaminated by carbon dioxide.[24 p162]

Work like this collaboration between Geissler and Plücker established a model for how scientists and glassblowers could work together to solve technical experimental problems — much as scientific glassblowers work today. Geissler went on to develop a successful business for the supply of both custom and standard chemical glassware. His *Kaliapparat* was widely used in experimental contexts ranging from organic analysis to plant physiology. The glassware revolution transformed physics and physiology as well as chemistry.[17 p68]

Professional scientific glassblowers assisted chemists in tackling one particularly tricky problem: identifying what their reactions made. Chemists today have ready access to formulas and molecular structures through a range of instrumental analytical methods. As the difficulty of reliable organic analysis indicates, it was far from straightforward for nineteenth-century chemists to know what they had made. This problem became acute when chemists began trying to synthesize specified natural targets. Successful target synthesis required chemists to demonstrate the identity of their synthetic products with the natural targets. Chemists solved this problem by working with glassblowers like Geissler to develop standard methods of melting- and boiling-point measurement. Nineteenth-century chemists made melting and boiling points into reliable indicators of purity and identity, and they did so using standard glassware. Glass and glassblowing are central to the early history of laboratory standardization.[9]

Recovering the History of Glassware

Ever since Steven Shapin published his essay "The Invisible Technician" in 1989, historians have been aware that technical expertise in the sciences is not the preserve of people called philosophers and scientists.[26] There is still a long way to go before we have a good account of the rise of technical expertise in any area of science.

The present volume, with its focus on tools and materials, provides a most welcome opportunity to highlight this important endeavor. This chapter has situated glass as a material medium and glassware produced by lampworking at the core of the development of modern chemistry, and the sciences generally. It therefore indicates that the history of glassblowing would be a good place to begin telling that story. Yet, despite some excellent studies in the history of physics, for example, we still know remarkably little about glassblowers like Geissler. In part, this state of affairs is a consequence of the difficulty of writing the history of glass. I therefore conclude with some reflections on the important question of how that history may be recovered. Many avenues are open and they extend far beyond the history of chemistry.

Its fragility is an impediment to historical studies of the material agency of glass in science, but it is not an insuperable obstacle. Historical reconstruction is a possible approach. Indeed, several studies within the history of chemistry have used reconstruction to provide a series of useful insights into what it was like to do organic analysis using Liebig's *Kaliapparat*. Those studies, however, focused on the pedagogically stable analytical procedure Liebig published some seven years after his original introduction of the *Kaliapparat*, and they also mobilized manipulative skills and material resources that were not available in 1830s Giessen. As a result, they are distanced in important ways from both Liebig's process of creative innovation and the work that made his new method of analysis stable in the context of large-scale teaching and collective research.[27]

My work has already begun to answer questions about why and how the glassware revolution happened and how glassblowers contributed to this development. This chapter has cited several of my essays that used documentary sources, including images and laboratory notebooks, to uncover both the context of Liebig's innovation and its consequences, within and beyond chemistry. Combining documentary sources with the evidence that is only accessible to skilled hands has the potential to illuminate our material and social understanding of how past science was done, and what made the knowledge it produced useful. In this approach, it is possible to ask not just what was done and how, but also by whom. In the end, the goal is to elucidate the limits of possibility of science in its own historical, technical, and social space. The history of glassware is a history of the synergy between head and hands. And this synergy is surely fundamental to understanding science's unrivalled ability to produce reliable knowledge about the natural world. That is why this historian of chemistry is learning to blow glass.

References

1. Armstrong I. Victorian glassworlds: glass culture and the imagination 1830–80. Oxford: Oxford University Press; 2010.
2. Berzelius JJ. *Chemische Operationen und Gerätschaften, nebst Erklärung chemischer Kunstworter, in alphabetischer Ordnung*, trans. Friedrich Wöhler. Dresden: Arnold; 1831.
3. Brock WH. Justus Liebig: the chemical gatekeeper. Cambridge: Cambridge University Press; 1997.
4. Chang H. Inventing temperature: measurement and scientific progress Oxford: Oxford University Press; 2007
5. Child E. The tools of the chemist: their ancestry and American evolution. New York: Reinhold; 1940.
6. Danger FP. The art of glass-blowing, or plain instructions for making the chemical and philosophical instruments which are formed of glass; such as barometers, thermometers, hydrometers, hour-glasses, funnels, syphons, tube vessels for chemical experiments, toys for recreative philosophy, & c. London: Bumpus and Griffin; 1831.
7. Faraday M. Chemical manipulation; being instructions to students in chemistry, on the methods of performing experiments of demonstration or of research, with accuracy and success. London: Phillips; 1827.
8. Gee B, Brock WH. The case of John Joseph Griffin: from artisan-chemist and author-instructor to business-leader. Ambix. 1991;38(1):29–62.
9. Jackson CM. Chemical identity crisis: glass and glassblowing in the identification of organic compounds. Annals of Science. 2015;72(2):187–205.
10. Jackson CM. Chemistry as the defining science: discipline and training in nineteenth-century chemical laboratories. Endeavour. 2011;35(2–3):55–62.
11. Jackson CM. The curious case of coniine: constructive synthesis and aromatic structure theory. In Klein U, Reinhardt C, editors. Objects of chemical inquiry. Science History Publications: Sagamore Beach, MA; 2014. pp. 61–101.
12. Jackson CM. Re-examining the research school: August Wilhelm Hofmann and the re-creation of a Liebigian research school in London. History of Science. 2006;44(3):281–319.

13. Jackson CM. Synthetical experiments and alkaloid analogues: Liebig, Hofmann and the origins of organic synthesis. Historical Studies in the Natural Sciences. 2014;44(4):319–63.
14. Jackson CM. Visible work: the role of students in the creation of Liebig's Giessen research school. Notes and Records of the Royal Society. 2008;62(1):31–49.
15. Jackson CM. The 'wonderful properties of glass': Liebig's *Kaliapparat* and the practice of chemistry in glass. Isis. 2015;106(1):43–69.
16. Johnson JA. The Kaiser's chemists: science and modernization in imperial Germany. Chapel Hill, NC: University of North Carolina Press; 1990.
17. Just L. Ueber die Möglichkeit, die unter gewöhnlichen Verhältnissen durch grüne, beleuchtete Pflanzen verarbeitete Kohlensäure durch Kohlenoxydgas zu ersetzen. Forschungen auf dem Gebiete der Agricultur-Physik, 1882;5:60–79.
18. Körner F. Anleitung zur Bearbeitung des Glases an der Lampe, und zur vollständigen Verfertigugn der, durch das Lampenfeuer darstellbaren, physikalischen und chemischen Instrumente und Apparate. Jena: Schmidt; 1831.
19. Lafond. Ueber die Kunst Glas zu blasen, mit Verbesserungen von Danger. Annalen der Chemie. 1833;7:298–313.
20. Liebig J. Ueber einen neuen Apparat zur Analyse organischer Körper, und über die Zusammensetzung einiger organischen Substanzen. Annalen der Physik und Chemie. 1831;97:1–47.
21. Liebig J. Anleitung zur Analyse organischer Körper. Braunschweig: Vieweg; 1837.
22. Morrell JB. The chemist breeders: the research schools of Liebig and Thomas Thomson. Ambix. 1972;19(1):1–46.
23. Morris PJT, editor. From classical to modern chemistry: the instrumental revolution. Cambridge: Royal Society of Chemistry; 2002.
24. Plücker J. Ueber das magnetische Verhalten der Gase. *Annalen der Physik und Chemie* 1851;84:161–80.
25. Reinhardt C, Travis T. Heinrich Caro and the creation of modern chemical industry. Berlin: Springer; 2000.
26. Shapin S. The invisible technician. American Scientist. 1989;77(6):554–63.
27. Usselman MC, Reinhart C, Foulser K, Rocke AJ. Restaging Liebig: a study in the replication of experiments. Annals of Science. 2005;62(1):1–55.

CHAPTER 1.2

THERMOMETERS

JOHN C. POWERS

Thermometers have been an integral part of the scientific study of heat and the properties of materials since the establishment of standard temperature scales in the early eighteenth century. In most scientific and industrial contexts today, thermometers and related temperature-measuring technologies — thermocouples, pyrometers, and others — have been, in the words of Bruno Latour, "black boxed:" These instruments and their readings are used as unproblematic components of research and process control.[17 p68] The current situation, however, belies the early history of the instrument, when fundamental questions, such as what property the thermometer actually measured and how one could calibrate instruments to read in agreement with each other, dominated discussions in thermometry. In effect, the understanding of what the thermometer measured changed as scientific practitioners used it to study the properties of heat and deployed it to improve practical applications of that knowledge.

There is debate about who made the first thermometer, but Galileo's air thermoscope in the 1590s was one of the early ones. The first liquid-in-glass thermometers originated in the 1640s in Florence, culminating in the work (1657–1660) of the *Accademia del Cimento*. The *Accademia*'s thermometers used spirit of wine (alcohol) as the thermometric fluid and were primarily used to measure the "changes of the air." Most had a fifty-degree scale, but were not calibrated following what would later become a standard practice — fixed points. As W. E. Knowles Middleton suggests, the Grand Duke of Tuscany had an "artisan" who made new thermometers by using the first thermometer as a standard.[23 p30–38] The 1661 arrival of a Florentine thermometer in London generated interest among several Royal Society members, including Robert Boyle, Robert Hooke, Edmund Halley, and eventually Isaac Newton, in fabricating and experimenting with their own instruments. From England, the instrument made it to the Netherlands and France.[23 p41–63]

As suggested above, the key problem of thermometry was making instruments that were comparable, that is, that indicated the same degrees of heat under the same conditions. In pursuit of this aim, seventeenth-century savants devised the idea of fixed points: thermally stable, empirical phenomena that would anchor a scale of degrees of heat. Many fixed points were discussed and used, such as solid ice, an ice and water bath, boiling water, "blood heat" (that is, body temperature), and cellars or caves, which seemed to maintain a constant heat.[8 Ch1] Research into the stability of these points proved that most were unreliable. For example, French academician, Philippe de la Hire (1640–1718), showed that although a bath of water and ice maintained a constant degree of heat, solid ice could get colder.[9] His colleague, Guillaume Amontons (1633–1705), who used air thermometers, demonstrated that water boiled at a degree of heat which varied according to the atmospheric pressure.[1; 2]

Daniel Gabriel Fahrenheit (1686–1736) was the first to construct thermometers that were perceived to be comparable to one another and calibrated according to a reliable scale. He first made spirit thermometers, but around 1714 began to use mercury as the thermometric fluid in order to extend the range of measurement for use in chemical analysis. He attempted to construct mercury and spirit thermometers that read in agreement at all degrees on his scale, but he was never completely successful in doing this. This project, however, led him to investigate problems in thermometry, such as fixed points and the expansion of thermometer glass. He adapted his scale from that of Ole Roemer (1644–1710), whom Fahrenheit visited in Copenhagen in 1708, but he improved on Roemer's calibration methods. His method, which dates back to Robert Hooke, was to use an ice and water bath as a single fixed point and graduate the thermometer tube based on the measured expansion of mercury over a range of heats.[30 p138–139, 170–173; 8 p66–79] However, in public presentations, Fahrenheit advocated a two-fixed-point method, using a water and ice bath and boiling water to set his scale. This became the standard method for calibrating mercury thermometers.[12]

Fahrenheit's patron, the Dutch professor of medicine and chemistry, Herman Boerhaave (1668–1738), devised the first widely used theory of the thermometer's action. For Boerhaave, the thermometer measured the presence of fire: a weightless, aether-like fluid, which permeated matter and space and was responsible for the observable phenomena related to heat, such as the sensation of warmth through friction and certain chemical reactions as well as the thermal expansion of solids and fluids. He suggested that every degree of increase on Fahrenheit's scale, as measured by the dilation of a thermometric fluid, represented the presence of a proportional amount of fire.[5 p78–106] In his chemistry courses, Boerhaave deployed Fahrenheit's thermometers as demonstration instruments, showing, for example, that the generation of heat or cold as various chemical species (in solution) were mixed. He also advocated for its use in controlling certain chemical reactions, such as fermentation.[5 p217–222; 25] His aim was not mathematical rigor; he argued rather that the thermometer was more reliable than the senses of the human chemist. In some cases, the human senses could be fooled, and in others they simply could not detect phenomena at extremely high or low degrees of heat.[25]

By the middle of the eighteenth century, natural philosophers and instrument makers, such as the Scottish physician, George Martine (1700–1741), advocated the use of two fixed points to anchor scales, because this allowed for thermometers to be comparable. In 1740 Martine published the first table with which one could convert readings between fifteen different scales.[20] This push for fixed points had the effect that extant scales were reworked to make them fit the two-fixed-point standard. For example, in 1730–1731, Antoine Reaumur (1683–1757), described in two somewhat convoluted *Memoires* the construction and calibration of an alcohol thermometer based on one fixed point and the expansion of the alcohol. Extrapolating from his description of the resultant scale, makers later in the century inferred that the boiling point of water would be 80 degrees on the "Reaumur scale," and began using this standard to construct Reaumur mercury thermometers.[23 p79–89; 13] By the time Anders Celsius (1701–1744) developed the 100-point scale, which now bears his name, two fixed points were a standard practice, although his original version used an inverted scale: 0° marked the boiling point of water, while 100° marked the freezing point.[23 p95–101]

Work to improve thermometer calibration undermined Boerhaave's and Fahrenheit's understanding of the thermometer. Fahrenheit's failure to get his spirit and mercury thermometers to read in concordance revealed an erroneous assumption: that the expansion of mercury and spirit of wine (alcohol) was linear with respect to heat. The standard method to test the calibration of thermometers was the method of mixing: two equal quantities of a liquid at different temperatures were mixed together, and the resulting, measured temperature should have been the mean of the two original temperatures. Deviation from the mean suggested a problem with the thermometer.[8 p60–64] Studies by the English mathematician Brooke Taylor (1685–1731) and the Swiss natural philosopher Jean-Andre Deluc (1727–1817) showed that fluids typically do not expand linearly when heated over a significant range of temperatures.[10; 28]

Investigations into thermometer calibration yielded significant insights into the nature of heat. For instance, to find the best thermometer for his chemical work Joseph Black (1728–1799) used the method of mixing to determine that the expansion of mercury was reasonably linear (for his purposes) from the freezing point to the boiling point of water. However, one outgrowth of this work was that when Black also tried this method with different liquids, and then with solid bodies immersed in liquid, he determined that different species of matter had different *capacities* for fire (or heat), meaning that they absorbed different amounts of heat in order to register the same increase on a thermometer. He interpreted this discovery within a chemical understanding of matter, specifically the attractions of chemical species via chemical affinity. Black's chemical theory of heat had one other consequence: an explanation for the fire, which seemed to be absorbed as a liquid boiled but was not detected by the thermometer. This *latent* fire (or heat) combined chemically with the liquid during the change of state to a vapor.[4; 14; 21 p194–199] Black's results suggested that the thermometer did not measure, as previously thought, the quantity of fire or heat, but rather its *intensity*.

In the 1780s, Antoine-Laurent Lavoisier (1743–1794) and Pierre-Simon Laplace (1749–1827) developed a new instrument, the calorimeter, to measure the amount of heat chemically combined in ponderable matter. This heat was measured by the weight of water melted from the ice which was packed around the reaction chamber. Thus, the latent heat of ice became the standard for measurement. As Jan Golinski pointed out, the balance, rather than the thermometer, became the prime instrument in the study of heat, although the thermometer still had a role to play in setting the experimental conditions and determination of heat capacity.[14 p201–202] The work of Lavoisier and Laplace also shaped the future of thermometry by positing a new material substance for heat, caloric, which could exist in two states: latent and sensible, the latter being detectable by the thermometer.[18]

Adherents of the developing caloric theory assumed that in the ideal temperature scale, the expansion of the thermometric fluid was proportional to the quantity of heat it absorbed as determined by heat capacity. This model, however, proved to be problematic. John Dalton (1766–1844), for example, noted in 1802 that the heat capacity of substances was not fixed, but rather seemed to increase with temperature.[3 p361; 8 p66–68] This finding suggested that thermometer scales were always arbitrary and depended upon the thermometric fluid used. Within this context, Dalton and the French chemist, Joseph Louis Gay-Lussac (1778–1850), working separately on the absorption of caloric and expansion gases, argued that all gases had reasonably the same thermal expansion with respect to heat.[19 Ch4]

This result suggested that air was the best thermometric fluid, a notion endorsed by no less an authority than Laplace.[8 p69] In pursuit of a better understanding of the thermal properties of gases, the First Class of the *Institut de France* sponsored three prize competitions (in 1812, 1815, and 1817) aimed at measuring the heat capacities of gases.[19 p134–150, 236–238] In response to these last two competitions, two French physicists, Alexis-Thérèse Petit (1791–1820) and Pierre Dulong (1788–1827), compared the thermal expansion of mercury and air thermometers over a wide range of temperatures, showing that the two began to diverge at temperatures over 300°C. They concluded that air yielded the most consistent thermal expansion and, thus, air thermometers represented the true temperature scale.[8; 11 p100–102; 3 p360–361]

In 1840, French physicist, Henri Victor Regnault (1810–1878), was commissioned by the French state to examine all of the laws and data related to the design of steam engines. This led him to conduct a thorough experimental examination of the principles of practical thermometry. He reaffirmed Petit and Dulong's divergence between air and mercury thermometers, but he also noted that this was caused, in part, by the expansion of the thermometer glass. Air thermometers performed better in this regard because the higher volumetric expansion of air as temperature increased minimized this effect. Similarly, Regnault asserted the ideal air thermometer was a constant-volume instrument, in which a column of mercury was used to measure the changing pressure of the air as the temperature changed. This method, he argued, was superior to the constant pressure air thermometer (in which the temperature was measured by the change in the volume of the air), since the expanding air would lose heat (and diminish the accuracy of the instrument) as it expanded out of the substance being measured.[8 p75–81; 3 p295–301; 26]

Regnault is considered to have reached the high point of practical thermometry in the nineteenth century, but he eschewed the question of what property the thermometer measured. William Thomson (later, Lord Kelvin, 1820–1907) and James Prescott Joule (1818–1889) addressed this question within the context of the new field of thermodynamics. Thomson and Joule defined temperature as follows: within a thermodynamic system, the proportion of heat gained or lost (the difference being converted into work) was equivalent to the temperature difference between the two states. Thus, defined in terms of the work generated by an ideal heat engine, Thomson argued that the "absolute temperature" could be calculated independently of the thermometric substance used in a thermometer.[8 p173–192; 29] They tested their idea of "absolute temperature" against real air thermometers, showing that the more air or gas thermometers deviated from the "absolute temperature" the further they deviated from the properties of an ideal gas. Thomson and Joule published a correction table, which related the scale of Regnault's air thermometers to the "absolute" temperature.[8 p192–197; 29] Additionally, they asserted that constant-volume hydrogen thermometers behaved closest to the absolute temperature scale; as such, these became the standard for calibrating to absolute temperature.[23 p114]

Although air and hydrogen thermometers were prized for their accuracy, they were not useful in all applications, especially those that involved very high and very low temperatures. Early high-temperature thermometers estimated temperature by measuring the expansion of metal bars or springs, such as Cromwell Mortimer's (1693–1752) instrument, which used a brass rod and was calibrated up to 900°F.[23; 24 Ch7] In an attempt to measure the heat of his kilns, potter Josiah Wedgwood (1730–1795) devised a pyrometer, which was based on the color and shrinkage of clay slugs after they had been fired, and with which he estimated temperatures up to an astounding (and incorrect) 5,000°F.[7; 8 p118–128] Modern pyrometers determine the temperature of an object by measuring its thermal radiation. In 1901 Ludwig Holborn and Ferdinand Kurlbaum designed an optical pyrometer, in which an operator looked through an eyepiece to match the color of an incandescent object with that of a heated wire in the instrument. More modern pyrometers measure the emissivity of heated objects at several wavelengths and relate the intensity of radiation to temperature via Planck's law.[22 p162–208]

For industrial applications, and in many materials laboratories, a common way to measure temperature is through electrical temperature-measuring devices, such as thermocouples. In 1822

Thomas Johann Seebeck (1770–1831) discovered that a circuit of two metals, in which the temperature at one juncture was raised, generated an electric current. By adding a galvanometer to measure the small current generated, later physicists, such as Claude Pouillet (1790–1868), were able to show that, for example, a bismuth-copper thermocouple generated a current that was reasonably proportional to the temperature differential of the two junctions.[23 p177–178] Thermocouples became invaluable tools in the development and industrial production of metals, especially steel, that require high temperature measurement.[16]

The engineer Carl Wilhelm Siemens (1823–1883) was the first to suggest using changes in electrical resistance in metals as a means for measuring temperature. In 1871, he delivered the Bakerian Lecture to the Royal Society of London on this topic and suggested that platinum be used as the metal for electrical-resistance thermometers.[27] Soon his firm, Siemens and Halske, began making "electric pyrometers," although these instruments were seen as having questionable reliability. Hugh Longbourne Callendar (1863–1930) worked on calibration techniques for the platinum resistance thermometers by comparing them to gas thermometers, finding that the resistance of platinum varies with temperature following a parabolic function.[6] Eventually, Callendar's thermometers became, with hydrogen gas thermometers, the standard-holders for temperature measurement. In 1899, a Committee on Electrical Standards headed by John William Strutt, Lord Rayleigh (1842–1919), accepted Callendar's interpolation formulas for bringing these instruments into agreement with gas thermometers. By the 1920s, platinum resistance thermometers had become the practical standard for scientific temperature measurement (hydrogen gas thermometers often being difficult to use), and by the 1980s, the United States National Institute of Standards and Technology (formerly the National Bureau of Standards) considered platinum resistance thermometers to be the standard thermometer for uses between 13.81 K and 630.74 K.[15; 19]

Since the introduction of reasonably accurate instruments in the eighteenth century, thermometers have been important tools in the scientific study of heat and the role of temperature, as a separate property from heat, in physical processes. More than this, thermometers have played a key role in establishing the thermal and thermodynamic properties of materials, which are essential to the scientists and engineers who manipulate matter for human aims. From brewing and pottery-making in the eighteenth century to metallurgy and semiconductor doping in the twentieth, the proper measurement and control of temperature using thermometers, pyrometers, or thermocouples has been a standard part of materials science processes.

References

1. Amontons G. Discours sur quelques propriétés de l'Air, & le moyen d'en connoitre le temperature dans tous les climats de la Terre. Mémoires de l'Académie royale des Sciences, année 1702. Paris; 1737. 216–43.
2. Amontons G. Le Themometre réduit à une mesure fixe & certaine, & le moyen d'y rapporter les observations faites avec les anciens Thermometres. Mémoires de l'Académie royale des Sciences, année 1713. Paris; 1739. 64–72.
3. Barnett MK. A brief history of thermometry. Journal of Chemical Education. 1941;18(8):358–64.
4. Black J. Lectures on the elements of chemistry, vol 1. Robison J, editor. Edinburgh; 1803.
5. Boerhaave H. Elements of chemistry, vol. 1. Dallowe T, trans. London; 1735.
6. Callendar HL. On the practical measurement of temperature: experiments made at the Cavendish Laboratory. Philosophical Transactions. Section A 1887;178:161–230.
7. Chaldecott JA. Josiah Wedgwood (1730–1795) — scientist. British Journal for the History of Science. 1975;8(1):1–16.
8. Chang H. Inventing temperature: measurement and scientific progress. Oxford: Oxford University Press; 2004.
9. De la Hire P. Experiences sur le Themometre. Mémoires de l'Académie royale des Sciences, année 1711. Paris; 1715. 66–79.
10. De Luc J-A. Researches sur les modifications de l'atmosphere. 2 vols. Geneva; 1772.
11. Dulong, P-L, Petit A-T. Researches sur la measure des températures et sur les lois de la communication de la chaleur. Annales de chimie et de physique. 2nd ser. 1817;7:113–54, 225–64.
12. Fahrenheit DG. Experimenta & Observationes de Congelatione aquae in vacuo factae. Philosophical Transactions. 1724;33(382):78–84.
13. Gauvin, J-F. The instrument that never was: inventing, manufacturing, and branding Reaumur's thermometer during the Enlightenment. Annals of Science. 2012;69(4):515–49.

14. Goliksi G. "Fit instruments:" thermometers in eighteenth-century chemistry. In Holmes FL, Levere TH, editors. Instruments and experimentation in the history of chemistry. Cambridge, MA: The MIT Press; 2000. Chapter 8.
15. Griffiths E. Methods of measuring temperature. London: Charles Griffin; 1918.
16. Heselwood W. C. and Manterfield D. Liquid steel temperature measurement: a review of the quick-immersion thermocouple method. Platinum Metals Review. 1957;1(4):110–18.
17. Latour B. Science in action: how to follow scientists and engineers through society. Cambridge, MA: Harvard University Press; 1987.
18. Lavoisier A-L, Laplace P-S. Memoir on heat, read to the Royal Academy of Sciences, 28 June 1783. Guerlac H, trans. New York: Neal Watson Academic Publishers; 1982.
19. Mangum BW. NBS Measurement services: platinum resistance thermometer calibrations. Washington, DC: US Government Printing Office; 1987.
20. Martine G. Essays medical and philosophical. London: A. Millar; 1741. 175–214.
21. McKie D, Heathcote NH de V. The discovery of specific and latent heats. London: Edward Arnold; 1935.
22. Michaleski L *et al.* Temperature measurement, 2nd ed. New York: Wiley; 2001.
23. Middleton WEK. A history of the thermometer and its uses in meteorology. Baltimore: Johns Hopkins University Press; 1966.
24. Mortimer C. A discourse concerning the usefulness of thermometers in chemical experiments. Philosophical Transactions of the Royal Society of London. 1746–1747;44:672–95.
25. Powers, JC. Measuring fire: Herman Boerhaave and the introduction of thermometry into chemistry. Osiris. 2014;29:158–77.
26. Regnault HV. Relations des expériences entreprises par ordre de Monsieur le Ministre des Travaux Publics, et sur la proposition de la Commission Centrale des Machines à Vapeur, pour déterminer les principals lois et les données numériques qui entrent dans le calcul des machines à vapeur. Mémoires de l'Académie Royale des Sciences de l'Institut de France. 1847;21:1–748.
27. Siemens C. W. On an electric resistance thermometer for measuring temperatures in inaccessible places. Proceedings of the Royal Society of London. 1871;19:443–45.
28. Taylor B. An account of an experiment, made to ascertain the proportion of the expansion of a liquor in the thermometer, with regard to degrees of heat. Philosophical Transaction of the Royal Society of London. 1723;32:291.
29. Thomson W. Mathematical and physical papers, vol. 1. Cambridge: Cambridge University Press; 1882.
30. Van der Star P, editor. Fahrenheit's letters to Leibniz and Boerhaave. Amsterdam: Rodopi; 1983.

CHAPTER 1.3

SIMPLE HEATING

CYRUS C. M. MODY

In the twentieth century, researchers learned to expose materials to tremendous extremes of temperature using lasers, highly specialized ovens, and other techniques. Yet materials researchers still required more modest means of raising temperature: Bunsen burners, hot plates, blowtorches, and so on. These techniques and their precursors go back centuries if not millennia. Indeed, controlled production and application of heat was one of the technologies that enabled the emergence of complex societies, if not our species itself. The editors have therefore chosen to place discussion of forms of heating with ancient roots in this section of the volume and label them "Simple Heating" in contrast to the techniques discussed in Part 3 under "Complex Heating." This is, however, something of a misnomer; some of these ancient techniques are not very simple, and all of them continue to evolve.

Prometheus to Ourselves

The myth of Prometheus shows that some societies, at least, have linked the control of heat — fire — with knowledge and civilization for a very long time. Indeed, strong evidence suggests that controlled use of fire for cooking coincided with the transition from *Homo erectus* to the immediate predecessors/closest relatives of *Homo sapiens*. The ability to cook food, to stay warm, to modify environments on a large scale with fire, and to stay up past dark conversing around a common hearth must all have played an important role in early human groups.

Presumably even before humans achieved full control of fire, they were curious about how it affected materials such as wood, stone, mud, and bone. Inspection of the remnants of wildfires must have revealed the interesting properties of, say, charcoal. At least 150,000 years ago (if not more), humans were apparently using fire to harden implements such as spears. The great leap in using fire to fashion materials, though, was surely the invention of pottery. Now, formless mud could be turned into useful ceramic forms capable of holding grains, oils, wine, and other valuable substances. For better or worse, the ability to store grains seems to have been a prerequisite for the formation of states; and certainly the ability to transport grains and liquids was a prerequisite for long-range trade.

More pertinent to this volume, experimenting with fire and mud co-evolved with new ways to produce and apply heat: from open fires to semi-enclosed pits to fully enclosed kilns capable of much hotter temperatures. With kilns came vitreous pottery and hence eventual familiarity with glass. Kilns may also have given rise to smelting: bits of rock left in the kiln or decorative stones on pots may have melted, leaving behind pure plugs of metal. Even semi-enclosed pits are capable of melting lead and — in extreme cases — copper. Hence, in most parts of the world, the first advanced metallurgy was practiced with copper and its alloys, especially bronze.

Clifford Conner's *People's History of Science* speculates that the earliest metallurgists may have been women, since women may have controlled kilns and pottery-making in some early societies.[2 p77] What is much more certain is that ideas about gender were incorporated into — and were inspirational for — innovation in heating techniques. For instance, in sub-Saharan Africa — where iron smelting was extraordinarily widespread from a very early period — archaeological, historical, and ethnographic evidence shows a close affinity between the practice of smelting and ideas about human reproduction.[9] Furnaces needed to be made fertile and impregnated in order to give birth to a bloom of iron; it takes relatively little

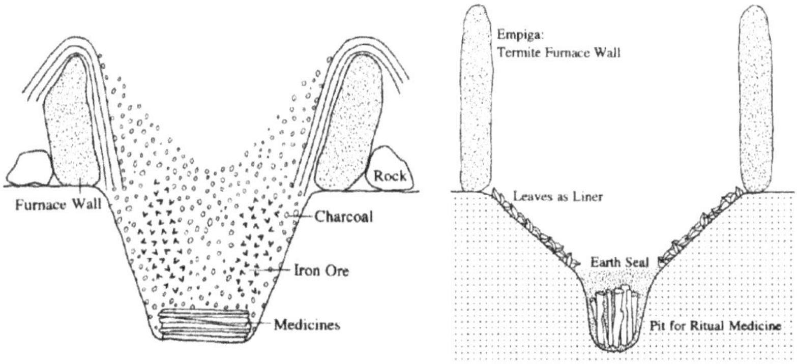

Fig. 1.3.1. Iron smelting furnaces of longstanding design from Burundi (left) and Tanzania (right). *Source*: Schmidt PR, Mapunda BB. Ideology and the archaeological record in Africa: Interpreting symbolism in iron smelting technology.[9]

imagination to see a certain similarity between furnace designs — despite their dizzying variety — and the human reproductive system (Figure 1.3.1).

Furnaces were often prepared with objects and rituals relating to reproduction, or decorated to resemble a woman giving birth. There were also often restrictions on who could approach the furnace (for example, no women at all and no men who had recently had intercourse), that also related to reproduction. These practices, reflective of cosmology and social order were inseparable from aspects modern eyes would see as "technological" — for instance, the shape and number of tuyères to deliver a blast to the furnace and the preheating of that blast to allow much hotter combustion. The multiplicity of ideas about cosmology, reproduction, gender, and social order must therefore have fed the multiplicity of furnace designs and hence the constant innovation in African metallurgy over more than two millennia.

The Alchemical Era

Whatever inspired them, design features such as tuyères and preheating are still fundamental to heating technology for materials research today. We also see a similar entanglement of cosmology, social order, and technology in a domain that is more obviously a predecessor of today's materials research: namely, alchemy.

As much as many scientists today picture alchemy as the domain of zealots and frauds, it was in fact closely linked to the state and to commerce in ways that prefigure the university-industry-government linkages that have typified materials research from the beginning. Those linkages were reflected in the tools they developed. For instance, many alchemists — such as Isaac Newton — were employed by royal mints, where they supervised much of the process of "transforming" mined ores into currency. For this, they refined techniques such as cupellation: the separation of precious from base metals by heating in a crucible and then oxidizing the base metal to form a slag — a technique still used today.

It is no accident that alchemists would be responsible for innovations in heating technology. Alchemists were fascinated with purification and heating was the most direct means they knew for smelting, distilling, and other purification techniques. They developed kilns, forges, and crucibles capable of delivering extraordinarily high temperatures as well as set-ups that delivered a gentler heat. Many of these techniques remain in use today, though often outside materials research.

A good example is the blowpipe torch. Although it is sometimes claimed that the blowpipe was an eighteenth-century (CE) invention, in fact — as noted by the prominent historian of alchemy William Newman — "the forced-air tube was used by jewelry makers and metal workers, even in ancient Egypt, for fusing and soldering metals."[6 p35] Blowpipes were being used for analysis — figuring out what a substance was made from — by the medieval period. Today the blowpipe is most familiar from glassmaking, but as a means of delivering heat for "materials research" it has a long history in alchemy.

Another example is the *bain marie*, which is now more often employed in cooking. As its name indicates, this apparatus is credited to Maria the Jewess, a circa first-century CE alchemist living somewhere in the eastern Mediterranean. She at least improved and possibly invented a number of heating technologies that are still in use in some form. She also understood — and, again, possibly coined — state-of-the-art alchemical concepts of her day.

Between them, Maria and Newton span some sixteen centuries of continuous innovation in alchemical apparatus, not least apparatus for heating: at various temperatures and rates, with direct or indirect heat, with varying degrees and rates of inlet and outlet of gas, reagents, and effluents. As historian Robert Anderson puts it, "each chemical process — fusion, cupellation, calcination, reverberation, cementation, distillation, digestion (and the bain-marie), sublimation, fierce distillation, and distillation per ascendum — needed a particular kind of furnace."[1 p17] A particularly ingenious solution to this need for variety was the alchemist Johann Becher's late seventeenth-century portable furnace made of eight parts that could be assembled in six different ways.

The same ideas about the furnace giving birth which informed African smelting also hovered around much alchemical innovation, as did references to the influence of the heavens and of various occult forces. Much of that language is obscure if not indecipherable to us today. It was long assumed that alchemists were speaking metaphorically or perhaps projecting their unconscious onto the material world. A more updated view is that they were speaking in coded language to make it difficult for competitors to pirate their procedures. But in at least some cases, their talk of, for instance, "Hermes' Tree" was descriptive rather than metaphorical: they were able to create chemical reactions resulting in ramifying crystalline growth, which indeed resembles a tree.[7]

From Chymistry to Chemistry

Alchemy did not end with the Scientific Revolution, but it did enter a new phase. Historians sometimes designate an interim period of "chymistry" — roughly the second half of the seventeenth century through the late-eighteenth century — which contained elements of both alchemy and modern chemistry. In this period, alchemy was

Fig. 1.3.2. The Alchemist's Experiment Takes Fire, Hendrik Heerschop, 1687.

increasingly subject to criticism, some of which related to techniques for heating. For instance, genre paintings of alchemists — a surprisingly popular topic — often cast them in a morally dubious light by depicting their experiments exploding or causing their homes to catch fire (Figure 1.3.2).

The central critique that pointed the way to modern chemistry, though, was that alchemists were too secretive and that they needed to open their methods and findings up to public debate. As sympathetic as we moderns might be to this point, we should remember that alchemy's critics were not acting entirely in good faith: the enormous personal fortunes needed to construct chemical experiments such as Robert Boyle's air-pump or Antoine Lavoisier's bomb calorimeter meant that what they really envisioned was elite private control over public chemical discourse, rather than a true public sphere for chemistry.

We also must keep in mind that the alchemists had good reasons for their secrecy — reasons that still inform materials research today. The basic architecture of a kiln, for instance, was relatively well known: anyone who wanted to try their hand at distillation could approximate it from knowledge that was reasonably public. However, building a good furnace that would achieve high heat and not melt or explode was quite difficult. The subtleties of building effective experiments were therefore considerably less public. Every alchemist either had to learn them from an adept or develop their own. For instance, almost every alchemist had their own recipe for lute, a sealant for vessels and liner for furnaces which often included, among many other possible ingredients, animal hair, flour, and manure.[3] This hindered theft of their practices in the early modern era and presents an obstacle to replication of their experiments today.

Whether due to these critiques or other reasons, the era of chymistry seems to have been associated with a slowdown in innovation in heating methods for materials research. This was a period of enormous *interest* in heat, of course. Experimentally, it saw Daniel Fahrenheit (and others) developing new ways of measuring temperature, Joseph Black demonstrating latent heat, Thomas Newcomen and James Watt building steam engines, all culminating in Lavoisier meticulously measuring the heat of chemical reactions with the bomb calorimeter. Theoretically, this was the period of debate over phlogiston, caloric, and other material carriers of heat. But the technologies of delivering heat to an experimental apparatus remained fairly static. This is a good example of Peter Galison's argument that revolutions in science are rarely total: a conceptual or experimental revolution often proceeds even as instruments stay the same, and vice versa.

In Industrialization's Wake

The middle of the nineteenth century saw a return to innovation in simple heating methods for materials research. The most famous of these advances is no doubt the Bunsen burner, invented in 1855 by Robert Bunsen's technician, Peter Desaga. Bunsen sought a hot, efficient flame with little contamination, and Desaga responded by adapting a Michael Faraday design in which fuel and oxidizer mixed prior to combustion. But Bunsen also wanted a small, light burner that could be batch-produced so that it could be used in his teaching laboratory. Desaga, followed by his son Carl, complied: the younger Desaga became an important manufacturer of laboratory equipment for both research and teaching, contributing to the spread of "Bunsen's" burner far and wide.

Thus, the Bunsen burner was invented within (and for use by) a new type of organization, the research university, which originated in Germany and spread in tandem with the Second Industrial Revolution. But the Bunsen burner was a product as much as a symptom of industrialization: it was only convenient to use in places where municipal gas lines — primarily used for lighting — could be tapped for laboratory use.

Indeed, modern metallurgy and later materials research owe a great deal to industrial heating innovations in the nineteenth century, particularly in the steel industry. The Bessemer process and later the Charles William (or Carl Wilhelm) Siemens open-hearth process made it possible to manufacture large amounts of high-quality steel with reproducible qualities. They did so via new methods for heating steel in the presence or absence of oxygen and nitrogen such that elements other than iron — particularly carbon — could be drawn out of the metal, added to it, or retained in just the proper ratio. Though these new kinds of furnaces were invented for mass production, they contributed to fundamental materials research concepts such as the phase diagram and Le Chatelier's principle.[4]

Precursors of both Bessemer's furnace and Siemens' open-hearth furnace go back many centuries. What set these nineteenth-century inventions apart was their scale: tons of steel could be produced in a matter of minutes (Bessemer) or hours (Siemens). That scale was necessary in the Second Industrial Revolution, which depended upon the rapid expansion of rail networks, the sprouting of steel-skeleton skyscrapers, and the launching of megaton steel ships. But by the beginning of the twentieth century, we start to see a parallel focus on new furnaces for the manufacture of high-value materials in rather small quantities. A notable example is Auguste Verneuil's 1902 furnace for growing synthetic ruby and other gemstones by blowing a hot gas laced with molecules of the desired crystal past a "seed;" as the seed grows it becomes a single crystal "boule."[8] In the twentieth century, high-purity crystals — most notably of silicon — would become increasingly important in both industry and materials research.

The final piece of the puzzle of "simple" heating was the application of electricity. In addition to his open-hearth process, Carl William Siemens was also a pioneer in electric arc heating — commonly used by the end of the nineteenth century for lighting and up to the present in spot welding and some production furnaces. Later in the nineteenth century, it was discovered that running an alternating current through an electromagnet could produce eddy currents — and hence heating — in a nearby conductor.[5] With induction heating, an object could be heated without contact, making possible twentieth-century advances in crystal growth such as the Czochralski process (where the seed is used to "draw" a single crystalline boule out of a melt) and the float-zone process (where heating melts one "zone" of a crystal, flushing out impurities as the zone shifts from one end of a solid crystal to the other).

The line between "simple" and "complex" heating is, of course, quite arbitrary. With induction heating we get very close to the techniques covered in the entry on complex heating, but one of the points of the present entry is that innovation in even seemingly simple techniques has been continuous for many centuries and hence has produced a large number of variations for different applications. Heating is one of the most fundamental means of altering material properties and hence has been central both in understanding the material world and in the production of new materials. Innovation in heating techniques has thus long been linked to knowledge-based industry, from the village smith to magnates such as Bessemer and Siemens.

References

1. Anderson RGW. The archaeology of chemistry. In Holmes FL, Levere TH, editors. Instruments and Experimentation in the History of Chemistry. Cambridge, MA: MIT Press; 2000. 5–34.
2. Conner CD. A People's History of Science: Miners, Midwives, and 'Low Mechanicks'. New York: Nation Books, 2005.
3. Kirkemo FM. Adam Lonicer's Kreuterbuch and 16th century distillation [master's thesis]. Trondheim, Norway: Norwegian University of Science and Technology; 2014.
4. Misa TJ. A Nation of Steel: The Making of Modern America, 1865–1925. Baltimore: Johns Hopkins University Press; 1995.
5. Mühlbauer A. History of Induction Heating and Melting. Essen: Vulkan Verlag; 2008.
6. Newman WR. Alchemy, assaying, and experiment. In Holmes FL, Levere TH, editors. Instruments and Experimentation in the History of Chemistry. Cambridge, MA: MIT Press; 2000. pp. 35–54.
7. Principe LM. Apparatus and reproducibility in alchemy. In Holmes FL, Levere TH, editors. Instruments and Experimentation in the History of Chemistry. Cambridge, MA: MIT Press; 2000. 55–74.
8. Scheel HJ. Historical aspects of crystal growth technology. Journal of Crystal Growth. 2000;211:1–12.
9. Schmidt PR, Mapunda BB. Ideology and the archaeological record in Africa: Interpreting symbolism in iron smelting technology. Journal of African Archaeology. 1997;16(1):73–102.

CHAPTER 1.4

SCALES AND BALANCES

JOSEPH D. MARTIN

In Ancient Egyptian mythology, entry into the afterlife was secure only after the weighing of the heart (Figure 1.4.1). The gods would place the deceased's heart on a scale opposite the feather of Maat, symbolizing the seven cardinal virtues of truth, justice, propriety, harmony, balance, reciprocity, and order.[4 p374] Those whose hearts were in equilibrium with Maat could pass into the afterlife; hearts heavy with misdeeds would be devoured by Ammut — a chimerical, crocodile-headed goddess — consigning the soul of the deceased to oblivion.[4; 10]

Although it concerns the fates of the dead, this story hints at the practices of the living in the ancient world, and the importance weighing held for them. Ancient civilizations curious about the nature of weight soon realized that two arms of equal length balanced on a fulcrum could be used to establish either that two objects weighed the same, or that one weighed more than the other. Balances of this type have been found dating to as early as the fourth millennium BCE. The moral consequence evident in the weighing of the heart hints at the import such balances held for civic life of the ancient world, guiding the equitable conduct of trade and commerce.[6] That same moral consequence carries through to the modern world, where the personification of Justice is commonly portrayed blindfolded, holding a set of scales.

Fig. 1.4.1. The weighing of the heart depicted in the Egyptian Book of the Dead, overseen by the jackal-headed Anubis, the god of embalming. *Credit*: Courtesy of the British Museum, via Wikimedia Commons.

Like many of the tools discussed in this volume, scales and balances make clever use of materials in order to better understand them. In the ancient world, that meant crafting stone weights to be used as the standards against which objects of unknown weight could be tested. This fundamental aspect of weighing changed little over the subsequent millennia. Even into the twentieth century, many scales and balances and their standard weights, although much refined in their construction and operation, would have been perfectly intelligible to an ancient Egyptian or Mesopotamian shopkeeper (Figure 1.4.2).

Over the next several millennia, improvements to weighing techniques came in the form of improved scales, but also in refinements to the systems that ensured the accuracy and precision of standard weights. The precision required for weighing of the sort that greases the wheels of day-to-day life in a settled society — for trade, assaying, and minting, for example — depended as much (or more so)

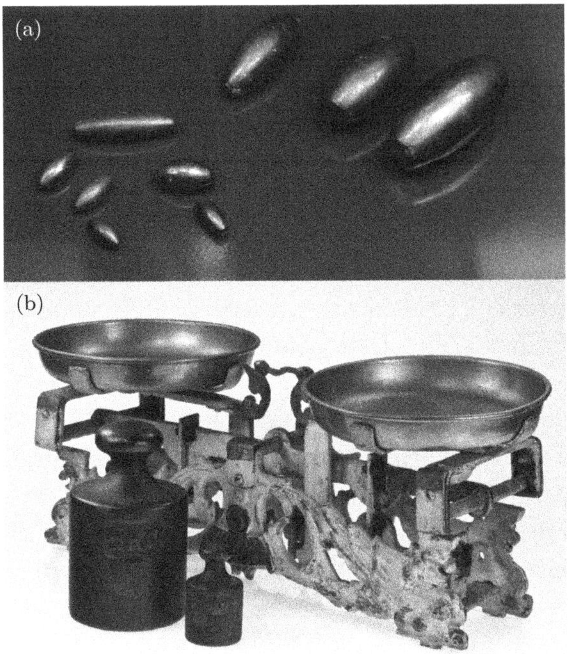

Fig. 1.4.2. (a) Standard hematite weighs from the Egyptian late period (ca. 500 BCE). (b) A nineteenth-century two-pan balance and brass weights. *Credit*: Courtesy of Wikimedia Commons.

on the robustness of standards as it did on the form of the scales set by them. As the following examples show, further refinements to weighing techniques were frequently inspired by the use of scales in investigations into the properties of matter.

Antoine Lavoisier and the Chemical Balance

Chemical problems in the late eighteenth century provided ample motivation to seek out more precise ways of weighing. Chemical investigations posed distinctive problems that called for precision balances. Assayers, whose job it was to determine the composition of metals, had long demanded precision scales, but they worked with a small class of substances whose properties were well known. Combined with the firm standards that were in place in much of Europe by the eighteenth century, this meant they received little trouble from standardized balances optimized for relatively small weights. But the research programs that emerged in the eighteenth century, in particular on the composition and properties of air, drove demand for increasingly sensitive scales for a wider variety of types of measurements.

Of special interest were scales that could hold heavy loads (on the order of kilograms) while also maintaining their sensitivity. Antoine Lavoisier (1743–1794), the virtuoso French natural philosopher, sought out scales that could manage containers big enough to hold considerable quantities of air, so that he might observe the results of chemical reactions on the weights of different airs. Precision weight measurements were crucial in the wide-ranging debate over the nature and existence of phlogiston, the hypothesized matter of fire. The precision balances Lavoisier commissioned permitted the measurements by which he noticed that many metals gain weight during calcination (burning), posing a problem for the notion that phlogiston was a substance with a finite weight.[9 p176] The suggestion that phlogiston might have negative weight slowed the bleeding of a mortally wounded theory, but Lavoisier's oxygen theory, flawed though it was, won favor over the phlogiston theory in large part on the basis of Lavoisier's precision measurements.[2]

Balances offering the precision sensitivity required by the eighteenth-century chemical laboratory had a limited commercial audience, and hence were unlikely to be developed in the absence of the demand provided by well-resourced experimenters. It is in Lavoisier's experimental practice that we begin to see something resembling the sort of networks characteristic of a modern chemical or materials laboratory. Lavoisier's laboratory was a site of practice not only for experienced instrument makers, but also for apprentices, so that the process of designing and refining instruments became intertwined with the training of their makers.[1] The social world that grew up around wealthy gentlemen natural philosophers therefore created a market for superlative precision that would not have existed otherwise.

Coulomb, Cavendish, and the Torsion Balance

Charles Augustin Coulomb (1736–1806) came to natural philosophy from a background in military engineering. His engineering training gave him both a sensitivity to the properties of materials and an appreciation for rigorous mathematical precision. While working to develop an extremely sensitive compass — one capable of detecting small local variations in the Earth's magnetic field — by suspending a magnetized needle from a silk thread, he became interested in what happens to thin wires when they are twisted.[3]

In the 1780s, Coulomb worked out that the force a twisted thread exerts against the direction in which it was twisted — what we would call the "reaction torque" and what he termed "the momentum of the force of torsion" — is proportional to the angle through which it has been rotated. After formulating the law governing the reaction torque of a wire, Coulomb realized that a sufficiently thin wire could be a sensitive tool for measuring extremely small forces.[5] The result was the torsion balance (Figure 1.4.3), which Coulomb used to verify the supposition that electrostatic charges exert forces that decrease with the square of the distance, and to articulate his eponymous law.

The British natural philosopher John Michell (1724–1793) independently developed a similar device. Michell's torsion balance made its way into the extensive instrument collection of his friend Henry Cavendish (1731–1810). Like Coulomb, Cavendish was interested in the nature of force laws. Isaac Newton had established the inverse-square relation for gravitational attraction, but his law of universal gravitation included a constant, G, the value of which remained difficult to establish experimentally. The precision offered by Michell's torsion balance permitted Cavendish to measure the gravitational attraction between two ordinary objects in the laboratory in his famous 1798 experiment, and to establish an accurate value for G.

The torsion balance and a related innovation from around the same time, the spring scale, first manufactured in Britain by the instrument maker Richard Salter in the 1770s, represented a transition in scale design. Whereas equal-arm balances relied on principles that had at least been informally understood for millennia, this new

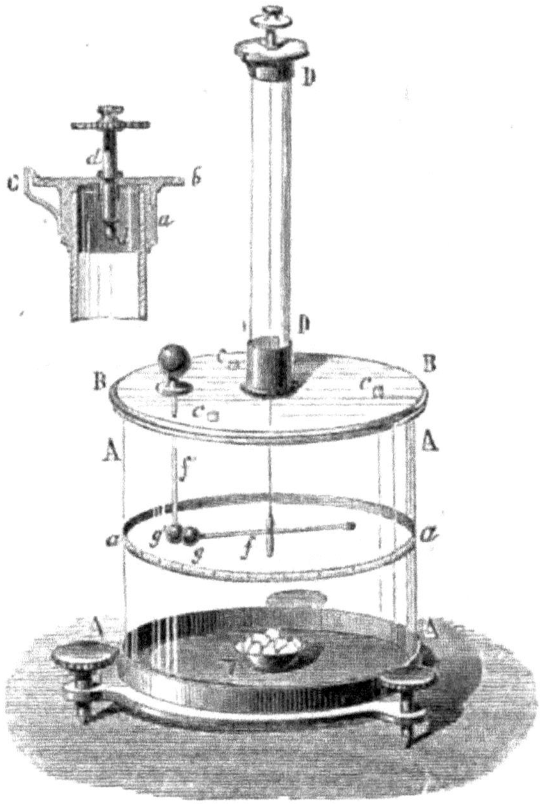

Fig. 1.4.3. Coulomb's torsion balance became a standard example in nineteenth-century scientific textbooks.[7 p519]

class of tools relied on identifying and formalizing a natural regularity and developing knowledge of how materials behave in accordance with it. In the case of the torsion balance, this was Coulomb's account of reaction torque; in the case of the spring scale, it was Hooke's law.

Precision Weighing as a Standard Practice

Scales and balances contributed to fundamental advances in physical and chemical science, but they also continued their function as standard pieces of laboratory apparatus. A precise, well-calibrated scale is necessary for preparatory work for a great many other tools, or for evaluating their results — for instance in preparing samples or in determining the results of an analysis. As in the past, ensuring the reliability of such instruments requires vigilantly maintained standards and carefully controlled systems of calibration and certification. Readers interested in standards more generally should consult Sharon Ku's contribution to this volume, but it will be worthwhile here to discuss weight standards in particular.

Up until the late-nineteenth century, most countries (and some smaller political units) maintained their own domestic weight standards. The British Imperial pound, for instance, was not quite one French livre. Local variation among standard weights created difficulties both for trade and for international scientific exchange, leading in the late-nineteenth century to the establishment of the SI (Système Internationale) system of units. On May 20, 1875, seventeen countries signed the Metre Convention, which established the meter as a standard international unit of length and the kilogram as an international unit of mass. Both would be maintained by artifact standards — physical objects defined has having a characteristic weight or length.

The kilogram was, until very recently, defined as the weight of a platinum–iridium cylinder — the International Prototype Kilogram (IPK; Figure 1.4.4) — kept in carefully controlled conditions at the International Bureau of Weights and Measures in Paris. Every SI scale in the world could trace its calibration back to this object.

But changes in how the metrological community defines and maintains standards have now caught up with even this most long-lived of artifact standards. The world's metrological community voted in November 2018 to start the process of redefining the kilogram in terms of a fundamental unit, Planck's constant — a change that went into effect on May 20, 2019.[8] What has changed? For most researchers in materials laboratories, very little. Their scales still work as before, and they talk about weight in the same terms.

The transition is deeper, however. Just like how the emergence of torsion balances and spring scales represented a new way of thinking about measuring weight through a deep understanding of physical law, redefining the kilogram in terms of Planck's constant was a way

Fig. 1.4.4. The International Prototype Kilogram, which until 2018 still defined the standard of weight in the SI system of units. *Credit*: Courtesy of the National Institute of Standards and Technology, via Wikimedia Commons.

of driving the girders of the mass standard into the firm bedrock of a fundamental physical quantity. For all the precision and control behind the manufacture and maintenance of the IPK, its real weight does drift ever so slightly over time, which means that the kilogram has been, on some level, a moving target since it was first manufactured in 1889.

It is worth mentioning in closing, however, that the current most precise measurement of Planck's constant relies on a Kibble balance. Like the torsion balance, the Kibble balance takes a subtle, delicate, but known force, in this case the electromagnetic force exerted by a current-carrying wire, and uses it to determine an unknown, counter-acting force. The cutting edge of precision weighing remains at the frontier of understanding material properties, and so the redefinition of the kilogram away from a material object will not make the properties of materials any less important for the practice of weighing.

References

1. Beretta M. Between the workshop and the laboratory: Lavoisier's network of instrument makers. Osiris. 2014;29:197–214.
2. Chang H. We have never been Whiggish (about phlogiston). Centaurus. 2009;51(4):239–64.
3. Heering P. Regular twists: replicating Coulomb's wire-torsion experiments. Physics in Perspective. 2006;8(1):52–63.
4. Karenga M. Maat, the moral ideal in ancient Egypt: a study in classical African ethics. New York: Routledge; 2004.
5. Martínez AA. Replication of Coulomb's torsion experiment. Archive for History of Exact Sciences. 2006;60(6):517–63.
6. Petruso KM. Early weights and weighing in Egypt and the Indus Valley. M Bulletin. 1981;79:44–51.
7. Privat-Deschanel A. Elementary treatise on natural philosophy. JD Everett, trans., London: Blackie; 1872.
8. Quinn T. From artefacts to atoms: the BIPM and the search for ultimate measurement standards. Oxford: Oxford University Press; 2011.
9. Stewart J. The reality of phlogiston in Great Britain. Hyle. 2012;18(3): 175–94.
10. Taylor JH, editor. Journey through the afterlife: the ancient Egyptian book of the dead. Cambridge, MA: Harvard University Press; 2010.

CHAPTER 1.5

LIGHT MICROSCOPY

HUBERTUS NEDERBRAGT

Collecting and Analyzing Objects

When considering the history of light microscopy, it is easy to forget that it is more than the history of microscopes and lenses. What we see with a microscope cannot be understood when we neglect the history of the manipulation of the material that we want to make visible.

In the eighteenth century, when natural history was the activity for which microscopy was used, collecting and classifying were the main purposes. Objects were magnified more or less as they had been found in order to reveal the wonders of nature. Although, in the microscopy handbooks of the first half of the nineteenth century, manipulation of objects received a prominent place, it was limited to cutting hard objects such as teeth, and covering objects with glass, cementing, and mounting. Objects were almost all of biological origin; crystals and minerals were considered interesting only because of their colors.[6]

Around 1850, the focus shifted to the use of the microscope for analytical purposes. The discussion of collecting/classification and analysis in the history of science is the theme of John Pickstone's *Ways of Knowing*.[5] Henry Clifton Sorby (1826–1908) published an investigation of the microscopic structure of crystals, both after *manufacturing* them by crystallization of salt solutions and after polishing very thin sections of rocks. His purpose was to show that geologists might benefit from the microscopical investigations of very small samples of rocks and minerals.[11] Sorby later moved to the study of the structure of metals by microscopical examination, using reflected light instead of transmitted light; for this method he was considered a pioneer.[8] The principles of the manipulation and study of crystals and metals have changed little since then: cutting, mounting, and polishing still are the main manipulations of objects under study. Further developments in these disciplines concern improvements of microscopes, lenses, and illumination. This kind of microscopy is still conducted with the aims of quality control of metals and identification of minerals for petrological purposes.

A comparable shift to analysis took place in the microscopical study of biological materials, of which the 1871 handbook of Charles Philippe Robin is a good example, although his emphasis was on collection.[7] He discussed several substances to be used for the conservation of objects, described a number of knives and microtomes for cutting thin slices of tissue, and gave a long list of chemicals and dyes, both of synthetic and plant origin, for specific staining of tissue elements, such as nerves with silver. His analytical interest is shown in chapters on forensic research, inspections of meat, and detection of falsifications in materials such as cacao and coffee.

The extent of manipulation to which objects may be exposed before the requested analysis takes place is well illustrated by following the protocol developed for the study of animal tissues. This protocol, using aniline dyes for better visualization of the details of tissues, gradually became a standard in the first decade of the twentieth century.

The Standard Procedure

First, a sample from animal tissues must be protected against biological degradation. This is obtained by fixation of the sample in formaldehyde.[3] In addition, the sample must be brought into a state that enables the cutting of very thin slices with a sharp knife (microtome); this is made possible by embedding and impregnating the tissue samples in paraffin.[9] Since paraffin dissolves in organic solvents only and formalin is an aqueous solution, the tissue slide has to be

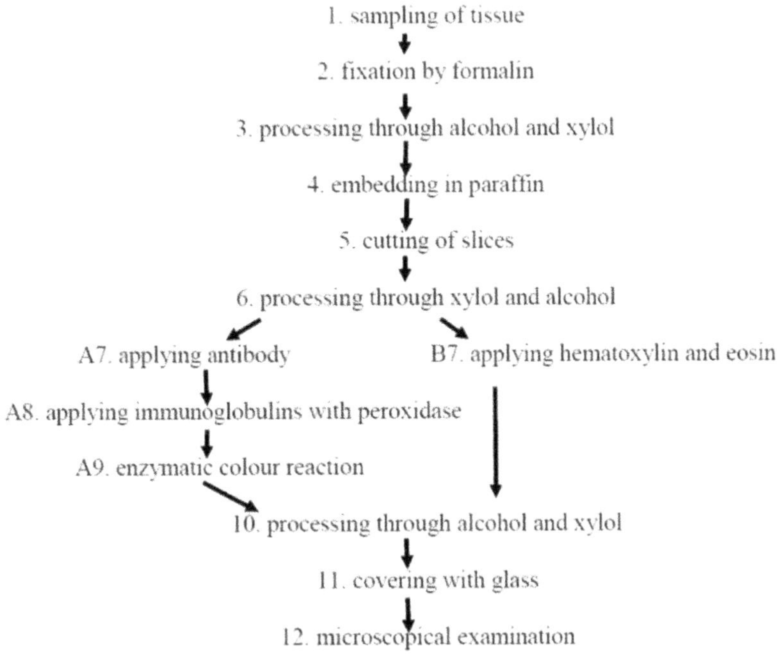

Fig. 1.5.1. Scheme of the main steps of the procedure to make a visual image in the microscope of a structure of a tissue.

dehydrated, using alcohol and xylol, and subsequently rehydrated by removing the solvents with water. Finally, since animal tissues do not have a color of their own, except for blood and pigments, the components of the tissues should be made visible by chemical dyes, for which as a standard eosin for proteins (pink) and hematoxylin (blue) for DNA of the nucleus are applied. This is called HE (hematoxylin-eosin) staining. The right side of the scheme of Figure 1.5.1 gives a summary of the procedure.

The standard procedure for histology and histopathology is the starting point for many methods that aim to elucidate the composition of animal tissues in more detail. An important application of HE staining is by pathologists to evaluate and diagnose the nature of diseases on the tissue level. Like in metallography and crystallography, the microscopical object evolved from being an aim in itself into material used for diagnostic purposes in the broadest sense.

Next, I will present an example of samples of animal tissues manipulated by (bio)chemical methods for more detailed microscopical evaluation; the example is taken from a study of mammary tumors (breast cancer) of dogs, but it may be considered a model for all biomedical research. The study described here was undertaken for a better understanding of the biological behavior of tumors by focusing on the composition of the extracellular surroundings of the tumor tissue, for which indications had been derived from biochemical investigations and cell culture studies. The localization of the proteins involved was expected to be revealed by chemical and (immuno)histochemical stainings of the tissue samples.

Microscopical Demonstration of the Presence and Composition of Glycosaminoglycans

In the second half of the twentieth century, biomedical scientists became interested in the biochemical composition of cellular and extracellular materials in tissues, and with that interest grew the need to be able to demonstrate these *in situ*. Since biochemistry isolated and revealed the characteristics of tissue proteins in the test tube, it was possible to synthesize special stains and dyes that could show the presence of these components *in situ*; that is, they could be localized very specifically in the tissue. I concentrate here on glycosaminoglycans (GAGs), formerly called mucopolysaccharides, a class of sugar-like chains, bound to proteins, to which sulfate molecules are attached. These sulfates give the molecules interesting functional properties, based on their negative charge. Certain dyes, on binding to these GAGs, change color, a process called metachromasia. Several dyes exhibit this property, but the more recent dye of choice is Alcian blue. The intensity of the binding of Alcian blue is dependent on the number of sulfate groups per GAG molecule. This is the basis for the staining method based on the Critical Electrolyte Concentration.[10] This method is explained by magnesium ions competing with Alcian blue for binding sites on GAGs; as the concentration of magnesium ions increases, more binding sites are blocked from access to Alcian blue. This makes it possible to estimate

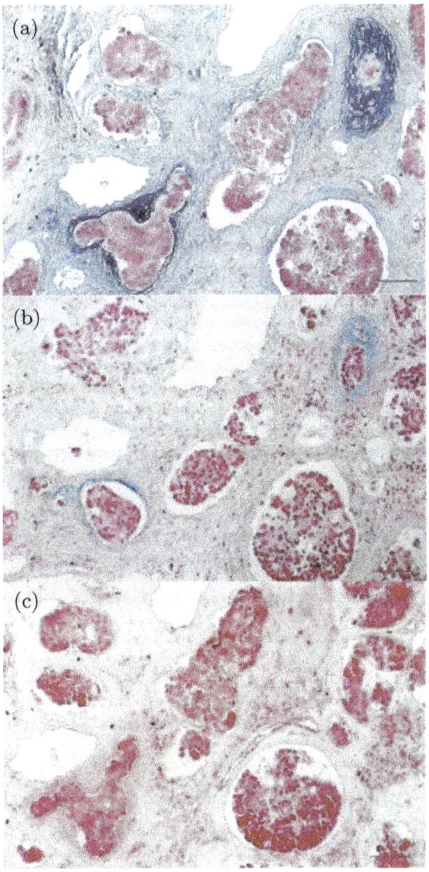

Fig. 1.5.2. Sulphated GAG (blue) accumulation around clusters of tumour cells, in serial slides of a breast carcinoma of a dog, stained with Alcian blue. Counter stain with neutral red. Magnification 200x. 2a (top): Alcian blue with 0.75 M magnesium chloride. 2b (middle): Alcian blue with 1 M magnesium chloride. 2c (bottom): Alcian blue with 1 M magnesium chloride after pretreatment with chondroitinase.[4] Reprinted with permission.

with the microscope the nature of the GAG molecules in a certain area of a tissue; GAGs with a low number of sulfate groups, such as dermatan sulfate, remain visible at low magnesium concentrations (Figure 1.5.2a), but become invisible at a high magnesium concentration that does not block the Alcian blue binding to GAGs, such as chondroitin sulfate which is highly sulfated (Figure 1.5.2b). A so-called counterstain — neutral red was used here — is a dye that gives a background color and an indication of the structure of the tissue as a whole. All these dyes are applied according to the same standard procedure as described above, in step B7 in the scheme of Figure 1.5.1.

To make a more decisive identification possible, an additional step may be introduced into step B7, the pre-treatment of the slide with an enzyme that specifically degrades a certain GAG. This is shown in Figure 1.5.2c, where the slide was incubated with the chondroitin sulfate–degrading enzyme chondroitinase, after which no Alcian blue staining was possible, indicating that the GAG staining intensely blue in Figure 1.5.2b must have been chondroitin sulfate in a highly sulfated form.

Microscopical Demonstration of the Presence of Specific Proteins

The use of metachromatic properties of tissues and cells has contributed to the identification of certain structures in tissues. But molecular and biochemical research of the cellular biology of developmental and tumorous processes in the last forty years has revealed the involvement of many specific proteins, including the GAG-bearing proteoglycans, in all kinds of biological processes. With the increase of the knowledge of the role of these proteins it also became necessary to detect and locate them in tissues with the light microscope. As soon as antibodies against these proteins were made available by immunological techniques, this localization could be accomplished. This created a new subfield of histochemistry, immunohistochemistry (IHC).[1]

Neglecting all the complexities of the immunological mechanisms in the body, the principle of IHC itself is simple: an antibody specifically binds to an antigen and it makes the antigen visible when it has a label attached to it, be it a fluorescent dye, a radioactive probe, or an enzyme that enables the development of a color when reacting with a certain substrate. These labels or their products may then be visible in the light microscope. Although this principle was developed in the 1930s, based on the pioneering work of Paul Ehrlich (1854–1905) early in the century, it took some decades before it was applied in histochemistry. On the one hand, this had to do with the slow and gradual progress in the improvement of reliability and stability of immunolabeling. On the other hand, it had to wait for the advancement of antibody production for laboratory research purposes. A well-equipped immuno-biochemical laboratory is necessary for the (commercial) production of these antibodies.

Once the antibody has been supplied and tested in pilot tests to see whether it does what it is supposed to do and in which concentration, the manipulation of the tissue samples can be started. The protocol for this method follows the outline of the scheme in Figure 1.5.1, but goes through the left branch. Sometimes it may be useful to pretreat the unstained tissue slice with seemingly rough manipulations, such as cooking in a microwave oven or a pressure cooker, to increase the exposure of the protein to the antibody.

The protein under investigation is versican, a proteoglycan that plays a role in the processes of tumorigenesis.[2] We start with the tissue slices on glass when they come out of the alcohol and are placed into water, at the moment in which, as a standard procedure, the hematoxylin-eosin staining should be performed. Instead of the aniline dyes, we now bring a drop of antibody solution onto the glass such that it covers the tissue slice. Since the antibody is directed against versican it will bind to versican by a strong interaction at all places where versican is present. Where it is not bound it will be washed away in a subsequent washing step. Suppose we have used an antibody derived from the cells of a mouse. In the next step, then, the tissue will be covered with a solution of mouse immunoglobulins. These are broad-spectrum antibodies from mice that will react with all mouse proteins they encounter; those are the specific antibodies that have bound versican in our case. These immunoglobulins have been labeled (by a chemical reaction) with the enzyme peroxidase which is chemically strongly coupled to the immunoglobulin. This enzyme is able to catalyze the process in which the colorless chemical compound diaminobenzidine is changed into a brown product.

Thus when we cover the tissue slice with a solution of this compound after it had been covered with the immunoglobulins and the peroxidase bound to it, a brown color will be made visible in all places where versican was originally present. Now the tissue slice is processed through alcohol and xylol again and covered with a protecting cover glass that is glued onto it, similar to the processing in the hematoxylin-eosin staining. This latter staining is always applied to a slice of tissue adjacent to that used for the versican staining, to be able to identify more clearly the structures of the tissue that contain versican. The slide can now be studied with the microscope; an example of how versican is revealed in the tissue by this method is given in Figure 1.5.3.

A later development that took place with the rise of molecular biology of DNA and RNA is the method with which messenger RNA molecules may be localized in tissues that are active in translating DNA to the synthesis of specific proteins in specific cells. This is made possible by a technique that is called *in situ* hybridization. Tissue samples (preferably, but not necessarily, frozen) are manipulated in comparable ways as in IHC and incubated with probes with a

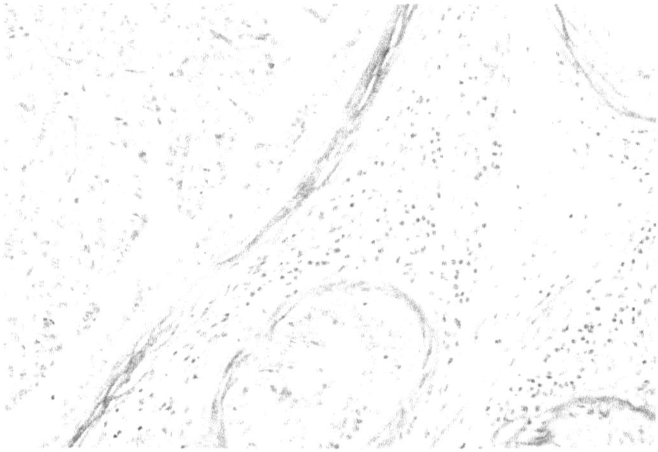

Fig. 1.5.3. Versican (brown; identifiable as the dark lines), made visible by the antibody 2B1, in a breast adenoma of a dog. Magnification 200×.[2] Reprinted with permission.

radioactive or fluorescent label, showing the actively and specifically synthesizing cells in a counterstained tissue.

Conclusions

In the history of the manipulation of objects for microscopical studies, several phases may be distinguished. The first phase was the period in which the emphasis was on collecting and classification. Objects were studied as they presented themselves. In the nineteenth century, scientists started to analyze objects with the microscope for purposes like quality control and diagnosis and the elucidation of details; for this, these objects had to be sectioned, cut, polished, and stained. With plant and animal material, methods for further manipulation developed, which was stimulated by the analysis of tissue compounds by physiology and (bio)chemistry and asked for means to detect those compounds *in situ*.

After all these manipulations, the structure of the sample, as seen through the microscope, does not look any more like it was when taken fresh from the plant or animal. It has become an artifact; yet we believe it to be an image of real life.

References

1. Childs GV. History of immunohistochemistry. In McManus LM, Mitchell RN, editors, Pathobiology of human disease. San Diego: Elsevier; 2014, 3775–96.
2. Erdélyi I, Nieskens DHM, van Dijk JE, Vass L, Nederbragt H. Immunohistochemical evaluation of versican, in relation to chondroitin sulphate, in canine mammary tumours. Histology and Histopathology. 2002;18(4):1067–80.
3. Fox CH, Johnson FB, Whiting J, Roller PP. Formaldehyde fixation. The Journal of Histochemistry and Cytochemistry. 1985;33(8):845–53.
4. Hinrichs U, Rutteman GR, Nederbragt H. Stromal accumulation of chondroitin sulphate in mammary tumours of dogs. British Journal of Cancer. 1999;80(9):1359–65.
5. Pickstone JV. Ways of knowing. A new history of science, technology and medicine. Manchester: Manchester University Press; 2000.
6. Quekett J. A practical treatise on the use of the microscope, including the different methods of preparing and examining animal, vegetable, and mineral structures, 2nd ed. London: Baillière; 1852.
7. Robin C. Traité du Microscope. Paris: Baillière; 1871, 950–60.

8. Rochow TG. Some incidental history of metallography and ASTM committee E4. In VanderVoort GF, Warmuth FJ, Purdy SM, Szirmae A, editors, Metallography: past, present, and future (75th anniversary volume). Philadelphia: American Society for Testing and Materials; 1993, 83–87.
9. Sanderson C, Emmanuel J, Emmanuel J, Campbell P. A historical review of paraffin and its development as an embedding medium. The Journal of Histotechnology. 1988;11(1):61–63.
10. Scott JE, Dorling J. Differential staining of acid glycosaminoglycans (mucopolysaccharides) by alcian blue in salt solutions. Histochemie. 1965; 5(3):221–33.
11. Sorby HC. On the microscopical structures of crystals, indicating the origin of crystals and rocks. Quarterly Journal of the Geological Society. 1858;4: 453–500.

CHAPTER 1.6

MANUALS, HANDBOOKS, AND RECIPES

JOANNA BEHRMAN

The history of science is intimately connected with the history of print. What many scholars call "print culture" has developed alongside science, supporting and shaping its development from before the early modern period until today. Print culture can be broadly described as the set of printed materials and practices associated with recording, dissemination, viewing, and so on of those materials.[5] This culture has structured the transmission of scientific knowledge from one person to another and among generations of practitioners. Scientific articles, for example, have been a crucial method of communication of professional scientific activity since the sixteenth century.[19] Research notebooks have also been important recording and interpretation tools, often revealing to historians the messiness of experimentation that published scientific articles erased.[23] Even taking notes is an act of print culture that shapes the practice of science.[16] The subjects of this article — manuals, handbooks, and recipes — are important aspects of scientific print culture from day-to-day scientific practice. Over the long term, they play a crucial role in bounding and regulating disciplines.

Manuals, handbooks, and recipes are all forms of print culture whose purpose lies largely in the dissemination of standardized information or in the replication of materials or processes. Scientists who need to know the mass density of aluminum, or how to apply a ceramic coating, consult these materials. These print materials are also social and disciplinary tools, and that is how I address them here. As the manufacture and manipulation of materials became a common part of laboratory practice through the twentieth century, manuals, handbooks, and recipes became essential means to regulate production, disseminate procedures, and enforce standards.

Traditionally, historians have regarded handbooks and manuals as maintainers of disciplines. Handbooks are usually considered distinct from textbooks in that classroom education is not their primary purpose, although the boundaries among genres are often blurred.[24; 42] For example, one philosophical explanation for the role of handbooks derives from the work of Ludwik Fleck, with similar arguments (partially drawn from Fleck) popularized by Thomas Kuhn. Fleck conceived of science as borne by "thought collectives," that is, groups of persons with similar philosophies and methodologies. A thought collective produces and consumes articles of print culture as a way to maintain and renew itself. Journal articles, for example, are used to push the boundaries of the discipline, and handbooks are where the knowledge of the discipline is consolidated.[17; 26; 41]

In this formulation, a handbook acts as a standardizing force in a scientific discipline and as a mark of its progress. This view of handbooks is fairly common among scientists and engineers as well. A former editor-in-chief of the *CRC Handbook of Chemistry and Physics* wrote that the growth of the handbook has "paralleled the growth of modern science," and certainly the handbook has visibly grown from its days as a pocket size manual in 1913 to a whopping 2,500+ pages in the recent ninety-eighth edition.[28; 37] The eleventh edition of the *Materials Handbook* proudly proclaimed its growth as measured by the number of materials described — it had surpassed 14,000.[11]

Handbooks, however, are more than mere collections of data. The interpretation and importance of information contained in a handbook needs to be understood in light of the larger body of scientific knowledge and practice. For instance, early editions of the *CRC Handbook* included a section on poison antidotes because many scientists pipetted by mouth.[14 p8] Moreover, scientific knowledge and practice have never existed in a vacuum separate from the broader society. Cost is a common concern for many materials scientists and engineers — it is possible to discern this concern in the *Materials Handbook* or the *Handbook of Engineering Materials* through their inclusion of information on manufacturing time, production cost, and durability of supplier-specific materials.[10; 11; 30] The study and use of materials is also dependent on their wider application, as the 1955 *Engineering Materials Handbook* illustrated with its contrast of materials suitable for high temperature use across long-term (industrial) or short-term (rocket) time spans.[29]

The use of, or contribution to, a manual could also be considered a mark of professional practice. Even ownership of a manual can confer a sense of community identity.[31] The status of a handbook depends on its ability to establish authority and trust. Simple ways of establishing authority include publication by a recognized professional group or respected individual, formatting in a manner which imitates another type of trusted print material, or testimonial by a trusted organization or individual.

However, contrary to Fleck's philosophical examination of scientific print culture, the handbook is by no means always a tool for increasing disciplinary solidarity. To illustrate this, the remainder of the chapter offers a historical example documenting the tenuousness of a manual's authority and role as an instrument of professionalization.

Rejecting Handbooks: A Historical Example

In 1901, Clifford Richardson, a former Inspector of Asphalts and Cements in Washington, DC, and then director of the public New York Testing Laboratory, sent identical samples of Portland cement to different chemists and found that the chemists widely disagreed on the composition of the material. Subsequent investigations by the New York Section of the Society for Chemical Industry and allied groups of the American Chemical Society determined that there were systemic problems with chemists' ability to perform analyses of various materials consistently and accurately. The poor results appeared to be due to the chemists' using unsuited

Fig. 1.6.1. Ceramics testing laboratory at the US Bureau of Standards. *Credit:* Courtesy of the Library of Congress LC-DIG-npcc-29234 DLC.

methodologies pulled from manuals, often conducting a particular procedure because it was the quickest, rather than the most appropriate or exact.[2–4; 22; 35]

At this time, national organizations and government agencies, such as the Chemistry Division at the Bureau of Standards (Figure 1.6.1), were producing manuals of standard chemical procedures as part of broader efforts at standardization.[15 p81–82] Industrial chemists blamed much of the widespread professional problems on these manuals, which chemists occasionally termed government "cook books."[12]

Language related to cooking was common in industrial chemistry of the early twentieth century. "Cooking," for example, was used to mean heating, and "cook" as a noun to refer colloquially to a synthesis or a procedure in general.[18; 20] But whereas use of the colloquial language of "cooking" or "a cook" did not carry negative connotations, reference to "cook books" usually did. The term came to be used to police boundaries between production and analytical chemists and among analytical chemists.

In the early twentieth century, disciplinary boundaries were being created within the larger field of industrial chemistry. Some industrial chemists coined the term "chemical engineer" in an effort to establish themselves as a separate and expert class as, or on the level with, management.[34] These chemists, many of whom worked on problems of scaling-up production in industry, specifically asserted a higher professional status than chemical technicians, who conducted much of the routine laboratory work and comprised the vast majority of all industrial chemists in the United States. Businesses tended to hold technicians in low regard, viewing them as interchangeable, semi-skilled laborers. Production chemists did not wish to be confused with poorly paid technicians, who were often stereotyped as (or were actually) female. Women comprised a significant portion of the chemical technician workforce, contributing to its perception as a feminized field. Lacking the licensing systems that exist for the legal and medical professions, new engineering groups in the United States developed other methods of enhancing and gatekeeping their profession, which simultaneously policed class, gender, and racial boundaries.[32 p51–90; 39]

The stereotyping of manuals helped to solidify professional prestige among individual or groups of industrial chemists through the exclusion of technicians, in part along gendered lines. Labeling manuals "cook books" carried an inescapable feminine connotation, evoking the idea of women's work in the domestic sphere. Pejorative usage of "cook book" reflected contemporaneous work specifically to channel women in chemistry away from managerial or engineering positions and into routine work as laboratory technicians. They were welcomed in these positions for the low wages they would draw, as well as for traits such as their "natural concentration on matters of minor import," as one chemist described it.[25] Although employers frequently typecast women into these positions, female scientists also took advantage of their skill, patience, and exhaustiveness to work in fields like botany or astronomy where they could classify specimens or photographic plates.[36] Through career advice books, and discriminatory hiring and educational practices, women were pushed into positions as laboratory technicians in which they would then be stigmatized as cook book users.[32 p170–173; 33; 13 p60–62]

Cook books, like those used in kitchens, carried with them the connotation of unknowledgeable work. The procedures in cook books were based on the empiricism of past successes,.without assuming a nuanced understanding of the underlying processes.[25; 27] Industrial chemists asserted that an ignorant technician would follow the directions of a handbook but would not understand why the procedure worked or how to deviate from it if necessary. If the directions were followed *exactly*, the procedure might be successful, but if there were contaminations or deviations from the directions, the technician would be at a loss. A supervisor of technicians needed a "know-how" that extended beyond the pages of the manual.[39] Naturally, male analytical chemists, supervisors of chemical laboratories, and the still-emerging chemical engineers volunteered their services as a way to assert their authority.

Although the term "cook book" began as a criticism of professional work, it came to be used almost entirely in discussions of proper scientific education. Various industrial chemists had argued that a reliance on manuals was indicative of a temperament more suited to subordinate occupations — and this disposition was being developed through terrible education. Students taught using prescriptive manuals akin to domestic cook books would become unable to think critically on their own and would have to be led by the hand through a task.[21] In society meetings, articles, and book reviews, industrial chemists asserted that cook book laboratories prepared students to function as routine workers, rather than as self-directed agents. They argued that students would not develop the self-reliance or even self-confidence that distinguished higher-ranked professionals from technicians.[22 p1632–1652; 8; 12]

"Cook book" as a rhetorical term spread to other fields during World War I (Figure 1.6.2), and by the 1920s "cookbook" had become a common adjective to describe laboratories at both the secondary school and college level in many scientific fields, in surveys

Fig. 1.6.2. US Army soldiers posed in cement laboratory, c. 1918. Laboratory manuals are visible under the table. *Credit*: Courtesy of the National Archives, photo no. 165-WW-125A-46.

of teaching, and in discussions of science education in general. This term remains to this day in critiques of "cookbook laboratories," meaning educational laboratories in which students follows a series of set procedures to achieve a known result.[1; 9]

Conclusion

Handbooks and manuals are where the knowledge and practices of a discipline accumulate. In their use, professional standards can be disseminated and enforced. Even in their rejection, as in the previous example, handbooks acted to construct disciplinary boundaries. The discipline of materials science has historically drawn from a number of fields, thus it is no surprise that the field has also produced a wide range of handbooks such as the Landolt-Börnstein, the CRC, and the ASM.[7] Although these publications have made a shift in recent years to a partly or largely online platform, their influence continues.[6; 38; 40] Research remains to be done to appreciate the full extent of how these diverse handbooks have affected work in materials science. Print culture may evolve and change with the move to different media, but it still shapes scientific practice.

References

1. Alozie NM, Grueber DJ, Dereski MO. Promoting 21st-century skills in the science classroom by adapting cookbook lab activities. The American Biology Teacher. 2012;74(7):485–89.
2. American Chemical Society New York Section. Meeting notes of 10 May 1901. In: ACS New York Section Meeting Notes 1892–1942. Box 1, Ledger 2 1899–1906. Chemical Heritage Foundation, Philadelphia. 84–85.
3. American Chemical Society New York Section. Meeting notes of 23 May 1902. In: ACS New York Section Meeting Notes 1892–1942. Box 1, Ledger 2 1899–1906. Chemical Heritage Foundation, Philadelphia. 115.
4. American Chemical Society New York Section. Meeting notes of 11 Nov 1904. In: ACS New York Section Meeting Notes 1892–1942. Box 1, Ledger 2 1899–1906. Chemical Heritage Foundation, Philadelphia. 271–72.
5. Apple RD, Downey GJ, Vaughn SL, editors. Science in print: essays on the history of science and the culture of print. Madison: University of Wisconsin Press; 2012.
6. ASM International. ASM Handbooks Online [Internet]. Materials Park, OH: ASM International; 2015 [updated 2015; cited 2017 Oct 24]. Available from: https://products.asminternational.org/hbk/index.jsp.
7. Bensaude-Vincent B. The construction of a discipline: materials science in the United States. Historical Studies in the Physical and Biological Sciences. 2001;31(2):223–48.
8. Bogert MT. Review of A course of practical organic chemistry by T. Slater Price. Journal of the American Chemical Society. 1908;30(6):1050.
9. Boyd-Kimball D, Miller KR. From cookbook to research: redesigning an advanced biochemistry laboratory. Journal of Chemical Education. 2018;95(1):62–67.
10. Brady GS. Materials handbook, 9th ed. New York: McGraw-Hill; 1963.
11. Brady GS, Clauser HR. Materials handbook, 11th ed. New York: McGraw-Hill; 1977.
12. Brogdon JS. The analyst versus the chemist. Journal of Industrial and Engineering Chemistry. 1912;4(9):685.

13. Bureau of Vocational Information. Women in chemistry: a study of professional opportunities. New York: Bureau of Vocational Information; 1922.
14. The Chemical Rubber Company. Handbook of chemistry and physics. Baltimore: Waverly Press; 1913.
15. Cochrane RC. Measures for progress: a history of the National Bureau of Standards. Washington DC: US Department of Commerce; 1966.
16. Daston L. Taking note(s). Isis. 2004;95(3):443–48.
17. Fleck, L. Genesis and development of a scientific fact. Translated by F. Bradley & T. J. Trenn. Chicago: University of Chicago Press; 1979.
18. Griffin ML. The comparative value of certain reagents for removing lime and magnesia from natural waters for industrial uses. Journal of the American Chemical Society. 1899;28(6):665–78.
19. Gross AG, Harmon JE, Reidy M. Communicating science: the scientific article from the 17th century to the present. New York: Oxford University Press; 2002.
20. Hamor, WA. Progress of the chemistry of papermaking. Journal of Industrial and Engineering Chemistry. 1912;4(4):307.
21. Hillebrand WF. Some thoughts on the present conditions of analytic chemistry. Journal of the American Chemical Society. 1905;27(3):304.
22. Hillebrand WF, Dudley CB, Stokes HN, Richardson C. Contribution from the committee on uniformity in technical analysis. Journal of the American Chemical Society. 1904;26(12):1644–53.
23. Holmes FL, Renn J, Rheinberger H, editors. Reworking the bench: research notebooks in the history of science. Dordrecht: Kluwer Academic Publishers; 2003.
24. Kaiser D. A tale of two textbooks: experiments in genre. Isis. 2012;103(1): 126–38.
25. Kugelmass IN. The criteria in the declaration of chemical independence in the United States. Science New Series. 1918;48(1251):610.
26. Kuhn TS. The structure of scientific revolutions, 2nd ed. Chicago: University of Chicago Press; 1970.
27. Lewis GN. Review of Electro-analysis by Edgar F. Smith. Science. 1908;27(696):689.
28. Lide DR. Preface to CRC Handbook of Chemistry and Physics facsimile first edition. In: The Chemical Rubber Company. Handbook of Chemistry and Physics. Baltimore: Waverly Press; 1913.
29. Mantell CL, editor. Engineering materials handbook. New York: McGraw-Hill; 1958.
30. Miner D, Seastone J, editors. Handbook of engineering materials. New York: John Wiley & Sons; 1955.
31. Mylander J. Early modern "How-To" books: impractical manuals and the construction of Englishness in the Atlantic world. Journal for Early Modern Cultural Studies. 2009;9(1):123–46.

32. Oldenziel R. Making technology masculine: men, women, and modern machines in America 1870–1945. Amsterdam: Amsterdam University Press; 1999.
33. Oldenziel R. Multiple-entry visas: gender and engineering in the US, 1870–1945. In: Canel A, Oldenziel R, Zachmann K, editors. Crossing boundaries, building bridges: comparing the history of women engineers 1870s–1990s. Amsterdam: Harwood Academic Publishers; 2000. 11–36.
34. Reynolds TS. Defining professional boundaries: chemical engineering in the early 20th century. Technology and Culture. 1986;27(4):694–716.
35. Richardson C. Uniformity in technical analysis. Journal of the Society of Chemical Industry. 1901;20(4):334–39.
36. Rossiter M. Women scientists in America: struggles and strategies to 1940. Baltimore: Johns Hopkins University Press; 1982.
37. Rumble JR, editor. CRC handbook of chemistry and physics, 98th ed. Boca Raton: CRC Press; 2017.
38. Rumble JR, editor. CRC handbook of chemistry and physics, 98th ed. [Internet]. Boca Raton: CRC Press; 2017 [updated 2017; cited 2017 Oct 24]. Available from: hbcponline.com.
39. Slaton A. "As near as practicable:" precision, ambiguity, and the social features of industrial quality control. Technology and Culture. 2001;42(1):51–80.
40. Springer Nature. Springer Materials [Internet]. Berlin: Springer-Verlag GmbH; 2016 [updated 2016; cited 2017 Oct 24]. Available from: http://materials.springer.com/.
41. Stuckey M, Heering P, Mamlok-Naaman R, Hofstein A, Eilks I. The philosophical works of Ludwik Fleck and their potential meaning for teaching and learning science. Science & Education. 2015;24:281–98.
42. Vicedo M. Introduction: the secret lives of textbooks. Isis. 2012;103(1):83–87.

CHAPTER 1.7

DISTILLATION

PHILLIP C. WANKAT

The most important nineteenth- and twentieth-century developments in large-scale industrial separations occurred in distillation. Although most of my research has been in adsorption, I reached this conclusion after reviewing the historical development of separation techniques such as the separation of ethanol from water, the separation of petroleum fractions, and the production of oxygen from air.[38] Distillation became and will remain the industrial workhorse because: 1) distillation is often the most economical separation process; 2) distillation is trusted; 3) distillation is understood well enough that most systems can be built without doing pilot plant studies; 4) unlike most separation techniques, distillation can usually separate binary mixtures completely, which means a component can be recovered with both high purity and high recovery; 5) construction materials can be selected to be compatible with any chemical mixture being separated; and 6) distillation is flexible and can be extended to azeotropic (constant boiling) mixtures such as isopropanol (a common solvent) and water (as well as mixtures with very small differences in boiling points) by adding an appropriate solvent. The resulting separation techniques are known as azeotropic distillation if a new azeotrope is formed or extractive distillation if there is no azeotrope.[26; 27] Of course, membrane separation experts may not agree with my conclusion.

Distillation separates compounds based on differences in volatility. Typically, compounds that boil at a lower temperature are more volatile and their concentration in the vapor will be greater than their concentration in the liquid. For example, consider the following methanol-water data at a pressure of 1.0 atm.[39] Methanol boils at a significantly lower temperature (64.5°C) than water (100°C). A methanol vapor mole fraction of 0.418 is in equilibrium with a methanol liquid mole fraction of 0.10 at a temperature of 87.7°C. The operating cost is the energy required to vaporize liquid and to condense vapor. Yet, even for an easy-to-separate mixture like methanol–water, the amount of separation achieved at equilibrium is modest. Fortunately, distillation columns are easily constructed, allowing the separation to be repeated with only minimal additional energy. These columns are constructed with countercurrent flow (vapor goes up the column and liquid goes down). When the vapor condenses, the energy is reused to vaporize more of the liquid. The repeated vaporizations and condensations concentrate the more volatile component in the vapor (for instance, methanol) and the less volatile component (for instance, water) is concentrated in the liquid. Since the vapor goes up the column, the product collected at the top of the column (the distillate) will be concentrated in methanol and the product taken at the bottom of the column (the bottoms) will be concentrated in water. Both the distillate and the bottoms product can be quite pure, and one is able to obtain high purity and high recovery of both products from a binary feed mixture. Although this might sound trivial, very few separation methods are capable of simultaneously producing high purity and high recovery of a desired product. The major downside for distillation is the relatively high energy use.

This short history focuses on large-scale, continuous distillation (that is, involving continuous feeding and withdrawal of products). The patent drawing in Figure 1.7.1 is a good representation of a continuous distillation system. After removing the partition and the side stream, we have a schematic drawing of a standard continuous distillation system. Batch distillation, a time dependent process in which a fixed amount of feed is distilled, is significantly older and has a long history that is reviewed in detail elsewhere.[9; 13; 14; 20] The histories of

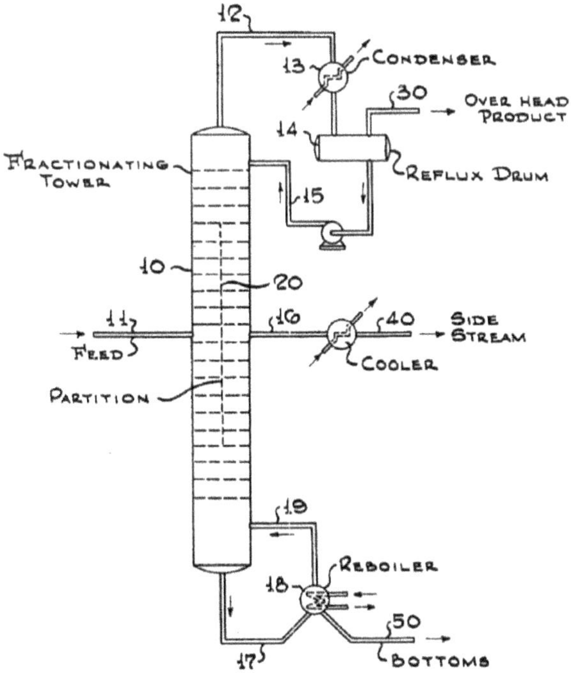

Fig. 1.7.1. Schematic of divided wall column from Wright patent.[41]

azeotropic and extractive distillation and distillation control are also reviewed elsewhere.[7; 27]

Characteristics of Distillation

Distillation, like other separation methods, often advances because of the desire to do a particular separation. Alcohol production, distillation of coal tar, air separation, petroleum distillation, and production of heavy water have all been driving forces for improving distillation.[9; 26]

A second characteristic of distillation is that change is often slow. Inventions usually take many years before they have significant impact. A good example is the slow uptake of tray columns or packed columns for large-scale distillation in, among others, the petroleum industry. Tray columns have vapor flowing upward between plates that contain liquid and have void spaces for vapor flow. The vapor bubbles through the liquid, forming froth with good contact between vapor and liquid. The liquid flows downwards in a separate channel (that is, a downcomer). An early precursor of tray columns was the continuously working vertical column that had bubble caps (typically bell-shaped devices developed to distribute the vapor while retaining the liquid on the tray), which were developed in 1813 by Jean-Baptiste Cellier-Blumenthal, "one of the most gifted designers of distillation columns."[20 p2] Although other types of trays have largely displaced bubble caps, they are still used over two hundred years later in applications that require relatively large amounts of liquid on the trays. In 1830 Aeneas Coffey developed a vertical sieve tray column for continuous distillation.[9; 10; 20] The development of columns was ignored in the petroleum industry, which, because of their large scale, are resistant to using new technology that may fail. Instead, through the 1920s, petroleum refineries used inefficient, directly heated, horizontal stills plus partial condensation to distil petroleum.[14; 26] Consultant Warren K. Lewis introduced vertical distillation columns and used the relatively new unit operations paradigm to convince the Standard Oil Company of New Jersey to install vertical columns.[14] The crude oil feed to the column was first preheated by condensing the low-boiling distillate product and further heated by heat exchange with the much hotter bottom product — methods that are still used.

From 1895, when he first liquefied air, to 1910, when the multi-effect column for air separation was developed, Carl von Linde, along with others, continually improved distillation processes for air separation.[20] Since oxygen, which is available in abundance in air, is needed for steel production, developing economies are interested in air separation. Despite large investments in research and development of alternative separation methods such as membrane gas permeation and pressure swing adsorption, cryogenic distillation, which operates at temperatures between the boiling point of nitrogen 77K and the boiling point of oxygen 90.6K, remains the method of choice for large-scale air separation units. Although quickly adopted for air separation, multi-effect columns have not been broadly employed outside of this niche. Another example of distillation technology used in a niche is the alembic still, which was invented by ibn Hayyan (aka Geber) in the late eighth or early ninth century, and is still used in some whisky distilleries (such as one visited by the author in Oban, Scotland). Applications of multi-effect columns, alembic stills, and bubble caps illustrate a third characteristic of distillation — technology that is not considered for most applications will often find other niches where it is used.

Air separation also illustrates well a fourth characteristic of distillation (and many other separation techniques): the sometimes critical role of trace components.[20] Argon, which is naturally present in air, boils between the boiling temperatures of nitrogen and oxygen. If nitrogen is withdrawn as the distillate and oxygen as the bottoms product, argon will concentrate in the middle of the column, and if the buildup is not controlled, argon will eventually contaminate the products. The buildup can be controlled by purging or by sending a side withdrawal to an argon purification column. When argon is desired, the concentration effect is useful since it results in a much more concentrated feed to the argon column. A severe problem occurs if the air contains trace amounts of light hydrocarbons such as ethylene or acetylene. These compounds will concentrate in the column bottom and, in the presence of oxygen, can form an explosive mixture. A constant purge at the proper column location will prevent the hydrocarbons from reaching the lower explosion limit of the mixture.

Current concerns about climate change have increased interest in energy reduction and revived attention to thermally coupled systems because they reduce both energy requirements and the amount of carbon dioxide released.[33] A current area of intense research is the use of divided-wall columns, which are practical configurations for thermally coupled systems. The divided wall column was patented by Richard O. Wright in 1949 as a way to decrease capital and operating expenses for the separation of three component mixtures (Fig. 1.7.1).[41] More than three decades later, the first industrial units were built.[18] Since then, some of the practical problems involving the control of vapor flow rates have been solved and growth in use has become explosive. As noted earlier, if the partition and the side stream are removed in Fig. 1.7.1, the result is a schematic of a standard distillation system. This illustrates a fifth characteristic of most distillation innovations: they add something new, but the basic distillation system is still recognizable.

Another twentieth-century development, reactive distillation, involves conducting a chemical reaction in a distillation column to remove reaction products. It was first developed by Arthur A. Backhaus, a prolific inventor who received a number of patents on the technique in the early 1920s.[4; 5] Although reactive distillations were studied for years, it was not until the work of Victor Agreda and Lee Partin, two young chemical engineers at Eastman Chemical only a few years out of graduate school, that their potential was realized. In 1980, Agreda and Partin started research to replace eleven unit operations with a single reactive distillation column for production of methyl acetate from the reaction of acetic acid with methanol. The first commercial plant was started in May 1983 with capital and energy costs reported to be one fifth of the costs of the standard design.[2; 11] Although many reactions are not candidates for reactive distillation, significant benefits are possible for reactions that can be run in a reactive distillation column. The principle demonstrated by this example is that combining functions in a single unit is difficult, but if it can be done successfully the rewards may be significant.

Trays and Packing

For over one hundred years, bubble cap trays, which are expensive, and sieve trays (sheets of metal with holes punched in them), which are inexpensive but have a narrow operating range, were the most popular trays. In the 1950s, valve trays were invented and were first commercialized by the Nutter company in 1951, with an improved design patented in 1955.[12; 25] To meet this new competition, Koch and Glitsch patented other designs of valve trays. Illustrations from the Koch patent are shown in Figure 1.7.2.[16] Valve trays cost more than sieve trays and have about the same efficiency and throughput, yet valve tray columns operate with good efficiency over a very wide range of vapor flow rates.[12] They started to be widely adopted in

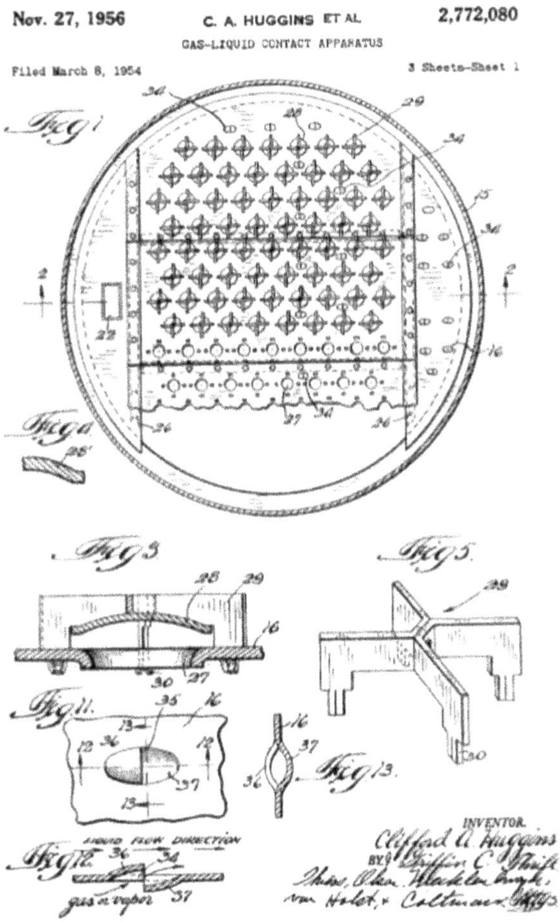

Fig. 1.7.2. Figures from Huggins and Thrift valve tray patent.[16]

the 1960s and are currently the most popular type of tray because of their flexibility.

Valve trays offer an interesting counterexample to the slow adoption rate of other distillation developments. They appear to have been relatively quickly adopted because: they fulfilled a known function better; the reasons they were better were easy to understand; they could be retrofitted into columns without changing the rest of the process; and they were invented by vendors who were adroit distillation equipment salespeople. However, sieve trays are the easiest trays to clean and remain the choice for highly fouling feeds such as those containing particulates. For example, sieve trays with large holes are used in beer stills (a distillation column with liquid feed introduced at the top of the column that usually uses direct injections of steam for heating) that receive the raw fermentation broth in biorefineries.

Instead of using plates or stages with downcomers, distillation columns can be packed with an inert material (the packing) that has a high surface area for contact of vapor and liquid. Random packing, which is dumped in the column and has no prearranged structure, was apparently first employed in 1820 and patented in 1847.[12] Because random packings can be made of a wide variety of materials, they are often the choice for corrosive environments. Fritz Raschig patented what became known as Raschig rings (cylinders with the radius equal to the cylinder height) in 1913. Since Raschig rings can be used in a prearranged structure, they can be thought of as a precursor of structured packings, which have a prearranged structure.[20] A major breakthrough in random packing occurred when thin-walled metal Pall rings were developed in 1958.[12] The Pall ring can be thought of as a Raschig ring with a slot that allows some liquid and vapor flow through the cylinder increasing mass transfer and reducing pressure drops.

Structured packings, commonly thought of as a modern development, were first employed in the 1930s. The UK Harwell Atomic Energy Research facility developed Spraypak for heavy-water separation, and the wire-mesh Stedman column and a perforated-metal sheet packing were developed by General Electric.[20] In the mid-1960s, the Sulzer company introduced a structured packing that worked very well, but was quite expensive.[12] Sulzer followed with a less expensive sheet metal packing, Mellapak, sold in the United States as Koch Flexipac.[12] In order for vapor to flow up the column the pressure in the column bottom must be greater than the pressure at the top. This pressure difference or pressure drop is lower in structured packings than in random packings and most tray designs. Vacuum distillation columns are commonly used for distilling thermally sensitive materials because operation at low pressure reduces the boiling temperatures in the column. Structured packings are often used for vacuum operation because the pressure, and hence the temperature, can be significantly lower in the bottom of the column, preventing thermal decomposition. On the other hand, structured packings often do not perform well at high pressures because as the density of the vapor approaches the density of the liquid maintaining countercurrent flow becomes difficult.

Books and Theories

Until the invention of the internet, printed materials were the main method for disseminating technical knowledge (see Joanna Behrman's contribution to this volume). In his *Handbook of Chemical Engineering*,[10] George Davis clearly developed the unit operations idea (not by this name) for distillation.[10; 13] Davis used spreadsheets with hand calculations for the mass and energy balances for distillation systems. He showed methods to reduce energy in distillation columns and discussed safety — topics that remain current. Arthur D. Little coined the term "unit operation" in 1915,[9] and unit operations were first treated comprehensively in William Walker, Warren Lewis, and William McAdams's textbook.[37] The schematics of continuous distillation columns in early textbooks,[31; 37] in the first edition of Perry's handbook,[28] and in patents (Figure 1.7.1) are similar to current schematics. Clark Shove Robinson's *The Elements of Fractional Distillation* was probably the first chemical engineering book to also become popular with a non-technical audience.[31] During the Prohibition Era in the United States, Robinson gave popular public lectures on distillation — many attendees were bootleggers.[40 p85]

Theoretical developments were also important in advancing distillation, but lagged behind practice, often by many years. Analysis of distillation requires thermodynamic (vapor-liquid equilibrium and enthalpy) data and correlations that were often not available in the 1800s. Ernest Sorel, who was interested in alcohol purification, pioneered theoretical analysis of continuous binary distillation in 1893 when he realized that a distillation column could be analyzed as a series of equilibrium stages.[34] Sorel's method is accurate if stages are close to equilibrium, but the trial-and-error calculation of every stage is laborious. Marcel Ponchon[29] and Paul Savarit[32] independently developed graphical procedures for Sorel's method. Warren K. Lewis[21] simplified Sorel's method by realizing that because in many cases vapor and liquid molar-flow rates are approximately constant, energy balances are often not required. In 1925 Warren McCabe and Ernest Thiele, Lewis's students in a graduate level course, converted Lewis's analysis to a graphical method.[23] Because the McCabe–Thiele plots clarified the physical reasons why column distillation works, Eger V. Murphree changed his 1925 paper on distillation efficiency to use the new McCabe–Thiele graphical method.[24] The McCabe–Thiele method appeared in textbooks only two years later.[20]

In the 1930s, stage-by-stage methods were adapted to multicomponent distillation by initially assuming either the product distributions or the temperatures.[22; 36] The latter choice agrees with current practice. However, neither method was totally satisfactory because obtaining convergence can be extraordinarily difficult.[15] Twenty-five years later, Neal Amundson and Arlene Pontinen demonstrated that distillation equations can be conveniently put into matrix form and solved on a computer.[3] The matrix analysis led to a number of computer methods for solving multicomponent distillation.[15] Commercial computer simulation programs with massive databases are now readily available for both continuous and batch distillation.

Equilibrium-staged computer models will produce designs that work well in about 80% of cases.[1 p25] Many engineers thought that the cases that did not work were either due to difficult convergence (a problem that has been solved) or low-stage efficiency. Twenty-seven years later, Ross Taylor and Rajamani Krishna demonstrated that a Maxwell–Stefan mass-transfer analysis could successfully model multicomponent systems that could not be modeled with an equilibrium approach.[35 p40] Many commercial simulators now include mass-transfer models as an option.

Current Advances and Research Funding

An argument can be made that distillation is a mature technology, that all currently important industrial distillation equipment was invented by the mid-1950s, and that additional major advances are unlikely. There are undoubtedly inventions that will become important in the future that are currently languishing in obscurity. For example, Higee, a centrifugal distillation system developed by Colin Ramshaw in the 1980s, shown in Figure 1.7.3, currently occupies a niche where small size or weight is important. However, there are a number of installations in China and Higee may eventually be much more widely adopted.[20]

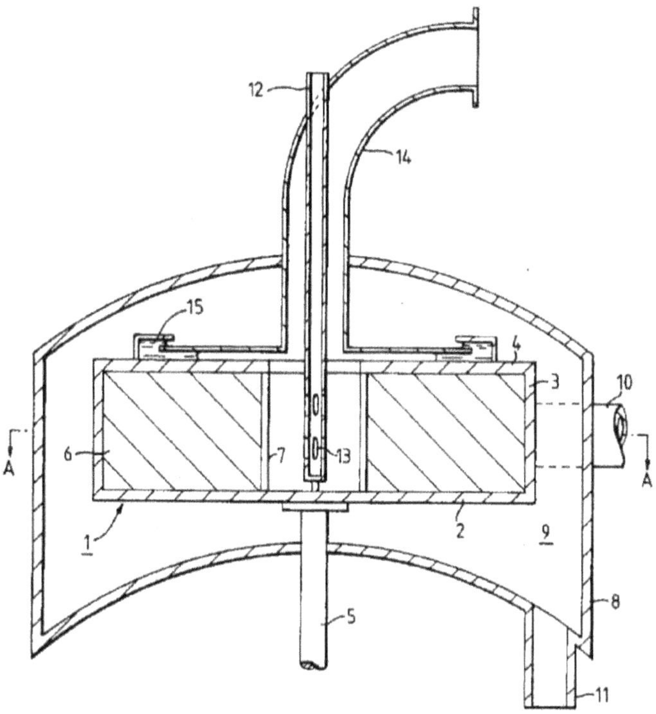

Fig. 1.7.3. Higee distillation system from Ramshaw and Mallinson patent.[30]

The author believes there will be new major advances; however, progress is slow because of the lack of funding for distillation research in the United States. Although distillation is technologically mature, it is by no means static; incremental improvements are still common, and more significant innovations do occasionally happen.[19] Since the market for distillation techniques is huge,[17] the impact potential of distillation research is relatively large, as shown by the historical episodes reviewed here.[38]

In the past, distillation has advanced when there was a great desire to do a particular separation, when techniques successful for other separations were adapted to a completely different separation, when someone questioned the basic tenets of distillation, and/or when a technology invented decades earlier became an industrial process. A distillation process that may be ready to meet these criteria is cyclic distillation, which essentially superimposes batch operation on a continuous counter-current process. Cyclic distillations do the same separations as standard continuous distillations with fewer stages or less packing and require less energy.[6; 8] Since difficulties with practically achieving and controlling cyclic distillations have reportedly been recently overcome and commercial systems are operating, cyclic distillation may become a major advance in the future.[6]

References

1. Adler S, Beaver E, Bryan, Rogers J, Robinson S, Russomanno C. Vision 2020: 1998 separations roadmap. New York: American Institute of Chemical Engineers, Center for Waste Reduction Technologies; 1998.
2. Agreda V, Partin L. Reactive distillation process for the production of methyl acetate. United States patent US4435595. 1984 Apr 26.
3. Amundson N, Pontinen A. Multicomponent distillation calculations on a large digital computer. Industrial and Engineering Chemistry. 1958;50(5): 730–36.
4. Backhaus AA. Continuous process for the manufacture of esters. United States patent US1400849. 1921 Dec 20.
5. Backhaus AA. Process of producing high-grade esters. United States patent US1454462A. 1923 May 8.
6. Bildea, CS. Kiss AA. Revive your columns with cyclic distillation. Chemical Engineering Progress. 2015;111(12):21–27.

7. Buckley P. History of distillation control. American Institute of Chemical Engineers Symposium Series. 1983;79(235):46–56.
8. Cannon M. Controlling cycling improves various processes. Industrial and Engineering Chemistry. 1961;53(8):629.
9. Davies J. Chemical engineering: how did it begin and develop? In Furter W, editor. History of chemical engineering. Advances in Chemistry, vol. 190. Washington, DC: American Chemical Society; 1980. 15–43.
10. Davis GE. A handbook of chemical engineering:illustrated with working examples, vol. 2. Manchester: Davis Bros.; 1902.
11. Doherty M, Malone M. Conceptual design of distillation systems. New York: McGraw-Hill; 2001.
12. Fair JR. Historical development of distillation equipment. American Institute of Chemical Engineers Symposium Series. 1983;79(235):1–14.
13. Freshwater D. George E. Davis, Norman Swindin, and the empirical tradition in chemical engineering. In Furter W, editor. History of chemical engineering. Advances in chemistry, vol. 190. Washington, DC: American Chemical Society; 1980. 97–111.
14. Gornowski E. The history of chemical engineering at Exxon. In Furter W, editor. History of chemical engineering. Advances in chemistry, vol. 190. Washington, DC: American Chemical Society; 1980. 303–11.
15. Holland C. History of the development of distillation computer models. American Institute of Chemical Engineers Symposium Series. 1983;79(235):15–38.
16. Huggins C, Thrift G. Gas-liquid contact apparatus. United States patent US2772080. 1954 Mar 8.
17. Humphrey J, Keller G. Separation process technology. New York: McGraw-Hill; 1997.
18. Kaibel G. Distillation columns with vertical partitions. Chemical Engineering & Technology. 1987;10(1):92–98.
19. Keller, G. Separations: new directions for an old field. American Institute of Chemical Engineers Monograph Series, no. 18, vol. 83. New York: American Institute of Chemical Engineers; 1987.
20. Kockmann N. History of distillation. In Gorak A, Sorensen E, editors. Distillation: fundamentals and principles. London: Elsevier; 2014. Chapter 1.
21. Lewis W. The efficiency and design of rectifying columns for binary mixtures. Industrial and engineering chemistry. 1922;14(6):492–96.
22. Lewis W, Matheson G. Studies in distillation: design of rectifying columns for natural and refinery gasoline. Industrial and Engineering Chemistry. 1932;24(5):494–98.
23. McCabe W, Thiele E. Graphical design of fractionating columns. Industrial and Engineering Chemistry. 1925;17(6):605–11.
24. Murphree E. Graphical rectifying column calculations. Industrial and Engineering Chemistry. 1925;17;960–64.
25. Nutter I. Valve mechanism for fluid and liquid contact apparatus. 1955. United States patent US2718900. 1955 Mar 3.

26. Othmer D. Distillation — some steps in its development. In Furter W, editor. A century of chemical engineering. New York: Plenum Press; 1982. 259–74.
27. Othmer D. Azeotropic and extractive distillation. American Institute of Chemical Engineers Symposium Series. 1983;79(235):90–117.
28. Perry J, editor. Chemical engineers' handbook. New York: McGraw-Hill; 1934.
29. Ponchon M. Etude graphique de la distillation fraction-née industrielle. La technique Moderne. 1922;13(1):20–53.
30. Ramshaw C, Mallinson R. Mass transfer apparatus and process. United States patent US4400275. 1983 Aug 23.
31. Robinson C. The elements of fractional distillation, 2nd ed. New York: McGraw-Hill; 1930.
32. Savarit P. Definition of distillation, simple discontinuous distillation, theory and operation of distillation columns and exhausting and concentrating columns for liquid and gaseous mixtures and graphical methods for their determination [in French]. Arts et Metiers. 1922;65:142,178,241,307.
33. Shah V, Agrawal R. Conceptual design of zeotropic distillation processes. In Gorak A, Sorensen E, editors. Distillation: fundamentals and principles. London: Elsevier; 2014. Chapter 7.
34. Sorel E. La Retification de l'Alcohol. Paris: Gauthier-Villars; 1893.
35. Taylor R, Krishna R. Multicomponent mass transfer. New York: Wiley; 1993.
36. Thiele E, Geddes R. Computation of distillation apparatus for hydrocarbon mixtures. Industrial and Engineering Chemistry. 1933;25(3):289–95.
37. Walker W, Lewis W, McAdams W. Principles of chemical engineering. New York: McGraw-Hill; 1923.
38. Wankat P. Separations: a short history and a cloudy crystal ball. Chemical Engineering Education. 2009;43(4):286–95.
39. Wankat P. Separation Process Engineering, 4th ed. New York: Prentice Hall, 2017.
40. Weber H. The improbable achievement: chemical engineering at M. I. T. In Furter W, editor. History of chemical engineering. Advances in chemistry, vol. 190. Washington, DC: American Chemical Society; 1980. 77–96.
41. Wright RO. Fractionation apparatus. United States patent US2471134. 1949 May 24.

CHAPTER 1.8

MIXING EQUIPMENT

PHILIPPE MARTIN

The fertilizer industry — especially the compound fertilizer industry — offers a revealing case for studying the history of mixing tools. This industry emerged in the nineteenth century and developed during the twentieth century. As it grew, it evolved to meet the challenge of providing large volumes of low-cost fertilizer for intensive agriculture. In response to demands from the agricultural industry, mixing equipment was mechanized, electrified, integrated, and specialized. The story of mixing and blending machines is therefore a story of technical development that takes general, multi-purpose tools and refines them for more specific applications. Many tools and techniques, including those involved in mixing, begin as generalized instruments and procedures shared among many activities. They gradually become more specialized in the course of their application to specific problems. The field of fertilizers exerted a powerful specializing effect on mixing apparatus, as it was implemented by specialists, from farmers themselves to fertilizer manufacturers and chemical engineers.

The following describes the main technical developments of fertilizer mixing equipment in the history of the fertilizer industry: manual mixing with non-specialized utensils, mechanical mixing, and chemical mixing in a specialized machine.

Non-Specialized Tools: Mixing "with a Man's Arm"

For much of the nineteenth century, fertilizer manufacturers and farmers used rudimentary manual techniques to prepare fertilizers. Practices differed according to whether the materials to be mixed were solid or liquid. Solid materials would first be weighed and passed through a sieve, then put in piles and returned to the sieve with a shovel one or more times before finally being sifted.[15 p333] Liquid materials would be deposited in a pit dug into the ground and would be agitated with a shovel, a rake, or a stick. The preparation of compound fertilizers, called "mixtures," consists of combining raw materials (organic or mineral). The procedure varies depending on whether the substances to be mixed are all solid, all liquid, or some combination of solids, liquids, and pastes. Superphosphate, for instance, is prepared on the floor of a plant by mixing solid phosphate ground with sulfuric acid with long wooden rakes.[11] The preparation of poudrette relied on a less complex procedure. In the 1850s, the company Baronnet & Cie began to produce fertilizers called "poudrette," which used feces from sewage as a precursor. Mixing dry feces with charred earth in a pit by means of a stick produced a fertilizer.

Even in manual mixing, mechanical operations were often performed before the preparation of the mixture itself. Any mixture with solid materials, such as minerals, calls for mechanical means to reduce them to powder prior to their use in a mixture. For crushing or grinding, the equipment used is neither specialized nor unique to the fertilizer industry. These were devices originally intended for the grinding of nuts, flour, or ores, and were powered by muscle power, water, horse carousels, or steam engines. From the 1860s, the pulverization of mineral materials began to be carried out with the English Carr disintegrator, a multi-purpose grinder (used for rock, ores, black bones, and other solids).[14]

These processes did not require specific know-how. They resulted from the appropriation of tools that were not specific to fertilizers and the reuse of "standard" techniques shared by a number of craft activities. The manipulation and manufacture of different materials often required different combinations of such shared techniques. The manufacture of compound fertilizer is part of a "technical group" consisting of grinding apparatus, screening apparatus and the mixing apparatus itself.[4 p15]

The actors in these fertilizer mixtures were manufacturers of compound fertilizers, but also farmers — often those running larger farms — who bought the raw materials and made their own mixes. These techniques lasted until the twentieth century for small plants or for the production of small quantities.[15 p333]

Electrification and Mechanization of Mixing

In the early twentieth century, electricity-powered mechanical mixing apparatus was increasingly used to avoid shoveling, which was expensive and inefficient.[15 p333] Mechanical mixers (Figure 1.8.1) were mainly discontinuous; that is, precisely weighed components would be introduced at predetermined intervals. When the mixture became homogeneous, the content of the mixer would be extracted.[10] For crushed solid products, conical or biconical rotary mixers were used. To mix a liquid or paste product with a solid, fixed trough mixers were generally employed. In these machines, a pallet device, carried by one or more shafts, ensures the mixing of the products and the unloading is effected by traps at the bottom.

In the interwar period, continuous compound fertilizer mixers became as efficient as those operating by a discontinuous process. The Raps mixer (Figure 1.8.2) was a combination of these two systems: it was an intermittent mixer operating on a batch basis and was therefore producing products as homogeneous as discontinuous mixers. It emptied automatically, by the simple maneuvering of a release fork and therefore offered the same advantages of speed and simplicity as a continuous mixer.[15 p334]

This mechanization took place as part of a larger trend toward rationalizing industrial processes through electric and mechanical means. In the 1920s, in France, the Compagnie de Saint-Gobain equipped its production units with mechanical shovels, which took

Fig. 1.8.1. Superphosphate vertical mixer. *Credit*: Album photo by Joseph Clément. Union française et Compagnies régionales unies de fabriques d'engrais et de produits chimiques. Paris: Jean Morax Imprimeur; 1941.

the fertilizer components directly into their respective compartments for mixing, and with rotating mixers operated by cranes.[1 p589–601] This increasingly expensive equipment required the greater investment that became possible in an increasingly concentrated industry. The use of this expensive and specialized mixing equipment was under the control of specialists, the fertilizer manufacturers, who gradually dispossessed farmers of autonomy in their fertilizer preparations.

Fig. 1.8.2. Raps mixer.[15 p335]

The Entry of Chemists: Chemically Combined Materials

At the same time that "mixed" compound fertilizers were being developed, so-called "complex fertilizers" emerged. In the United States, the term "mixed fertilizers" is generally synonymous with "compound fertilizer." In some European countries compound fertilizers are further subdivided into "mixed fertilizers" and "complex fertilizers," and I use that distinction here.[7 p248] The distinction turns on whether chemical reactions are integral to the mixing process. This issue became relevant in France in the 1920s when the possibility arose of producing the nitrogen component of fertilizers from synthetic ammonia.[18] In 1930, the French chemist Camille Matignon articulated his view of the chemists' and industrialists' shared task: "The current trend in fertilizer industry, is to eliminate from them all inert substances they contain, often expensive by their origin and transport, null or insignificant interest in plants. Thus it is oriented in search of fertilizer with high-analysis materials, formed from phosphoric acid itself, by its union with ammonia or with potassium hydroxide and ammonia."[12] This research led to the chemical processes of "ammoniation" of phosphates — that is, the combination of ammonia and phosphate — by phosphoric acid or nitric acid: ammonium phosphate and ammonium nitrate.

Caking and decreasing the phosphate titre during storage are serious problems in manufacturing powdered ternary fertilizers. The remedy would come from granulation.[17 p105–106] The first work on the granulation of fertilizers was undertaken by a public organization in the United States, the Bureau of Soils in Washington, DC, in 1922.[6] A number of the synthetic products were readily fusible, and Bureau of Soils researchers found that they could be converted into granular form by heating the material to a temperature just above the melting point and spraying the molten product into the top of a tower through which a stream of air is passed in an upward direction. The sprayed molten products congeal in their fall through the tower to form spherical particles of uniform size and shape.

The trend in this period was toward combining mixing and granulation operations. The new compound fertilizer forms required many treatments and sometimes included mixing components that had previously been granulated separately. The separate granulation of all the materials in a mixture would add greatly to its cost. In 1935 a study was initiated to address this question in the Fertilizer Research Division of the Bureau of Chemistry and Soil, the successor to the Bureau of Soils. This study sought methods for the granulation of complete mixtures in a single operation.[6] The method that showed the greatest promise of commercial application was granulation by rotary drying. This process was applicable to the granulation of either fusible or infusible materials and mixtures. It could be carried on simultaneously with ammoniation, and the heat developed by the reaction with ammonia could be used to aid in the granulation of the mixture. The Fertilizer Research Division laboratory tested the process with a rotary drum. When necessary, external heat was applied to the drum by means of a gas burner. The dryer was fitted with an electric heater of variable capacity and a fan for blowing the heated air through the apparatus, whereby a wide range of temperature and airflow could be obtained as desired. The steps in the rotary drying process of granulation that this research developed consisted of adjusting the moisture content of the mixture as necessary by the addition of water or steam up to the optimum for the mixture to be granulated, raising the temperature of the mass to 60–80°C by ammoniation or otherwise, rolling the mixture in a drum until granulation occured, and then drying the granulated product. At the same time, work on granulation was carried out in Berlin (Germany) by an American company, the Dorr-Company.[13 p238]

Compound fertilizer preparation, by the 1920s, was no longer a simple mechanical job, but a complex chemical engineering operation. As such, its economics depended on large-scale operation accessible only to large chemical groups.[7 p5–6]

Mixtures in Integrated and Automated Fluid Networks: Granulators, Highly Specialized Equipment

In the 1950s and 1960s, fertilizer mixers evolved into more automated systems incorporating all chemical processes leading to granular compound fertilizers. Production capacities were increased to meet the demands of intensive agriculture; this favored granulation techniques that were suited to highly mechanized agriculture. Many granulation processes were developed in this period, but we will focus on two American processes for the production of diammonium phosphate (DAP): the Dorr-Oliver process and the continuous "ammoniator-granulator" process. The Dorr-Oliver process (Figure 1.8.3) was developed by the American company Dorr-Oliver; the continuous "ammoniator-granulator" process was developed by the Tennessee Valley Authority (TVA).[16] As in mechanical equipment, grinding was necessary upstream of these chemical mixing processes, and screening downstream. However, these techniques mostly involved circulation of fluids.

In the Dorr-Oliver process, the ammoniation of phosphate — in fact phosphoric acid — is made first, in several stages, in reservoirs independent of the granulation (Figure 1.8.3).[13 p238] The resulting slurry is then sent to the granulator to obtain the granules, which

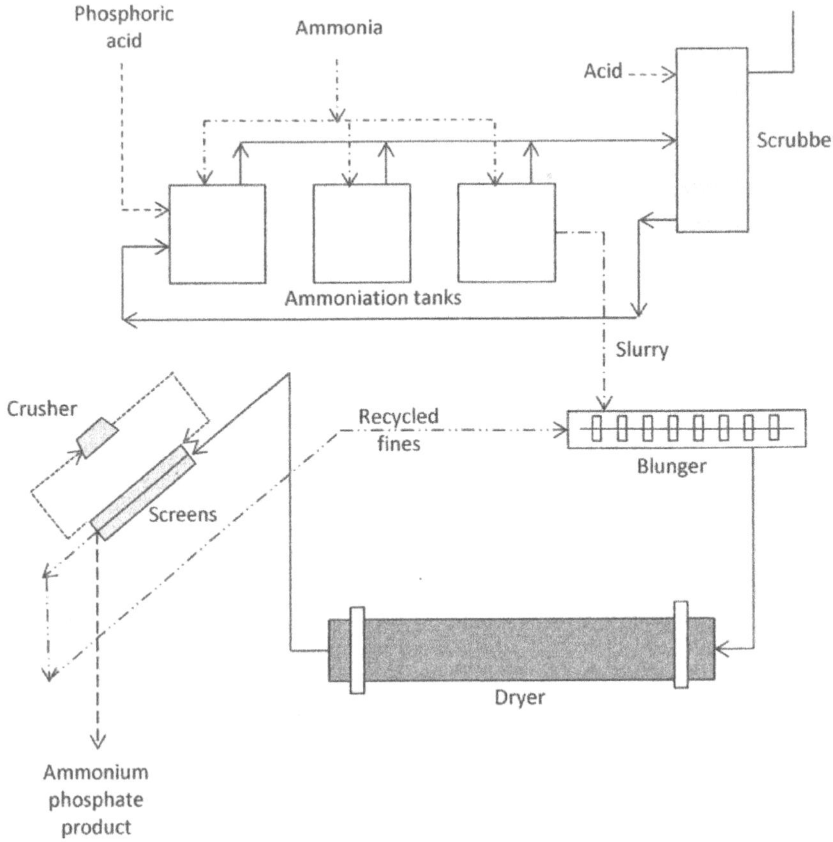

Fig. 1.8.3. Dorr-Oliver process for ammonium phosphate. Figure inspired by A. V. Slack's *Chemistry and Technology of Fertilizers.*[17 p115]

are dried and screened. In the continuous "ammoniator-granulator" process, ammonia and phosphate undergo pretreatment in a pre-neutralizer (Figure 1.8.4). But, unlike the Dorr-Oliver process, the ammoniation is only partial. The resulting material is sent to the granulator, or "ammoniator-granulator," in which ammonia continues to be injected. This second method is more efficient than the first. Granulation was demonstrated in the TVA pilot plant in 1953; by 1962, 164 plants in the United States — probably about two thirds of all US fertilizers plants — used the TVA process.[7 p253]

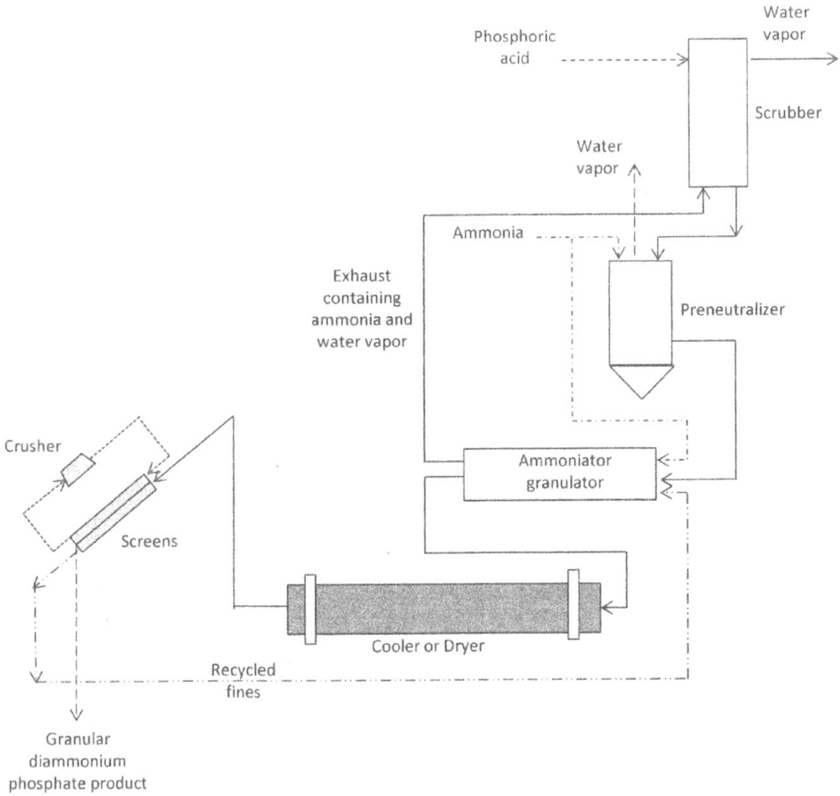

Fig. 1.8.4. TVA process for diammonium phosphate. Figure inspired by A. V. Slack's Chemistry and Technology of Fertilizers.[17 p116]

The invention of a continuous "ammoniator-granulator" by the TVA had a significant effect on the development of granulation processes in the United States. The continuous ammoniator was originally developed for more efficient ammoniation of superphosphate as compared with the batch mixers that were in common use. However, the consumption of granular fertilizers was growing more in Europe than in the United States. In the 1950s and 1960s, chemical engineering companies began to launch their own processes in the market. In the 1950s, granulation was more prevalent in Europe than in the United States. In Britain, perhaps 90% of fertilizer was granulated — the percentage was similarly high in France, Germany, and other continental countries — compared with 15% in the United States.[3]

New optimization process innovations began in the late 1960s. Early granulation techniques, particularly in England, were of the so-called "wet and dry" type. Enough water was added to a dry mix to cause particles to stick together, and the resulting granules were dried. This type of granulation is still widely used in Europe, where steam is often applied to help plasticize the mix, but it has fallen out of favor in the United States. The granules are relatively weak because of the small amount of liquid phase involved and drying to remove the added water increases process cost. To address these shortcomings, no water is added — instead, raw-material solutions are left more dilute than they otherwise would be.[17 p109] On the other hand, granulation in a drum was gradually replaced by granulation on a tray, ensuring greater regularity of granules.[9] Lastly, measures were taken to capture dust and toxic fumes and reintroduce them into the production process in order to limit environmental nuisances.

Conclusion

The evolution of mixing equipment is well highlighted by the case of the fertilizer industry. Reasons of costs, production volumes, and mechanization of agriculture have led the fertilizer industry to innovate, on the one hand, in its technical mixing processes, from manual mixing to chemical mixing, and on the other hand, in its products, from mixed compound fertilizer to granular compound fertilizer. Thus tools evolved alongside an emerging idea of industrial standardization: industry strove to realize the dream of being able to store fertilizers indefinitely in bulk and to handle them with conventional equipment such as that used in the grain industry.[3]

To meet these challenges, early industrialists proposed process innovations marked by chemists' influence. Originally a mechanical problem, the manufacture of compound fertilizers transitioned, catalyzed by the input of chemical expertise, into a chemical engineering issue. By examining the evolution of the tools used to manufacture compound and granulated fertilizers, we find innovation criteria that are described more generally by Gilbert Simondon in his technical objects genetics: we see a passage from linear to rotary (from the granulation tower to the rotating drum), vertical to horizontal (from the vertical drum to the horizontal plate), but also an evolution that develops synergies of function and self-regulation, such as the use of exothermic ammonization reactions to drive granulation.[2 p136–137; 8 p156–164]

References

1. Daviet JP. Un destin international: la Compagnie de Saint-Gobain de 1830 à 1939. Paris: Editions des archives contemporaines; 1988.
2. Deforge Y. Technologie et génétique de l'objet industriel. Paris: Maloine; 1985.
3. Fertilizer granulation. Agric. Food Chem. 1955;3(2):97–98.
4. Gille B. Histoire des techniques. Paris: Gallimard; 1978.
5. Gray A. N. Phosphates and superphosphates. New York: Interscience publishers inc; 1944 (first edition 1930).
6. Hardesty JO, Ross WH. Factors Affecting Granulation of Fertilizer Mixtures. Ind. Eng. Chem. 1938;30(6):668–72.
7. Hignette TP. Fertilizer Manual. Dordrecht: Kluwer Academic Publishers; 1985.
8. Jacomy B. L'âge du plip, Chroniques de l'innovation technique. Paris: Coll. Science ouverte, Seuil; 2002.
9. Les engrais complexes. Informations chimie. 1966;numéro spécial(43):117–51.
10. Les manutentions mécaniques dans l'industrie des engrais chimiques. L'industrie chimique. 1935;(252):11–21.
11. L'excavation mécanique des fosses à superphosphate, système Wenk, Le Génie Civil. 1913;(20):400–401.
12. Matignon C. Les progrès de la grande industrie chimique depuis 50 ans. Le Génie civil. 1930;Numéro spécial cinquantenaire:80–88.
13. Nielson FC. Manual of Fertilizer Processing. New York: Marcel Dekker, INC; 1987.
14. Oppermann C. A. Broyeur universel de Carr. Portefeuille économique des machines. 1869;(165):69–70.
15. Pluvinage C. Industrie et commerce des engrais. Engrais azotés et organiques. Paris: J.-B. Baillière et fils; 1926.

16. Sheridan RC. Chemical Fertilizers in Southern Agriculture. Agricultural History. 1979;53(1)[Southern Agriculture Since the Civil War: A Symposium (Jan., 1979)]:308–18.
17. Slack AV. Chemistry and Technology of Fertilizers. New York: John Wiley & Sons. Interscience Publishers; 1967.
18. Travis AS. The synthetic nitrogen industry in World War I, its emergence and expansion. New York: Springer; 2015.

PART II

INVISIBLE AND INFRASTRUCTURAL TOOLS

INVISIBLE AND INFRASTRUCTURAL TOOLS: INTRODUCTION

CYRUS C. M. MODY

What do you need in order to be able to do materials research? That is one of the animating questions of this volume as a whole. In Sections 3 and 4, we offer some relatively self-evident answers: tools for making and characterizing materials. In Section 1 we surveyed tools that are perhaps only self-evident upon second glance: that is, the kinds of tools that are so taken for granted that it takes some effort to perceive how indispensable they are. In this section, we will examine tools that are not self-evidently tools at all. Yet no one can deny that they are things — material objects, practices, modes of communications — that are indispensable to materials research.

To understand the kinds of invisible tools that I mean, think about the proverbial "toolkit" that most householders have for doing minor maintenance and repair. Some of the tools in the toolkit are canonical. Heidegger, for instance, elevated the humble hammer to iconic status in his famous discussion of ready-to-hand versus present-at-hand tools.[6] The analogue in materials research is probably the microscope, which has received similar philosophical attention.[5; 12; 15] But the family toolkit usually also contains things like pencils and work gloves that might not automatically be labelled "tools" but ought to qualify as such. The toolkit also incorporates — though it might not materially "contain" — more diffuse, infrastructural aids to action. A meter stick, for instance, is a fairly obvious tool, but one that only works if everyone agrees what a centimeter is. Nails are found in many toolboxes; but a nail is only useful if there's some agreement — and extension of trust — that the nail is made of the right materials and won't bend or break too easily or quickly rust and disintegrate. And the hammer, the nail, and the meter stick are all only useful if it is possible to acquire them: a whole infrastructure of catalogues and hardware stores and manufacturers stands behind the toolbox.

Each of these less obvious tools has an analogue in materials research, most of which are covered in this section. For pencils, think information display and recording devices. For work gloves, think safety equipment and procedures. For the meter stick, think international standards. For the nail, think mechanical testing and climate exposure. And for manufacturers and catalogues think — well, think manufacturers and catalogues. Once you know how to translate between them, the materials researcher's toolkit is not so different from the householders'. Both contain obvious and non-obvious, high- and low-tech, and tangible and intangible tools.

To understand how materials researchers make materials and knowledge about materials, you need to pay attention to both members of all those pairs. The obvious, tangible, high-tech tools are the kinds of things materials researchers apply for funding to buy, or invent, and/or found companies to sell. Those are often the tools they describe most explicitly in their research articles or in their pedagogy. Few research articles question what a meter is, though perhaps there is little reason they should. But it is also the case that few research articles describe in detail how their data were displayed — and maybe they should, since data display is a contingent achievement rather than a given.[2; 9] Non-obvious, intangible, and/or low-tech tools matter for the reliability and dissemination of knowledge about materials. Using such tools needs to be learned, and is never self-evident. Because such tools are often invisible, disagreement about them may be invisible too — one of the many sources of friction in collaborative research.[8] Quite simply, the tools presented in Sections 3 and 4 would tell us nothing without the tools examined in this section.

A Case Study: Conducting Polymers

A brief case study may help readers to understand how crucial these invisible and infrastructural tools are for materials research. Consider the case of research on organic conductors and conducting polymers in the 1960s and 1970s — work that has led to commercial products (organic light-emitting diodes or OLEDS) and the 2000 Nobel Prize in Chemistry being awarded to Alan Heeger, Alan MacDiarmid, and Hideki Shirakawa.[10 p85–91] This ought to be one of the canonical examples of "materials research" since it was the source of one of the few Nobel Prizes, and one of the most important commercial successes, to come out of the Materials Research Laboratory program in the United States — probably the most important institutional stimulant to the formation of a discipline of materials science. So if an overview of this case shows that invisible and infrastructural tools mattered, then there is good reason to believe that they matter across materials research.

The roots of Heeger, MacDiarmid, and Shirakawa's Nobel Prize lie in a thought experiment in physics. Quantum mechanics theorists had long tried to calculate how electrons would behave if their movement was constrained to two dimensions, or even one, but in the late 1960s there was still no physical system with which their calculations could be checked. Alan Heeger, a physicist at the University of Pennsylvania, decided to investigate organic conductors (carbon-based materials that conduct electricity relatively freely) as possible analogues for such hypothesized "one-dimensional metals." For reasons that are not entirely clear, organic conductors had never attracted much interest — historical study of this field would no doubt reveal much about the assumptions and aesthetics guiding more conventional fields in chemistry, metallurgy, and the like. In that sense, organic conductors resembled amorphous semiconductors in posing a challenge to the crystalline systems then dominant in materials science.[7] Yet organic conductors were not completely unknown and Heeger became especially interested in one class — the charge-transfer salts.

In particular, in a salt called tetrathiafulvalene-tetracyano-quinodimethane (TTF-TCNQ), electrons moved linearly through stacks of TTF and TCNQ groups (that is, roughly in one dimension). Chemists at DuPont and SUNY Buffalo had published syntheses of several charge-transfer salts in the open literature, which meant that Heeger's postdoctoral fellow, Anthony Garito, could use those published recipes to make his own. That also meant that a number of other lab groups became interested in TTF-TCNQ at roughly the same time, and Heeger soon found himself racing them to elaborate on its material properties. One of his and Garito's claims — that they had observed a superconducting transition in TTF-TCNQ — was immediately controversial and, ultimately, shown to be incorrect. But in the meantime the controversy about TTF-TCNQ brought sustained attention to organic conductors for the first time. Research on the topic spread to many universities and corporations, and talks about TTF-TCNQ became a regular staple at physics and chemistry conferences.

Some researchers were, like Heeger, mostly interested in the material's relevance to somewhat esoteric disciplinary questions. Others, however, thought there might be technological applications. IBM and Xerox, for instance, were interested because it seemed possible that organic conductors could provide a way around Xerox's patents on the use of amorphous selenium for the photodetectors in its copiers. That application never panned out, but at IBM it spawned interest in adapting single molecules of TTF-TCNQ for use as discrete microelectronics components such as diodes. Despite the inability of IBM (or anyone) to synthesize such molecular electronic devices, the company did secure a number of patents for these ideas.

Meanwhile, Heeger was somewhat stung by his experience with charge-transfer salts and cast about for a new material system to investigate related phenomena. Eventually he heard about polysulfur nitride, or $(SN)_x$. Margot Becke-Goehring had published a synthesis of this material in the 1950s but the synthesis was complex enough that Heeger's group was ill-equipped to carry it out (as were, in fact, many chemists).[3] Thus, Heeger needed to find someone who could make samples that he could then characterize — presumably a chemist and preferably someone nearby. As it happened, he had some contact with the University of Pennsylvania's chemists through their mutual participation in Penn's Laboratory for Research on the Structure of Matter (LRSM). This was the Penn laboratory funded through the Advanced Research Projects Agency's Interdisciplinary Laboratory program, which eventually became the Materials Research Laboratory program. In fact, Penn's was one of the first three labs funded through that initiative, largely because its combination of chemistry, physics, and metallurgy was seen as an ideal model for the new field of materials science more generally.

Thus, at the coffee break during a talk at the LRSM, Heeger was able to strike up a conversation with Alan MacDiarmid, a Penn chemist specializing in the chemistry of silicon. MacDiarmid, it turned out, had synthesized sulphur nitride — polysulfur nitride's monomer — for his master's thesis. Heeger and MacDiarmid began meeting regularly on weekend mornings in MacDiarmid's office to discuss conducting polymers: how to make them, how to characterize them, and what could be learned from them. Soon, a metaphorical pipeline ran from MacDiarmid's lab to Heeger's lab: one synthesizing, the other characterizing. Their co-authored publications drew the attention of the same people who had been interested in charge-transfer salts, plus new people interested in conducting polymers and then some because conducting polymers held greater promise of technological feasibility. A conducting polymers research community emerged quickly and Heeger and MacDiarmid became well-known leaders in it.

On that basis, MacDiarmid was invited to take a sabbatical in 1975 at Kyoto University. During that time, he gave a talk at Tokyo Institute of Technology and met a chemist there, Shirakawa, who was working on a mysterious silvery substance, polyacetylene. Intrigued, when MacDiarmid returned to the United States he contacted his grant officer at the Office of Naval Research, Kenneth Wynne, to request funding to bring Shirakawa to Philadelphia as a postdoc.

Despite some initial reluctance, Wynne complied. During the year Shirakawa spent at Penn, he discovered that doping polyacetylene with bromine and other elements could vary its conductivity by eleven orders of magnitude! With that, conducting polymers research exploded. As a review in 1984 put it:

> In contrast to the CT [charge-transfer] salts, the widespread fascination with conducting polymers is clearly motivated by the technological potential. However, as in the case of CT salts, these materials have also presented a strong scientific challenge. This challenge has attracted a diverse community of chemists, physicists, and materials scientists who have not previously been active in polymer science.[4]

What had begun as a simple thought experiment is today a multibillion-dollar market for applications such as cellular telephone displays and lighting.

Consider, then, where the invisible and infrastructural tools appear in this case. First, we should notice that all the actors in this story were paid, professional scientists working in academic, corporate, and government organizations. A vast organizational infrastructure of administration is necessary to make sure such people get paid, have workspaces and equipment, have access to publications, can order materials, and so on.[13] In this case, it was especially important that funding and administrative support existed for Heeger and MacDiarmid to hire key personnel (Garito, Shirakawa) at opportune times.

Second, we should notice how important the flow of communication was in this case, and therefore how important the infrastructures channeling that flow were. Heeger, Garito, and MacDiarmid needed others' published syntheses to make TTF-TCNQ and $(SN)_x$ in their own labs; that means they needed bibliographic tools to find out about those published syntheses, and then they needed the published syntheses themselves (found in journals located in or ordered from libraries). They also needed unpublished but formal channels such as conferences to tell the world about their results and to learn what others were doing. And they needed informal channels to enrol allies and pass along crucial tacit knowledge.

The role of Penn's LRSM is particularly interesting in this regard. As Brit Shields' entry on laboratory buildings shows, the LRSM was designed to facilitate such informal interactions by, for instance, grouping researchers along interdisciplinary corridors near shared instrumentation where they would naturally run into each other and, presumably, chat. (At least, that was the ideal. As usual with laboratory architecture, it did not quite work that way in practice.) Alongside that physical infrastructure was an institutional one — regular colloquia and so forth — that brought people like Heeger and MacDiarmid together. If you looked only to the most tangible notion of "tools in materials research," it might seem that the LRSM contributed little to this case; but if you expand to include less tangible tools, then its contribution becomes clearer.

Once they knew what they wanted to synthesize, and had the people and knowledge to synthesize it, Heeger and MacDiarmid's groups had to actually make the stuff. And for that they needed to order reagents — trusting that the reagents were what the manufacturers said they were — and various apparatus for manipulating those reagents. They probably also made some of that apparatus themselves (MacDiarmid was a reasonably skilled glassblower) or had it made by Penn's technical staff. They needed safety equipment, though they didn't always get or even want it: MacDiarmid, like many chemists of his generation, eventually contracted myelodysplastic syndrome, a cancer associated with exposure to common laboratory chemicals such as benzene. Interestingly, they needed tools for carrying their samples from MacDiarmid's lab to Heeger's — storage and transport tools are not included in this volume but their development has been essential to the expansion of collaborative research in the sciences. And then, once the samples got to Heeger's lab, they needed data output and recording devices to make a permanent record of the characteristics of those samples, without which the whole enterprise would have been pointless. That data then went into journal articles and conference presentations, which required their own specialized tools (such as, in those days, typewriters and overhead projector transparencies).[11] The data also went into patents filed by Heeger, MacDiarmid, and others, which required its own vast infrastructure of laws, lawyers, patent inspectors, and courts to enforce patent rights.[1; 14]

A Final Thought

This example shows that bringing these invisible and infrastructural tools into our historical accounts gives us something much more than an elaboration of ancillary details; it can reveal the processes that are essential to driving materials research. Thinking about the type of ready-to-hand tools that researchers rarely engage with consciously, but nevertheless use continually, focuses our attention on the ways in which different types of tools are coordinated within research programs. As you encounter the following descriptions of invisible and infrastructural tools, therefore, think not only about the importance they hold in their own right, but also about how they work in concert with other tools, and other varieties of tools, described in this volume. Infrastructures, after all, are large-scale systems designed to coordinate diverse constituent parts. And we often fail to see invisible tools precisely because they integrate so seamlessly with other, more visible instruments and procedures.

References

1. Arapostathis S, Gooday G. Patently contestable: electrical technologies and inventor identities on trial in Britain. Cambridge, MA: MIT Press; 2013.
2. Coopmans C, Vertesi J, Lynch M, Woolgar S. Representation in scientific practice revisited. Cambridge, MA: MIT Press; 2014.
3. Goehring M. Sulphur nitride and its derivatives. Quarterly Reviews, Chemical Society. 1956;10:437–50.
4. Green RL, Street GB. Conducting organic molecules. Science. 1984; 226(4675):651–56.
5. Hacking I. Representing and intervening: introductory topics in the philosophy of natural science. Cambridge, UK: Cambridge University Press; 1983.
6. Heidegger M. Sein und Zeit. Tübingen: Max Niemeyer Verlag; 1927.

7. Hoddeson L, Garrett P. The man who saw tomorrow: the life and inventions of Stanford R. Ovshinsky. Cambridge, MA: MIT Press; 2018.
8. Jordan K, Lynch M. The sociology of a genetic engineering technique: ritual and rationality in the performance of the 'plasmid prep'. In Clarke AE, Fujimura JH, editors. The Right Tools for the Job: At Work in Twentieth-Century Life Sciences. Princeton: Princeton University Press; 1992. pp. 77–114.
9. Lynch M, Woolgar S, editors. Representation in scientific practice. Cambridge, MA: MIT Press; 1990.
10. Mody CCM. The long arm of Moore's Law: microelectronics and American science. Cambridge, MA: MIT Press; 2017.
11. Ochs E, Jacoby S. Down to the wire: the cultural clock of physicists and the discourse of consensus. Language in Society. 1997;26(4):479–505.
12. Rasmussen N. Picture control: the electron microscope and the transformation of biology in America, 1940–1960. Stanford: Stanford University Press; 1997.
13. Stephan P. How economics shapes science. Cambridge, MA: Harvard University Press; 2012.
14. Swanson KW. The emergence of the professional patent practitioner. Technology and Culture. 2009;50(3):519–48.
15. Wilson C. The invisible world: early modern philosophy and the invention of the microscope. Princeton: Princeton University Press; 1995.

CHAPTER 2.1

BUILDINGS

BRIT SHIELDS

Studies of the places in which scientific inquiry happens, be it in the laboratory or the field, have generated insight into the practices, values, and cultures of scientific communities.[4; 6] The spaces of inquiry influence scientific practice. The architectural design of a laboratory, for example, influences the flow of researchers and the placement of equipment, and thus has profound consequences on how much — or how little — such researchers interact.[1] The same insight applies to a laboratory's placement within its environment. A remotely located laboratory can perpetuate disciplinary isolation; a strategically placed laboratory can foster serendipitous encounters, potentially leading to fruitful collaborations. At the same time, the demands of scientific practice influence the spaces of inquiry. The physical buildings that house scientific laboratories must be calibrated to accommodate tools that are sensitive to light, heat, vibration, electricity, dust, and other environmental conditions. The built environment serves as a sometimes invisible, yet nonetheless influential scientific tool. Like any other tool, it can be adapted over time to meet the evolving needs of its users and can have dramatic consequences on the research trajectories of its occupants.

The buildings that house the research facilities, laboratories, and offices for materials science research and training on US university campuses played just such a pivotal role in the development of the field. The history of the Interdisciplinary Laboratory (IDL) program sponsored by the US Department of Defense's (DoD) Advanced Research Projects Agency (ARPA) is an illustrative example of exactly how the built environment shaped the scientific practice of materials science researchers within the US university system. The program was designed to foster interdisciplinary research among materials-related disciplines while increasing the populations of researchers and graduate students in those fields, with the goal of encouraging a scientific approach to the technical problems posed by materials bottlenecks in fields like aerospace, electronics, and telecommunications, which were strategically valuable during the Cold War. Architecture was seen as one essential part of the solution. Carefully planned laboratory buildings, it was thought, would encourage interdisciplinary collaboration, be it in the form of consulting on research projects, sharing central facilities, or advising graduate work in various fields. The logic behind ARPA's strategy also took buildings to be vital scientific tools in and of themselves, designed to meet the exquisite demands of carefully calibrated and sensitive scientific equipment.

Historical Context

During the Cold War, and particularly following the launch of the Soviet satellite *Sputnik* in 1957, US federal agencies recognized the importance of technology and science for national security and economic prosperity. Under President Dwight D. Eisenhower's White House Office of Science and Technology, the Coordinating Committee on Materials Research and Development was assembled to assess how best to achieve the national technological and scientific goals in the space race and other sectors. The committee recognized the needs both for increased research efforts in materials and for an increase in the numbers of PhDs trained in materials science–related disciplines. The committee recommended that the federal government financially support a program of interdisciplinary laboratories at US universities.[9 p100] The Federal Council for Science and Technology accepted the recommendation and ARPA was tasked with creating the program.

Some studies on the growth of materials science have argued that, prior to 1960, most university departments would not, on their own, "justify" the title of a materials science department.[10] Rather, the research and training taking place on campuses within materials science were spread among traditional academic departments, including physics, chemistry, electrical engineering, and metallurgical engineering. ARPA's task was to develop a number of IDLs meant to foster the interdiscipline of materials science by grouping the disparate researchers and students across a campus into one building. The thought was that placing all of the materials-related researchers in one building, along with state-of-the-art shared facilities and laboratories, would result in heightened levels of collaboration and development in the field. Following ARPA's call for applications, three universities were selected to house the first IDL buildings. Contracts were signed with the University of Pennsylvania, Cornell University and Northwestern University. Over the following years, ARPA developed materials science IDLs at nine additional universities. The Atomic Energy Commission developed similar laboratories at three universities: the University of California, Berkeley, the University of Iowa, and the University of Illinois. NASA also developed a materials science laboratory. The IDL program was later renamed the Materials Research Laboratories program and was transferred from ARPA to the National Science Foundation in 1972. During this time of organizational change, the buildings served as a point of continuity for the materials science programs.[8]

Interdisciplinary Laboratory Buildings

The ARPA IDL contracts in 1960 directly spelled out the goal of the program: "of furthering and understanding factors which influence properties of materials." This would be achieved by offering the financial support to increase personnel and construct the necessary facilities.[10 p37] The IDL program held dual goals in research and education; its laboratories were meant to increase the volume of research within materials science, particularly along interdisciplinary lines, and also to expand the numbers of doctoral students within materials science–related disciplines.[7 p11–16]

The productive interdisciplinary collaborations that ARPA envisioned were to pull researchers in materials-related fields, such as physics, chemistry, and engineering, into closer proximity by placing them within the same laboratory building. The IDLs would offer shared central facilities — operated by trained technicians — which were available to all of the laboratory's faculty, postdocs, and graduate students, regardless of department affiliation. With common facilities and laboratories, the IDLs were meant to create a unique research environment in materials science by physically housing scientists and engineers within one building. Critically, this would allow for a more interdisciplinary approach to training as well. Rather than having doctoral students and postdocs separated by departmental structures and physically spread over campus, they would have ready access to other departments' faculty, students, and research facilities, which otherwise might have been unknown or inaccessible to them. The buildings were funded in part by ARPA through a building-use fee, which was then often supplemented by the host university and private industry.[10 p37]

University of Pennsylvania's Laboratory for Research on the Structure of Matter

One of the first IDL buildings to be constructed under the ARPA program was the Laboratory for Research on the Structure of Matter (LRSM) at the University of Pennsylvania. The four-story building opened in 1965, following years of negotiations over the best way to design a building intended to foster interdisciplinary research and training. The design of the building reflected the ideological commitment to interdisciplinarity, implementing a race-track scheme of offices and private laboratories on the outside perimeter of the building with "central facilities" and common laboratories occupying the center of the building, whenever possible.[2 p27]

In the year the building opened, the LRSM contained the following ten central research centers available to all of the laboratory's researchers: Analytical Chemistry, Electrical Standards, Electron Microscopy, High Magnetic Field, Magnetic Resonance, Materials Processing, Metallography, Microwave Spectroscopy, Spectroscopy, X-Ray Diffraction.[3 piv–xii] Each facility was developed under the guidance of an associated faculty member and operated by a laboratory assistant or supervisor. LRSM researchers received annual reports describing the capabilities of each research center as well as the relevant research projects of the laboratory. Such communications were meant "to encourage the widest possible use of these centers by acquainting everyone with their capabilities."[3 piv]

The IDL achieved its goal in housing the laboratories and offices of researchers who had previously been disparately located on campus. The academic affiliations of the LRSM faculty, postdocs, and graduate students included the School of Metallurgical Engineering, the solid state group from the physics department, and the chemistry department.[2 p28] Within the LRSM, the researchers were not assigned space based on their departmental affiliation. Rather, the researchers were placed according to proximity to the facilities their research demanded. Thus, for example, the chemist Robin M. Hochstrasser and the metallurgist Geoffrey R. Belton were given offices nearby as they both had specialties in spectroscopy.[2 p33] The building also represented collaborations with private industry, which ranged from visiting lecturers through a visiting scholars series to capital grants and equipment procurement. The General Electronic Company, for example, provided the equipment for the high magnetic field laboratory. By 1965, the LRSM had secured more than $2 million from industry.[11] In addition to attention paid to the layout of the floors, the building also demanded particular specifications to allow for the type of research being conducted within its walls. For example, the LRSM was air-conditioned and designed to have sealed windows, creating a "dust-free" environment, and was placed on a sunk foundation of bedrock with machinery installed on the roof to minimize potential vibrations.[2 p34]

In 1972, when the IDL program was transferred from the purview of the DoD to the NSF, it was renamed the Materials Research Laboratories program. Over the following decades, the research program of the LRSM grew, particularly in the 1990s in the areas of nanotechnology and biotechnology, and in the early 2000s with the establishment of the NSF-funded Nano/Bio Interface Center.[2 p37] The research efforts in nanotechnology ultimately put new demands on the building that had not been foreseen in the 1960s. The LRSM researchers started using parts of the adjacent Edison Building, a "temporary" facility constructed in 1965. In a 2004 feasibility study of the LRSM and the Edison buildings, done in collaboration Kling Architects, it was determined that the buildings were inadequate for the current and projected future needs of the researchers. For example, vibrations caused by the nearby street as well as mechanical equipment within the aging LRSM meant that highly sensitive equipment, such as scanning electron microscopes, could only be used on certain floors.[5 p30] It was ultimately decided that the Edison building be demolished, while the LRSM building remain and a new building be constructed.

The solution took the form of the Singh Center for Nanotechnology, which opened in 2013, constructed on the adjoining lot to the LRSM to house the growing nanotechnology research centers on campus (Figure 2.1.1). As was the case with the LRSM, fostering interdisciplinary research serves as a guiding principle in the ideology of the Singh Center. The building houses faculty from both the College of Arts and Science and the School of Engineering and Applied Science. The nanotechnology building itself serves as a critical tool to its occupants, as it meets numerous specifications to accommodate the extremely sensitive equipment and facilities it houses, including a 10,000 square foot cleanroom. Such provisions include the placement of the elevators at the ends of the building to avoid electromagnetic interference with the laboratories. This technological demand informed the design of the building and led the architectural firm Weiss/Manfredi to place a central staircase at the main entrance to the building, featuring a series of landings with tables and chairs meant to foster collaboration.

Fig. 2.1.1. The University of Pennsylvania's Laboratory for Research on the Structure of Matter (left) and the Singh Center for Nanotechnology (right). *Credit*: Photo by the author.

Conclusion

The built environment in which scientific study and training occurs can have profound impact on scientific practice; similarly, the housing of finely calibrated equipment imposes demands on the architecture of laboratory spaces. The study of such spaces can provide insight into the ideology of scientific research and training. The buildings housing materials science research on US university campuses as part of the ARPA IDL program were specifically designed to foster interdisciplinary collaboration, provide space to train PhDs in materials science–related disciplines, and house sensitive machinery. The IDL buildings allowed for researchers to organize themselves around shared facilities rather than departmental affiliation. This ideology also impacted the training of doctoral students and postdocs who were encouraged to participate in interdisciplinary research within the laboratory. More recently, similar considerations for both the physical demands of the building to house highly sensitive equipment as well as the interdisciplinary nature of the researchers have informed the design of nanotechnology buildings.

References

1. Allen T, Fustfeld A. Research laboratory architecture and the structuring of communications. R&D Management. 1975;5:153–64.
2. Choi H, Shields B. A place for materials science: laboratory buildings and interdisciplinary research and the University of Pennsylvania. Minerva. 2015;53(1):21–42.
3. Fifth Annual Technical Report, Laboratory for Research on the Structure of Matter, University of Pennsylvania, 1965.
4. Galison P, Thompson E, editors. The architecture of science. Cambridge, MA: MIT Press; 1999.
5. Kling Architects, LRSM nanotech feasibility study, May 21, 2004.
6. Leslie S. A different kind of beauty: scientific and architectural style in I. M. Pei's Mesa Laboratory and Louis Kahn's Salk Institute. Historical Studies in the Natural Sciences. 2001;38(1):173–221.
7. Martin J. What's in a name change? solid state physics, condensed matter physics, and materials science. Physics in Perspective. 2015;17(1):11–16.
8. Mody C, Choi H. From materials science to nanotechnology: Interdisciplinary center programs at Cornell University, 1960–2000. Historical Studies in the Natural Sciences. 2013;43(2):121–61.
9. National Research Council. Materials and man's needs. Washington, DC: National Academies Press; 1975.
10. Schwartz L. Materials research laboratories: reviewing the first twenty-five years. In Psaras P, Langford H, editors. Advancing materials research. Washington, DC: National Academies Press; 1987.
11. University of Pennsylvania Archives, UPF 8.55 Box 144 Folder 2, University of Pennsylvania, News Bureau, October 4, 1965.

CHAPTER 2.2

SAFETY EQUIPMENT

AMY E. SLATON

In following the history of safety in materials science, as in other cases of scientific and technological labor, we can start from the understanding that safety emerges in communities. The safe operation of scientific spaces involves prescriptions by the experts and regulators, who dictate what shall count as a safe space and the conduct that accords with those prescriptions on the part of researchers and technicians working in research labs, personnel in fabrication settings, or instructors and students in the classrooms where preparation for life in these workplaces takes place.[8 p9] Decisions about how best to make the work of laboratories safe cover a great many issues: material arrangements abound, including objects such as specialized storage and disposal containers or ventilation and fire suppression systems, and entire infrastructures for electrical, gas, or water supply and waste disposal. These are accompanied by complex arrays of protective personal apparel — goggles and gloves, lab coats, and hazmat suits — which surround the student's, scientist's, technician's, or custodian's body. All of these artifacts are paired with formal protocols and tacit best practices regarding, say, the order in which material handling steps are to be taken, or the appropriate response to spills, explosions, or injury. Responsibility and accountability are distributed across virtually every sort of personnel, including safety officers, principal investigators, equipment repair people, technicians, and assembly and custodial staff. In short, the history of safety in

Fig. 2.2.1. Laboratory fume hood, designed to allow manual work with hazardous materials and direct observation, being delivered to university research laboratory, 2018. Purpose-built wooden cabinets intended for chemical research and education replaced experimenters' use of preexisting hearths and chimneys in the late nineteenth century. The mass-production and commercial distribution of laboratory safety equipment including hoods and ventilation systems accelerated through the twentieth century, supporting the standardization of safety procedures in materials research and production facilities. *Credit*: Photo by the author.

materials science has involved a combination of material arrangements and ideas about optimal human behaviors (Figure 2.2.1).

In seems obvious that those responsible for the creation and enforcement of safety rules, including policy makers, lab directors, or instructors, repeatedly assert the necessity for safety compliance to everyone involved, stressing the bodily risks to self and others likely to result from non-compliance. But what does "compliance" actually include? Clearly, the degree to which a heated specimen can burn skin if accidentally touched with a bare hand, or the toxicity of two chemicals that produce fumes when inadvertently combined, must be well understood from a physical perspective. At the same time, however, judgments must be made about what risks are tolerable (for those immediately within the setting being controlled, and in proximity to it); what resources can be marshalled to ensure safety; and what personnel can reasonably be expected to do, and reasonably be expected to notice or predict, in either routine or emergency situations. In other words, danger and safety reside not in the strength of the acid being handled, in the vulnerability of human skin to acid, in the acid handler's behavior, or in the relative availability of acid-proof gloves, but in the meeting of all four of those factors.

With this sort of outlook, science-studies scholars, interested in both the knowledge and practices that make up scientific research, have helped us see that science safety is similarly layered.[2; 14; 15; 18] The pursuit of safety, we have learned, is similar to other modern work settings in reflecting both institutional and interpersonal distributions of blame, responsibility, and authority as well as ideologies of labor, skill, and competence in scientific practice. For example, what counts as safety in a particular setting is whatever counts as an appropriate limit to bodily suffering for those at risk; that is, safety signals the upward limit of acceptable pain or injury. Naturally, accidental death in a scientific workplace almost universally represents an absence of safety. But consider researcher or technician fatigue, repetitive strain injury, irritated eyes or skin, and other maladies, including chronic disabilities and even lethal illnesses that derive from, say, long-term laboratory exposures (consider here the deaths of Marie Curie and Rick Smalley, who both died of cancers that can be plausibly linked to their exposure to carcinogens in the laboratory). The presence of these psychological and bodily experiences is not necessarily seen to represent an absence of safety. It may instead signal a tireless or brave researcher, or simply someone doing their job as expected.[4] Doing one's job in spite of such difficulties may also be a choice made by insecure workers who feel that they have no choice but to tolerate some injury and discomfort.[9] In other words, far from a set of obvious and advancing technical developments, what has counted as laboratory safety has a great deal to tell us about the construction of valued, reliable, and compliant scientific actors, and vice versa.

Materials science, a field made up of techniques and knowledge drawn from multiple disciplines (chemistry, physics, engineering, and others) takes its safety practices from many sources. In the training and operations literature of materials science, we find discussion of what is required for the safe handling of chemicals, the use of power sources, encounters with testing machinery, and diverse other technical practices that constitute materials science, whether undertaken in classrooms, in factory assembly processes, or as part of a research enterprise.[1] Safety practices are deployed in both non-synthesis-based and synthesis-based investigations: that is, processes involving matter that is not chemically altered in handling and processes in which matter is altered.

Across instructional, manufacturing, and research settings where materials science is performed, we find ongoing testing and inspection of existing materials. When non–synthesis based, this sort of quality-control or investigative activity may involve comparison or measurement without any alteration of the materials or parts under study, or alternatively, the acts of heating, decomposing, stretching, or otherwise deforming a specimen to ascertain its character. Such work involves risks, including the handling of hazardous materials, the operation of large machines (strain and compression testing instruments can exert tons of force and are themselves heavy), and proximity to dangerous power sources such as high-voltage electricity or nuclear energy. But the second type of materials-science operation, involving synthesis, is perceived by subject matter and lab safety experts to carry additional personal and property risks.[11] Here, unknown conditions may derive from the mixing of existing substances or the exposure of novel materials to deforming forces. Outcomes are not predictable and thus determining what safety protocols are needed must involve some estimation, by definition adding to the danger of the situation. More complex safety protocols and oversight, as embodied in standards issued in the United States by the Occupational Safety and Health Administration (OSHA), often accompany synthesis projects in materials science.[3]

It would be misleading, however, to always associate safer labor with the application of existing knowledge in materials science and riskier labor with the pursuit of new knowledge, as the above distinction might imply. That is, we cannot presume that routine tasks in material science necessarily bear little bodily risk whereas experimentation demands greater concern. For example, computational materials science involves few bodily risks and is normally undertaken by advanced researchers. At the same time, even the most repetitive testing or inspection of nanoscale or radioactive materials may expose workers to significant dangers. A heightened level of danger is also seen in the work of setting up, cleaning, and maintaining the facilities and equipment used by those involved in materials testing. As Sims writes, in these operations the "usual physical barriers and roles that separate people from dangerous parts are often by-passed."[15 p337] This indirect relationship of ascribed skill level and risk is a telling feature of laboratory work; it alerts us to the contingent distribution of risks not simply among persons with different responsibilities in the lab (after all, it is sometimes high-ranking researchers who maintain or adapt equipment), but between lab personnel and the suppliers of apparatus and materials, or indeed between people and things.

Although the scientifically advanced work of researchers and more routine work of technicians and shop-floor workers do not automatically involve less and more safety challenges respectively, the risks that different groups experience in materials laboratories or production operations are nonetheless profoundly shaped by social differentiations. The historical study of safety in academic and industrial laboratories or in fabrication settings (factory shop floors or cleanrooms, sometimes called "fabs") illuminates these differentiations. These places of science are stratified labor settings beset by inequities along lines of job and wage security, or mobility and autonomy, all of which in the United States map in turn onto divisions of class, race, gender, and (dis)ability.[6; 16; 17] By extension, the processes of ascribing identities in the lab or fab (along lines of occupation or race, of credential or gender), distributing opportunities (for continued employment or occupational advancement), and sustaining individual productivity or institutional operations may all be seen as inseparable from the definition and enactment of lab safety. In this brief essay, I cannot explore many such socio-historical episodes, but I can draw attention to their diversity and ubiquity by establishing a basic typology of *safeties* in materials science. We might start with two particularly telling safeties: one safety deriving from knowing and following rules about what to do and what not do in the materials science workplace and the other deriving from an absence of knowledge, a strategic ability to recognize one's own ignorance about what is safe to do and not do.

The first sort of safety involves rules that reach their prospective users through science education, starting as early as elementary school — training programs in work settings, direct instruction in the lab or on the shop floor, observation of workplace norms, and other encounters with best practices that inculcate compliance. Protective clothing is to be donned in a particular sequence, glassware or gauges to be selected according to guidelines, machines powered up and down according to instructions. Proper entry to the cleanroom involves minutely detailed bodily movements and material arrangements, the violation of any one of which renders the others ineffective. Signage, labels, tags, and emergency kits, and instructions pervade the modern lab and point the way to correct conduct (Figure 2.2.2). Not only does such direction convey the nature of safe action, it also conveys the value and necessity of discipline and self-discipline. Action and attitude combine. Countless lab and fab safety courses and training videos include the message that safety must be taken seriously, that remembering to act in a particular way is as important as knowing how to act, and that danger is found in nature and nature cannot be denied — as a classic 1947 poster for Oak Ridge employees puts it, "Radiation need not be feared but it must be treated with respect!"[13]

The National Research Council (NRC), which has taken on a high-level oversight role in US science safety, describes progressive developments in laboratory safety over recent decades.[10] The NRC traces advancement from a "first epoch" in which chemical and material dangers were seen to inhere in "chains of events"

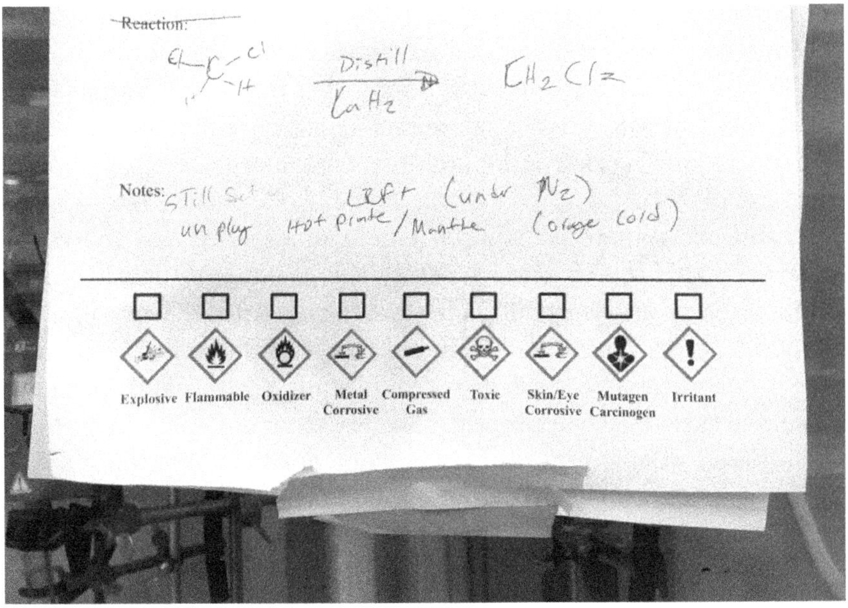

Fig. 2.2.2. Laboratory safety signage in a university teaching laboratory, 2017. This pre-printed paper form, to be temporarily taped to laboratory equipment, requires users to recognize and indicate hazardous conditions that could be present during a particular experiment. Hazards are represented on the form as both icons and text, possibly on the assumption that students may not yet be familiar with icons. As in the use of other lab safety protocols, experimenters or technicians are here given responsibility for reporting significant risks associated with specific materials or processes, but not for determining new or unexpected categories of hazard. *Credit*: Photo by the author.

precipitated at the hands of individuals (not pinpointed but presumably prior to the mid-twentieth century) through the current epoch of more systems- and culture-focused approaches to safety. Today's "culture of compliance" casts safety as a concern best distributed across levels of laboratory expertise and responsibility, and promotes reflection and analysis about such distributions so that institutional safety standards might be constantly improved. The authors of the NRC report that synthesizes these conclusions clearly recognize that institutional choices regarding safety matter on both material and ideational levels: what training and equipment is to be provided by an organization; what conditions shall count as danger and safety; which worker conducts shall be rewarded and which punished.

This is true across safeties as I am using that term, but this particular sort of safety relies on worker memory and repetition, with supervisor and worker accountability customarily assured through the use of security systems, log books, lock-out-tag-out apparatus, and other material stand-ins for human voice and movement (Figure 2.2.3). In other words, trust is a measured quantity in the workplaces of science. But note that along with the use of those stand-ins, every individual's sense of responsibility and accountability is nonetheless to be cultivated (Figure 2.2.3). Each person is meant to take on the appropriate level of anxiety about danger, care for self and others, and other compliant positions regarding safety,

Fig. 2.2.3. Laboratory technician using atomic absorption spectrophotometer, in the Electric Storage Battery Company, Philadelphia, undated. "Wash bottles" with integral spouts, seen in foreground, enable one-handed use, while reducing splashing and controlling strength and direction of flow. Simple material innovations of this kind, as much as complex apparatus and protocols, distribute responsibility for safety and blame for injury in the laboratory. *Credit*: Photo courtesy of the Science History Institute.

as well as investment in ongoing success as a student, researcher, technician, custodian, or manager. Success in any of these areas produces both safety and safe actors. Nor are these ideologies distinct from those involving actors outside the lab. "Rational decision making tools" of the type developed in laboratory settings also shape recommendations for the public assessment of environmental and health risks of newly engineered nanomaterials.[12; 18]

By contrast, the second sort of safety in our typology entails the veneration of not knowing, implicating ignorance as agency. Here, laboratory or factory personnel are expected to detect what they do not know and to seek "missing" information. Materials Safety Data Sheets (MSDSs) embody this kind of safety. First appearing in standardized formats in the 1940s, and mandated in the United States by OSHA and the Environmental Protection Agency as a stronger regulatory climate took hold through the 1970s and 1980s, published collections of MSDSs hold detailed information from chemical or material producers, distributors, and importers for the use of consumers in research and manufacturing settings.[5] Listing compositional and behavioral features of a given material, the sheets are vital for synthesis operations as they aid in the prediction of material reactions, but they also inform users of potential material toxicity through exposure, or of fire and explosion risks. Hygiene and first-aid recommendations appear as well. The efficacy of MSDSs relies on and produces a blend of routine *and* inquiry on the part of the employee. The sheets are useful only with existing funds of chemical knowledge and concern that will lead the user to then ask, "what more do I need to know?"

In some sense, therefore, the kind of safety that relies on identifying and correcting one's ignorance might be associated with practitioners who possess more extensive scientific knowledge, higher-level positions in the materials science workplace, and greater occupational autonomy. But safety protocols dictate that every worker, whether custodian or lab director, in order to remain safe, must incessantly question whether or not they know what they need to know about safety. Yet at the same time, "not knowing" is not normally credited as an asset among those engaged in seemingly routine operations. Countless training materials convey that the lower-level lab worker or neophyte must learn to dread and prevent gaps in knowledge and instead seek to be replete with memories and routine behaviors. Crucially, such cognitive gaps are the very sites from which open-ended inquiry (that is, discovery) derives; to deny the utility of students', technicians', assembly workers', or junior researchers' ignorance is to deny the scientific value of their curiosity. There is an interesting point here about scientific occupations and mobility: stratifications of expertise and responsibility that shape work in the laboratory may appear at first glance to be hierarchies with unlimited possibilities for growing one's knowledge and attaining greater scientific facility. Knowledge accumulates for the eager participant in scientific investigation; curiosity is rewarded. Yet, looking closely at procedures such as those surrounding scientific safety reveals a much more complex and perhaps inflexible set of occupational relations.

Other safeties exist, each bearing such consequential sociabilities, and the above examples suggest that to detect these we must move beyond the idea that safety involves only or primarily a downward flow of understanding from authorities toward workers who are more or less knowledgeable, more or less compliant. That is, we can reject the unidirectional image of dangers being ascertained by regulatory, managerial, or scientific experts; passed along in the form of safety practices, material arrangements, and a "culture of compliance;" and ultimately embraced or ignored by individual students, technicians, manufacturing employees, or researchers. Instead, we can see that the reverse process is also underway. Individuals are imagined by those in authority and by one another to be of variable competence, achievement, and deserved opportunity or security in the lab; "safety" is devised as the measurement and expression of those relative endowments; and finally, dangers thereby become reasonable. Whether we are considering the amount of knowledge presumed by the content of a material safety data sheet, the warnings posted by makers or users on electronic equipment, or instructions for clean room entry, we can detect prevailing ideas of who can and must do what to make science safe.

The dangers of scientific work are in all of these ways naturalized as both necessary and good for knowledge, personal security, institutional health, or, where science serves commercial interests, economic expansion. If, through our study of safety, we can somehow grasp that both of these sequences are happening at the same time in the places where materials science is undertaken, we may attain a fuller picture than we have had until now of the character of scientific activity as a set of social and labor relations.

References

1. Department of Materials Science and Engineering, Chemical Hygiene Plan. DMSE, MIT 2016. Available at: https://dmse.mit.edu/sites/default/files/documents/DMSE_CHP.pdf [Accessed January 2, 2017].
2. Doing P. Velvet revolution at the synchrotron. Cambridge, MA: MIT Press, 2009.
3. Fairfax RE. Letter to Kenneth Coppola, February 5, 2004. Available at: https://www.osha.gov/pls/oshaweb/owadisp.show_document?p_table=INTERPRETATIONS&p_id=24782 [Accessed January 4, 2017].
4. Herzig R. Suffering for science. New Brunswick, NJ: Rutgers University Press; 2006.
5. Kaplan SA. Development of material safety data sheets. Paper presented at ACS National Meeting, New York, NY, 1986.
6. Keefe J, Potosky D. Technical dissonance: conflicting portraits of technicians. In Barley SR, Orr JE, editors. Between craft and science: technical work in U.S. settings. Ithaca: Cornell University Press; 1997. 53–81.
7. Materials Science and Engineering, Lab Safety. Available at: https://drexel.edu/materials/research/safety [Accessed January 2, 2017].
8. Mody CCM. A little dirt never hurt anyone: knowledge-making and contamination in materials science. Social Studies of Science. 2001;31(1):7–36.
9. Mosby K. Technician skill and identity in a commercial laboratory. Unpublished manuscript, March 7, 2012.
10. National Research Council. Safe science: promoting a culture of safety in academic chemical research. Washington DC: National Academies Press; 2014.
11. National Research Council Committee on Prudent Practices in the Laboratory. Prudent practices in the laboratory. Washington, DC: National Academies Press; 2011. 2–3.
12. Nel A, Zhao Y, Madler L. Editorial: environmental health and safety considerations for nanotechnology. Accounts of Chemical Research 2013;46(3): 605–872.

13. Oak Ridge National Laboratories. Health physics posters. Available at https://www.orau.org/ptp/collection/hpposters/hpposters/htm [Accessed January 4, 2017].
14. Silbey S, Ewick P. The architecture of authority: the place of law in the space of science. In Sarat A, Douglas L, Umphrey M, editors. The place of law. Ann Arbor: University of Michigan Press; 2003.
15. Sims B. Safe science: material and social order in laboratory work. Social Studies of Science. 2005;35(3):333–66.
16. Slaton AE. Body? What body? Considering ability and disability in STEM disciplines. American Society for Engineering Education Annual Conference Proceedings 2013. AC 2013-7526.
17. Slaton AE. Race, rigor and selectivity in U.S. engineering: the history of an occupational color-line. Cambridge, MA: Harvard University Press; 2010.
18. Viseu A. Caring for nanotechnology? being an integrated social scientist. Social Studies of Science. 2015;45(5):642–64.

CHAPTER 2.3

VACUUM CHAMBERS, PUMPS, GAUGES, AND SYSTEMS

CYRUS C. M. MODY

We often think of scientists as working directly with their experimental materials. The image of the chemist staring upward into the beaker is, after all, one of the most pervasive icons of the scientist. Yet in many fields, experimental materials (and the conditions necessary for their existence) are a threat to scientists, and/or scientists (and the conditions necessary for their existence) are a threat to experimental materials. In such cases, scientists must essentially build remote laboratories — micro-environments in which they can only interact with their materials at a distance. Vacuum is one of the oldest and most important such micro-environments in the history of laboratory science.

Pre-1950

Some historians have argued that the scientific method began with the vacuum chamber.[8] From the late-seventeenth through the mid-twentieth centuries, vacuum experiments were usually conducted in bell jars or sealed glass envelopes (see Figure 4.0.1 in the introduction to part 4 of this volume). Hence, many experimentalists depended on artisans who could form glass into complicated shapes; quite a few scientists could even do decent glasswork themselves (see Catherine Jackson's contribution to this volume). As late as the 1980s, many universities and laboratories employed full-time glassblowers (and a handful do even today), though that practice became less common from the 1960s onward.

Vacuum chambers and other glass-enclosed micro-environments have been used for research on many different topics, but investigations of materials have been a consistent thread. For instance, Joseph Priestley's synthesis of (what came to be called) oxygen and other "airs" took place in a pneumatic trough consisting of a glass envelope resting in a water bath. In the late nineteenth century, sealed glass vacuum environments became the basic apparatus of electron physics and related fields. Despite its name, electron physics was not just about electrons — electron physicists' experimental apparatus was also used to make significant contributions in materials research. Irving Langmuir's Nobel Prize in Chemistry, for instance, was awarded for his study of surface reactions on tungsten filaments in a glass envelope.[7] Similarly, the field-emission microscope, invented by Erwin Müller in 1936, offered near-atomic resolution images of tungsten and a few other materials inside an evacuated glass envelope.[4]

Gradual Movement toward Ultrahigh Vacuum

Glass vacuum chambers were ideal in the "string and sealing wax" era. They were cheap, quick, and easy to make. Properly sealed, a sample could stay pristine for decades. And glass, of course, allows the experimentalist to see what is going on inside the apparatus. But by the 1950s, glass's limits were being reached. In most cases, a glass chamber required the experimentalist to waste time breaking vacuum in order to replace the sample or tinker with the equipment. In addition, glass was ill suited to extremely low pressures. That did not matter so long as scientists had no desire or ability to generate or measure ultrahigh vacuum (UHV; that is, ultra-low pressure) conditions. But by mid-century, scientists did have the means, motive, and opportunity to move toward UHV.

The first piece of the puzzle was the achievement of lower vacuum. Until the late nineteenth century, vacuum was generated mechanically — enclose some of the gas in a chamber separately in a piston or airlock; expel that gas into the outside environment; seal the airlock to the outside again and re-open it to the rest of the chamber (thereby leaving the chamber at a lower overall pressure); and repeat. In the late nineteenth and early twentieth century, though, numerous techniques emerged for coaxing gas out of a vacuum chamber by adsorbing gas molecules with cooled charcoal, mercury, low-pressure oils, and other media. Innovation of these diffusion and molecular pumps was greatly aided by a long-standing collaboration between Wolfgang Gaede (a professor in Karlsruhe) and the Leybold Company in Cologne.[1 p20–22] After World War II, ion pumps became the final-stage means of removing the last trace gases from a chamber by ionizing them and accelerating them into the surface of a cathode. Ion pumps were invented at Varian Associates, one of the early important firms in what later became Silicon Valley.[3 p182–186]

Ultrahigh vacuum would be useless, however, without some way to measure how "ultra" it was. One of the basic reasons to conduct experiments in vacuum is to very precisely control the environment that a sample (especially the surface of the sample) resides in. But there is little reason to control the environment more precisely than you can measure it. Curiously, in retrospect it appears that some of the diffusion pumps of the 1930s could generate vacuum two orders of magnitude lower than the ionization gauges of the day could measure — it was just that no one knew so at the time.[6 p34] Improvements to ionization gauges in the 1950s pushed their sensitivity down to the 10^{-11} Torr region (atmospheric pressure is 760 Torr) and made possible an explosion of activity in surface science and other fields that required ultrahigh vacuum.

A Community Forms

That explosion of activity soon led to the creation of institutions to support the growing vacuum community. In 1953, a Committee on Vacuum Techniques was formed in the United States, largely by scientists and engineers in industry.[2 p4] In 1957, the CVT became the American Vacuum Society (AVS), a full-fledged scientific professional society with increasing membership from government and universities and, after 1965, its own peer-reviewed journal. Nor was the United States alone; in 1958 an International Organization for Vacuum Science and Technology (IUVSTA) formed in Belgium as an umbrella organization for various national vacuum societies. This became the International Union for Vacuum Science, Technique and Applications in 1962.

The AVS and IUVSTA were needed in part because their meetings allowed researchers and equipment vendors to network and find out about each other's requirements and offerings. Once the frontiers of ultrahigh vacuum opened in the early 1950s, the toolkit of vacuum research first needed to be reconstructed and then expanded. Once one could generate and measure pressures of 10^{-11} Torr, sealed glass envelopes had to be replaced by stainless steel chambers with airlocks, portals, gauges, pumps, flanges, manipulators, and, of course, experimental apparatus and materials.

Whereas experimentalists could make a glass envelope themselves (or ask a technician to make one), a fully-kitted UHV system was much too complex for homemade equipment. Instead, experimentalists increasingly bought systems from vendors. Some continued to make their own instruments, which they placed inside their UHV systems; others simply waited until commercial versions of those instruments were available before buying them. From the 1960s onward, the number of such instruments proliferated rapidly — several of these are profiled in both the "Making" and "Characterization" sections of this volume. Thus, experimentalists needed AVS and IUVSTA conferences to hear about the newest instruments and to check out the ones that had matured enough that vendors had commercial versions on display in the exhibit hall.

The Physical Plant

The AVS and IUVSTA meetings were also useful places to learn less-obvious tricks for dealing with vacuum's vicissitudes. Obtaining and maintaining UHV conditions is an arduous process for which most materials are unsuitable. UHV systems first need to be "baked out" at high temperature in order to pry volatile compounds off the chamber walls before the system can be pumped down. Then, once vacuum is achieved, all the materials inside the chamber must be inert — they must not "outgas" molecules, which would ruin the vacuum. Every UHV researcher has had experiments ruined by poor choices of materials for either their samples or their equipment. AVS meetings were thus good places to exchange gossip about which materials would work in UHV conditions. Once a vacuum-stable material was identified, word-of-mouth would ensure that it quickly became an entrenched favorite. An interesting example is Viton, a brand of rubber marketed by DuPont, which UHV researchers enthusiastically adopted because it provided the qualities of ordinary rubber without the outgassing.

Over time, UHV researchers became more and more adept at working in vacuum and had more technological options for how to manipulate and characterize samples in vacuum. Thus, after about 1980, vacuum systems began to grow in size and complexity. Vendors developed multi-chamber systems in which multiple specimen preparation and characterization techniques could be housed and samples could be moved from one technique to another without breaking vacuum. To get a sense of what the growth in vacuum systems entailed, let us return to the conceit of this volume and imagine what we would see if we walked into a materials laboratory. Figure 4.0.1 shows the kind of glass-envelope experiment we would have seen during most of the twentieth century. Experiments in the glass envelope days were small and light enough to be easily portable. Many scientists continued to display them on their bookshelves as relics long after experiments with those devices were over. In contrast, Figure 2.3.1 shows a typical post-1980 UHV system weighing hundreds of pounds and occupying a footprint approaching that of a small car.

As Figure 2.3.1 indicates, though, if you wandered into a post-1980 lab that owned a UHV system, you might only be able to glimpse a small piece of that system due to the yards of aluminum foil wrapped around it. Foil is the most distinctive visual and material characteristic — in a superficial sense — of labs that own a UHV system. The foil is used to distribute heat when "baking" and to protect the system from airborne contamination in between bake-outs.

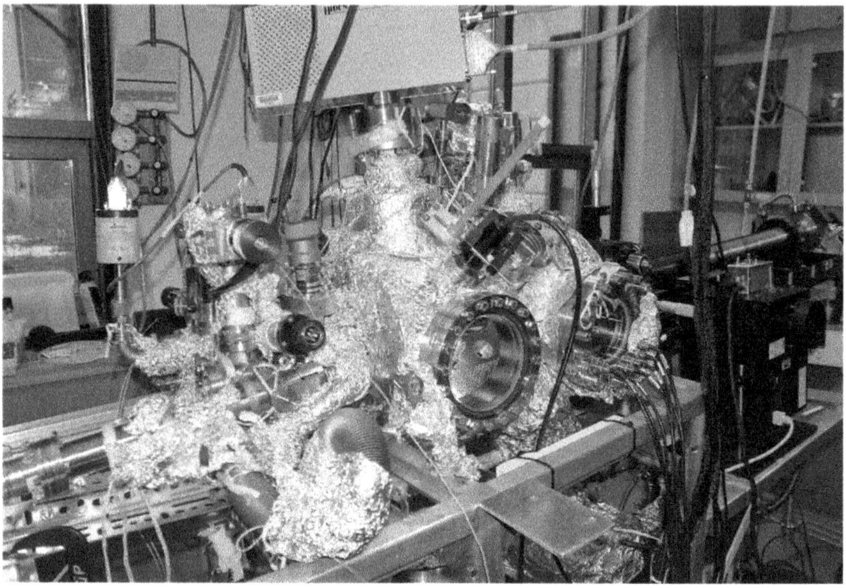

Fig. 2.3.1. Modern, multi-chamber UHV apparatus. Note the tin foil. Also note the flanges, where different segments of the system connect together, or where the system can be opened to the outside. Note too the portals and manipulators offering the experimenter visual and tactile access to the system without breaking vacuum. *Credit*: Photo courtesy of the Cynthia Friend Lab, Harvard University.

If you happened to wander into a lab during bake-out of its UHV system — well, bad luck, because you were not going to see any materials research being done any time soon. It often took several days to bring a system from atmospheric pressure down to UHV conditions. For many researchers, that meant a highly variable work pattern — conducting experiments twenty-four hours per day when the system *was* working, and reading and doing theoretical work for the long stretches when it was not.

Why Vacuum?

At this point we must ask, why were researchers so interested in experimenting with materials at very low pressures? An answer can be inferred from the order in which the AVS added divisions catering to specialized interest groups: vacuum metallurgy (1961); thin films (1964); surface science (1968); vacuum technology (1970); electronic materials and processing (1979); fusion technology (1980) — tellingly, renamed plasma science and technology in 1986; nanometer-scale science and technology (1992); magnetic interfaces and nanostructure (1999); and, interestingly, biomaterial interfaces (2003).

The vacuum technology division was the original core of the AVS. Its members were largely industrial engineers who built vacuum systems for everything from particle accelerators to iceberg-lettuce shippers.[5] Gradually, though, the AVS was taken over by people who wanted to do research and development in vacuum, rather than build vacuum systems for other people to use. Metallurgists were the first because techniques such as sputtering and vacuum welding were developing rapidly in the 1950s as new ways to modify metal surfaces.

Surface science was next because UHV conditions were necessary for the discovery and/or study of a variety of surface phenomena such as surface reconstructions in semiconductors. Surface scientists were important sources of, and markets for, UHV technology and (especially) the specimen preparation and characterization tools that went into UHV chambers. Surface science was perhaps the quintessential Cold War field, with close ties to the microelectronics industry (surface effects are important in integrated circuits), and to the civilian and military space and missile programs (which had an interest in what happens to solid surfaces exposed to high vacuum). In the 1970s, surface scientists were also deeply involved in the boom in environmental research (since surface reactions are important in catalytic converters), which also led to the creation of AVS's fusion technology division. The thin films and electronic materials and processing divisions were were closely linked to the surface science division, as their memberships were often employed by the same companies (for example, IBM or Bell Labs). These divisions used vacuum techniques such as sputtering to grow thin films and other components of integrated circuits. Whereas surface scientists traditionally focused on the atomic scale (ten-billionths of a meter) and thin films/electronic materials researchers on the micron scale (millionths of a meter), by the 1990s all three groups had converged on the intermediate nanoscale (one to 100 billionths of a meter) — hence the 1992 founding of AVS's nanometer-scale division. UHV equipment remains an important enabling technology for nanoscience and nanotechnology.

Over time, life scientists have also been enrolled into nanoscience research — hence the AVS's 2003 founding of a biomaterials division. As other contributions to this volume show, the inclusion of biomaterials alongside traditional metals and semiconductors has typified materials research generally since the late 1990s. Vacuum research is a particularly notable case, though, because vacuum conditions are generally inimical to living beings, and organic compounds were long thought of as antithetical to maintaining a good vacuum. Nevertheless, vacuum can provide the highly controlled conditions necessary for study of the structures and phenomena that are of interest to life scientists today.

Conclusion

The creation and study of vacuum was essential to the invention of science in the seventeenth century. For almost three hundred years, though, the technology for creating and maintaining vacuum environments stayed roughly the same, even as the science done in vacuum changed dramatically. The middle of the twentieth century, however, saw a convergence of innovations in vacuum pumps, gauges, and chambers leading to an ultrahigh vacuum revolution. The UHV revolution might have meant little, however, had it not been for a variety of postwar industries (especially microelectronics) and state initiatives (especially in aerospace fields) that needed to know how materials behave in vacuum, and/or that needed delicate, tiny surface structures that could only be manufactured in ultrahigh vacuum. Thus, UHV was one of the key tools of materials research in that it helped push our knowledge of materials from the bulk out to the surface.

References

1. Hablanian MH. Comments on the history of vacuum pumps. Journal of Vacuum Science and Technology A. 1984;2(2):118–25.
2. Lafferty JM. History of the American Vacuum Society and the International Union for Vacuum Science, Technique, and Applications. Journal of Vacuum Science and Technology A. 1984;2(2):104–9.
3. Lécuyer C. Making Silicon Valley: innovation and the growth of high tech, 1930–1970. Cambridge: MIT Press; 2006.
4. Melmed AJ. Recollections of Erwin Müller's laboratory: the development of FIM (1951-1956). Applied Surface Science. 1996;94/95:17–25.
5. Petrick GM. 'Like ribbons of green and gold': industrializing lettuce and the quest for quality in the Salinas Valley, 1920-1965. Agricultural History. 2006;80(3):269–95.
6. Redhead PA. The measurement of vacuum pressure. Journal of Vacuum Science and Technology A. 1984;2(2):132–38.
7. Reich LS. The making of American industrial research: science and business at GE and Bell, 1876–1926. Cambridge: Cambridge University Press; 1985.
8. Shapin S, Schaffer S. Leviathan and the air-pump: Hobbes, Boyle, and the experimental life. Princeton: Princeton University Press; 1985.

CHAPTER 2.4

CONTAMINATION CONTROL

DAN HOLBROOK

A contaminant can be defined as anything that interferes with or frustrates imposed order. In the words of Mary Douglas, contamination is "matter out of place" and "essentially disorder."[11 p2; 35] More specifically, a contaminant is "any unwanted particulate, gaseous, liquid, solid, dissolved matter, or radiation within any environment."[32 p3] The technologies of contamination control are relatively simple and few. The most important is controlled ventilation, with filtration an integral element. Another is the use of materials that do not contribute contaminants (particles or vapors) to either the environment or to processes. Lastly, cleaning technologies, both mundane and advanced, contribute to contamination control.

Controlling Contamination

The desire to control contaminants emerged in several industries, for reasons of both worker health and product and process integrity. Since the mid-nineteenth century, chemistry and metallurgy have played increasingly important roles in determining the characteristics of materials, and were put to use by an increasing range of industries.[1] A recent article outlining the emergence of materials science as a discipline asserts that the core of that discipline is the "composition, structure, and processing of materials."[2 p223] Neither the chemistry nor the atomic and crystalline structures of materials can be understood without successful attempts to control their composition; that in turn is only achievable by controlling contamination.[23]

Ventilation

Systems and technologies designed to ventilate spaces certainly have their histories, but coverage is uneven. The most basic ventilation technology, the chimney, has not yet found its serious historian; the best accounts indicate its appearance in the West sometime in the twelfth century.[6] Prior to that, smokes and fumes vented through simple holes in roofs and walls. Interior structures designed to help capture smoke and fumes — smokeboxes and hoods — seem to predate common use of the chimney, but their functionality certainly increased when paired with the increased draft a chimney provided.[30 pxxvii] Depictions of alchemists' laboratories clearly show hoods (Figure 2.4.1). Hoods are common to laboratories, and serve both to evacuate hazardous vapors and dusts, and to protect samples and other workpieces from the environment of the lab. With

Fig. 2.4.1. A sixteenth-century woodcut by Hans Weiditz depicting an alchemist's workshop. Note the hood on the left-hand side.

the additions of doors, filters, shrouds, and other methods of controlling the flow of air, the hood remains a vital part of laboratory contamination control.[33 p6; 4 p37–38]

Complex mechanized ventilation systems emerge from a variety of activities and industries. Air conditioning, which controls air temperature, humidity, and dirt content, and then moves it in the directions and amounts desired, developed in the late nineteenth century, solving processing issues in a variety of industries whose products demanded cleaner conditions.[8 p119–130; 19 p2–5] The knowledge and skills required to engineer ventilating systems — early on the captive of a few firms and a few men within those firms — spread and soon became routinized.[7]

Some facility types demanded especially complex ventilation systems. A surgical operating room (OR), for example, would need to be designed to thwart the presence and movement of biological agents.[14 p177–178] Ventilation needs also dictated control of air within larger structures. To continue with the surgery example, the rooms surrounding an OR needed protection from contamination generated in the OR; designing and constructing ventilation systems that maintained lower overall air pressure in the OR ensured one-way flows of airs within multi-room complexes. Much effort in the mid-twentieth century, especially in England and the United States, went into controlling not only the contamination present on surfaces and in the air, but also that generated by the human occupants of a space.[14 p178] Laboratories are an equally important example. Indeed, ventilation needs drove building design for laboratories, as reflected in the fact that ventilation accounts for up to half of the costs of designing and building laboratory spaces.[33 p3; 24 p65]

Filtration

Filtration is central to modern contamination control. High-efficiency particulate air (HEPA) and ultra-low particulate air (ULPA) filters resulted from the same set of industrial needs that account for most contamination-control technologies. Dusts, along with their cousins mists, smokes, clouds, and fumes, presented obstacles to a number of the desired goals of engineers, scientists, and manufacturers. Military needs during both World Wars added pressure and funding to the search for better filtration, and set university and industry labs to work on these problems.[22 p39]

Advances in gas mask filters, and, later, filters able to make larger spaces relatively free of particulate and gas/fume contamination, resulted from research conducted most prominently at the Harvard School of Public Health's (HSPH) Department of Ventilation and Illumination, with financial help from a range of relevant industries.[9; 28; 29] Early analytical and laboratory work at the school included investigations into dust and disease among manganese workers, in jute mills, and in the lead, zinc, and gas industries.[29 p105–111]

Interactions between HSPH and industry were reinforced by cooperation between dust researchers and government agencies. The US Bureau of Mines, like HSPH founded largely in reaction to late-nineteenth-century concerns with the health and welfare of industrial workers and urban residents, had been researching dust since at least 1913.[12 p80; 27 p10–11; 18] The Bureau had also been involved with the Army's attempts to develop improved gas masks during World War I, based on its experiments with developing respirators for use in mines.[3] Anxiety in the run-up to World War II brought renewed energy to this research; the risk of battlefield gas necessitated the safeguarding of the air in operational buildings on the front, where whole rooms needed the same protection that gas masks provided. A relatively simple adaptation of gas-mask filter paper folded into "deep pleats separated by a spacer panel and sealed inside a rectangular frame using rubber cement" and integrated into a building ventilation system handled the particulate elements of chemical, biological, and radiation aerosols.[5 p2–3] The Manhattan Project presented particularly difficult filtration problems, largely solved by adopting these so-called "absolute" filters. Declassified after the war, high-efficiency filters made their way rapidly into a wide range of industrial and research settings.

HSPH once again was a primary actor. Its Air Cleaning Conference, held almost every year starting in 1951, constituted the major avenue for the dissemination of information on air cleaning techniques, equipment, problems, and research questions, initially for Atomic Energy Commission (AEC) personnel, then for its contractors, and, before the 1950s were done, for a wider industrial audience.[13] The later development of so-called ULPA (Ultra Low Penetration Air) filters, which to be classified as such must remove 99.999% of 0.1–0.2 μm particles, was motivated largely by the further miniaturization of various technologically important devices, particularly semiconductor electronics. These filters do not differ in concept or basic construction from HEPA filters, but use a filter medium constructed using smaller fibers.[19]

Construction Materials

The construction of spaces where contamination is an issue must take into account the materials composing the floors, ceilings, and walls, as well as the materials making up fixtures and furniture. The selection criteria are straightforward; the materials must not themselves contribute contaminants, and they must be easily and thoroughly cleanable. Cleanability also involves construction details — coved corners, fixture leg design — that both frustrate dirt accumulation and allow access with brooms, mops, wipes, and so forth.

Historically, these materials and construction techniques originated in efforts to control contamination in surgical operating rooms and food processing facilities, including restaurants. Some of the materials were of course old and well-known, such as marble and glazed tiling. Others were of more recent vintage, such as enameled sheet metal (although enameling is of course also an ancient technology), which had found applications in food facilities.[16 p96–97; 20 p18–20,25–26; 10] Stainless steel, plastic laminates, vinyls, and epoxies have also found their applications in contamination-control situations.[14]

Fabrics: Wipers and Uniforms

Though thoroughly prosaic, wiping and wipers are a major part of maintaining clean working conditions. Wiping (taken broadly to include mopping), forms the front line in contamination control.[15]

The increasing need for clean conditions pushed the development of specialized textiles for use in wiping and mopping, along with systems of use. Simple wadding and cotton rags produced far too many particles; nylons and other fabricated materials promised less detritus and were soon adopted. Microfiber, non-woven textiles, usually adapted to specific cleaning environments, became common by the end of the twentieth century.[30] Likewise, mopping systems using special fiber heads became the norm for highly clean systems. So-called "critical" and "precision cleaning" put an emphasis on "the cleaner and the cleaning equipment [being] chosen to work together and address the particular cleaning application."[17 p39] Prevention of contamination also involved the uniforming of personnel, again using specialized lint- and particle-free textiles. Perhaps most important in cleaning is the proper behavior of individuals involved. Workers and cleaners alike, if contamination is to be controlled, must adhere to strict limits on their behaviors.[17]

Conclusion

The tools and equipment used to control contamination in laboratories are, as outlined above, neither complicated nor, except in the case of very high standard cleanrooms, extraordinary. Nonetheless, they are crucial to almost any attempt to control the composition of materials. Semiconductor materials provide a prime example of the importance of contamination control. Semiconductor device performance depends almost completely on the composition of materials; purity levels as high as one part per ten billion (1:10,000,000,000) are regularly attained in silicon and germanium. The addition of other materials, dopants, alters their electronic characteristics; for these characteristics to be governable, the dopants also must be of extraordinary purity. With other materials, as with semiconductors, tailoring characteristics to specific applications demands controlling their structure and composition — the essence of materials science. In the laboratory and on the manufacturing floor, contamination control is fundamental to that work.

References

1. Bensaude-Vincent B. The concept of materials in historical perspective. NTM Zeitschrift für Geschichte der Wissenschaften, Technik und Medizin. 2011;19(1):107–23.
2. Bensaude-Vincent B. The construction of a discipline: materials science in the United States. Historical Studies in the Physical and Biological Sciences. 2001;31(2):223–48.
3. Brophy LP. Origin of the chemical corps. Military Affairs. 1956;20(46): 217–26.
4. Buie, John. Evolution of fume hoods. Lab Manager. 2011;December 1:36–37.
5. Burchsted CA, Kahn JE, Fuller AB. Nuclear air cleaning handbook: design, construction, and testing of high-efficiency air cleaning systems for nuclear application, 4th ed. ERDA 76-21. Oak Ridge, TN: Oak Ridge National Laboratory; 2003.
6. Butler OR. Smoke gets in your eye: the development of the house chimney. Ultimate History Project. Available from: http://www.ultimatehistory project.com/chimneys.html.
7. Carpenter RC. Heating and ventilating buildings: an elementary treatise. New York: J. Wiley & Sons; 1895.
8. Cooper G. Air-conditioning America: engineers and the controlled environment, 1900–1960. Baltimore: Johns Hopkins University Press; 1998.
9. Curran JA. Founders of the Harvard School of Public Health, with biographical notes, 1909–1946. New York: Josiah Macy, Jr. Foundation; 1970.
10. Dickens EHF, editor. Porcelain enamel in the building industry. Washington, DC: Building Research Institute, National Academy of Sciences, National Research Council; 1953.
11. Douglas M. Purity and danger: an analysis of concepts of pollution and taboo. London: Ark Paperbacks; 1984.
12. Drinker P, Hatch TF. Industrial dust: hygienic significance, measurement and control. New York and London: McGraw-Hill; 1936.
13. First MW. A brief history of the air cleaning conferences. In Proceedings of the 23rd DOE/NRC nuclear air cleaning conference, Buffalo, NY, 25–28 July 1994. 1995:3–15.
14. Holbrook D. Controlling contamination: the origins of clean room technology. History & Technology. 2009;25(3):173–91.
15. Holbrook D. Discipline and polish: on wipers and wiping. Maintainers; Hoboken, NJ2016. Available from: http://themaintainers.org/s/Discipline-and-Polish-Holbrook.pdf
16. Jakle JA, Sculle KA. Fast food: roadside restaurants in the automobile age. Baltimore: Johns Hopkins University Press; 1999.
17. Kanegsberg B, Kanegsberg E. Handbook for critical cleaning. Boca Raton: CRC Press; 2001.
18. Katz EN. The white death: silicosis on the Witwatersrand gold mines 1886–1910. Johannesburg: Witwatersrand University Press; 1994.

19. Klocke SD, Whyte W. High efficiency air filtration. In Whyte W, editor. Cleanroom design, 2nd ed. Chichester, UK: John Wiley and Sons; 1999.
20. Langdon P. Orange roofs, golden arches: the architecture of American chain restaurants. New York: Knopf; 1986.
21. Legget RF, Hutcheon NB. The design of research laboratories. Part I: A general assessment. Part II: Air conditioning and conditioned rooms. Ottawa: National Research Council of Canada; 1966.
22. Maines R. Asbestos and fire: technological trade-offs and the body at risk. New Brunswick: Rutgers University Press; 2005.
23. Mody CCM. A little dirt never hurt anyone: knowledge-making and contamination in materials science. Social Studies of Science. 2001;31(1):7–36.
24. National Research Council Committee on Design, Construction, and Renovation of Laboratory Facilities. Laboratory design, construction, and renovation: participants, process, and product. Washington, DC: National Academy Press; 2000.
25. New marvels of food factories. Popular Science. July 1934:20–21.
26. Parker ME. Food-plant sanitation. New York: McGraw-Hill; 1948.
27. Powell FW. The Bureau of Mines: its history, activities and organization. New York: D. Appleton and Company; 1922.
28. Sellers CC. Hazards of the job: from industrial disease to environmental health science. Chapel Hill: University of North Carolina Press; 1997.
29. Shattuck GC. Industrial medicine at Harvard. 1954.
30. Shuffrey LA. The English fireplace: a history of the development of the chimney, chimney-piece and firegrate with their accessories, from the earliest times to the beginning of the sixth century. London: B. T. Batsford; 1912.
31. Siegerman H. Wiping surfaces clean. Amherst, NH: Vicon Publishing; 2004.
32. Sivinski HD, editor. Contamination control principles. Albuquerque: Sandia Corporation; 1967.
33. Smith HL, editor. Buildings for research: an architectural record book. New York: F. W. Dodge Corporation; 1958.

CHAPTER 2.5

TRADE CATALOGS

MAX MOELLER

Trade catalogs are marketing tools created by businesses for use in a particular time and place to advertise products and make money. But since all products are made out of materials, it stands to reason that old trade catalogs may have something to contribute to the history of materials research. Even more, trade catalogs were at once a way that researchers might find materials or products that were useful for their work and the vehicle through which the products of their work were often disseminated. So the research value of trade catalogs should not be overlooked.

Originally, trade catalogs were produced "for the trade only," meaning they were shared or directed only to those involved in the supply side of an industry.[7] A catalog of goods, often accompanied by a price list, would be supplied by manufacturers to retailers in the hopes of securing an outlet for new products. This simple marketing tool evolved to take advantage of different opportunities over time. Single-page broadsides hawking wares would be tacked up in public spaces to attract the attention of passersby. The development of a national mail-distribution system allowed manufacturers to mail catalogs directly to consumers. Special-issue catalogs were created for, and distributed at, important events such as World's Fairs. And a premium system was developed so loyal, repeat customers could select rewards based on a point system. The print catalog format was stretched and pulled in every conceivable direction in order to share information and take full advantage of the medium's potential to maximize profits.

Today, many of these marketing efforts have moved online. Yet in spite of the internet, we are still awash in trade catalogs. They are mailed to our homes and offices. They are given away at stores or handed out on the street. They are omnipresent, and yet as little valued as ever — much like the materials they were so often used to distribute.[10] Given this oversight, what can we learn about the historical use of trade catalogs within the field of materials research?

Trade catalogs were used widely and appropriately as tools in the field of materials research to acquire standard materials, supplies, and instruments for laboratory use. Synthesizing materials could be a tricky, elaborate, and extensive process. A company or laboratory that specialized in making certain types of materials or instruments could find a market among other companies or laboratories that needed those materials or instruments for other tasks. Trade catalogs, which listed the essential stuff of laboratory life that could be bought off the shelf, therefore reflect the intricate web of dependencies that build up as materials research became a more industrial enterprise through the mid-twentieth century.

But in materials research and elsewhere, trade catalogs were taken only at face value. They served a purpose and that was that. There was no obvious reason to collect or preserve such ephemeral items for the future, so most were trashed or recycled as junk mail without thought. As a result, old trade catalogs have become rare. But some libraries recognize the research value inherent in such publications and collect them for future use.[12] An exploration of trade catalog collections can offer additional insights beyond a narrow focus on their use within a particular industry. So, what can we learn from trade catalogs with regard to materials research? Consider nylon, the world's first truly synthetic fiber. DuPont synthesized a polymer into a textile fiber in the lab for the first time in 1935.[3] Nylon was formally announced to the public on October 27, 1938. DuPont focused its initial marketing of nylon solely on hosiery, samples of which were displayed at the New York World's Fair in 1939. This built tremendous anticipation before nylon hosiery first went on sale to the public in May 1940. Because nylon was a completely new material, it captured the public imagination like no other fiber has, before or since. It is hard to overestimate its appeal — whereas today we are awash in synthetic materials and mass-produced products, these were novelties in the early twentieth century and carried the kind of cachet we now reserve for the natural and the handmade. But understanding why it appealed historically is different from understanding the particulars of how it was first sold both within the textile industry and to consumers. The devil is in the details, and the adoption of nylon by manufacturers and retailers can be traced in trade catalogs to a great extent.

"What's New in Nylon" is an unusually detailed catalog found in the Hagley Library's collection.[2] Although technically it does not fit within the parameters of a trade catalog as defined above, it was categorized and filed as a trade catalog at the DuPont Experimental Station library before eventually being gifted to Hagley, where it is also shelved as a trade catalog, which makes a larger point: you never know what exactly you will find when researching within trade catalog collections in the library! "What's New in Nylon" (Figure 2.5.1) dates from December 1941 and specifies where the new nylon products could be purchased in Wilmington, DE, and Philadelphia, PA. It lists the different types of products nylon was used to create, the names of manufacturers employed to produce these products, the price of such items, and the retail stores where they were available for sale.

Whereas Wikipedia might tell you that nylon was a sensation and proved extremely popular with customers lined up for blocks to get their hands on nylon stockings, trade catalogs like "What's New in Nylon" can tell you specifically that 51 gauge nylon stockings were being manufactured by Oakbrook and sold for $1.65 each at the Famous Maid retail shops. You can also learn that 51 gauge nylon stockings were available for ten cents less at Lerner's Shops. Blouses, brassieres, brushes, fishing line, girdles, gloves, neckties, raincoats, and shower curtains are just a sampling of the products detailed in this catalog, a mere eighteen months after nylon hosiery was first offered for sale. The wealth of information provided offers an unusual

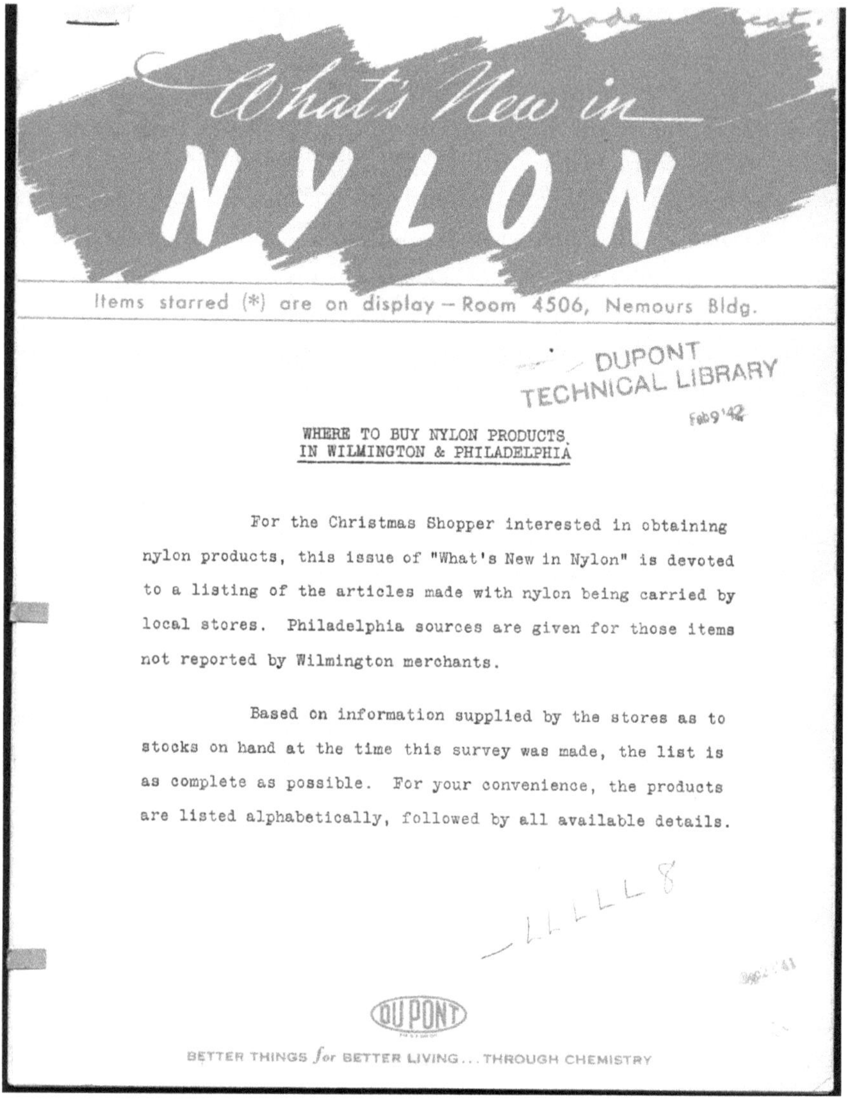

Trade cat.

What's New in
NYLON

Items starred (*) are on display — Room 4506, Nemours Bldg.

WHERE TO BUY NYLON PRODUCTS
IN WILMINGTON & PHILADELPHIA

For the Christmas Shopper interested in obtaining nylon products, this issue of "What's New in Nylon" is devoted to a listing of the articles made with nylon being carried by local stores. Philadelphia sources are given for those items not reported by Wilmington merchants.

Based on information supplied by the stores as to stocks on hand at the time this survey was made, the list is as complete as possible. For your convenience, the products are listed alphabetically, followed by all available details.

DUPONT

BETTER THINGS for BETTER LIVING . . . THROUGH CHEMISTRY

Fig. 2.5.1. The catalog "What's New in Nylon?," a piece of direct-to-consumer marketing with a production gloss that contrasts the functional aesthetic feel of a trade catalog, emphasizing the origins of a new product in industrial materials research.

snapshot of the market at this moment in time. Researchers could have a field day charting price variations, distribution patterns, business networks, and point-of-sale locations. The example of how nylon was marketed as a consumer product through catalogs helps us see how materials in general became consumer products, some targeted at the general public, some targeted at specialist laboratories.

Trade catalogs can reveal other kinds of surprises as well. The "Kendex … Guaranteed" trade catalog (Figure 2.5.2), produced by Kendex Corporation in 1951, functions as a sales manual that details not only its nylon hosiery but also shares business details in order to attract prospective independent sales dealers.[5] It discusses the many advantages of working with Kendex, detailed conditions for a product guaranty, and multiple photographic illustrations of interior and exterior working conditions at the Kendex production facility in Babylon, NY. The catalog also contains a wealth of additional items "laid in" to promote its hosiery line. Those items contain additional information on purchasing, more advertising matter, supply order forms for retailers, a booklet on how to make money selling hosiery, a typewritten pitch letter, a sample card depicting the range of hosiery colors available, and a packaged full sample of Kendex first quality full fashion nylon hosiery, style no. 251 (15 denier, 51 gauge). It is rare to find such a wealth of primary-source printed and sample materials.

A similar bonanza awaits researchers looking to study the nylon output of the Kentcraft Hosiery Company. "Guaranteed Socks, worn by millions coast to coast" was produced in 1955, and the catalog goes to great lengths touting the wonders of nylon for its full line of men's hosiery.[6] The catalog also contains multiple items laid in, such as advice on how to become self-employed, order forms for display samples, customer testimonials, a list of premium gifts available for varying levels of sales accomplishments, and one wrapped sample pair of "miracle socks" made of 100% DuPont nylon.

These sources can tell us a great deal about how information about materials and the consumer products made from them traveled in the mid-twentieth century, and how that in turn influenced the industrial production of materials. It is tempting to think of

Fig. 2.5.2. A Kendex Corporation trade catalog highlighting the industrial production process.

industrial laboratories identifying a demand for a new material and deploying eager researchers to tackle the problem. But just as often, laboratories at places like DuPont conducted research into new materials without a preconceived need in mind because the company was confident that it could create a new market, in part via tools such as trade catalogs. In that sense, trade catalogs are "tools for materials research" insofar as they are tools for justifying and commercializing the type of materials development industrial researchers were conducting, and nylon offers a particularly potent example of this practice.

Nylon appears repeatedly in catalogs across industries: safety, work clothing, shoes, fishing, sports, carpeting, and others. It is also one of the few new materials that achieved name recognition, so nylon added value both to products like shoelaces or jacket linings and to the advertising and catalog titles that prominently bore its name. Want to study outdated office technology? Why not try "Data Processing Nylon Post Binders, Indexes and Accessories," by the Wilson Jones Company, from 1970.[14] "Control the paper explosion!" of office reports and electronic data processing record keeping by purchasing sturdy nylon post binders, available in three lengths, two styles and multiple colors. "The special nylon is formulated to withstand impact shocks, temperature changes, even frequent misuse." The catalog provides great detail on the construction of binders, available accessories, and the intended uses of this product.

In addition to consumer products, the benefits of nylon were quickly appreciated for industrial applications. J. E. Rhoads and Sons had its origins as a tannery in the 1720s. Gradually, the company shifted focus and produced leather power transmission belts for industrial use. Trade catalogs for this firm suggest that the company quickly recognized the benefits of nylon, incorporating the new fiber into its belt production processes beginning in the 1950s. Unlike the traditional material of leather, its brand of Texalon nylon belting provided greater strength and longer life, maintained continuous belt tension, eliminated edge fray, and ultimately reduced production costs.[4] Multiple Rhoads catalogs from the 1950s through the 1970s, available at Hagley, demonstrate the advantages and applications of

Texalon belts, hinting that nylon's success often came at the expense of traditional materials. Rhoads managed to make a successful business transition, moving from leather to a new and improved synthetic material. Other businesses were not always so flexible, however, and trade catalogs can be used to study the range of effects new materials had when introduced into industrial processes.

Catalog No. 22 (1971) for the Reliable Nylon Sling Company touts the same strength and continuous tension benefits of nylon as discussed in the example above, this time while promoting airlift slings. "When the lift is Special, and the load is Valuable, they specify Reliable Nylon Slings."[9] The company also manufactured nylon cargo nets, small tarpaulins, canvas bags, and tool rolls.

More thorough research shows that nylon can be found readily, even when it is not recorded as a search term in the library's online catalog. My perusal of trade catalogs for industries that are likely to use nylon paid off handsomely, as multiple examples of the use and promotion of nylon surfaced. The "Big Smith Price List Number 435" effective May 4, 1970, shows that 100% nylon was being used as a fabric material for long-line quilted utility coats, as well as for the inner lining of overalls and uniform jackets.[11] "Aramsco Safety Supplies for the Hazardous Environment," circa 1988, records several applications of nylon, such as Sinco Retract-A-Matic Lifelines, yellow PVC/Nylon weave bib overalls and jackets, and even brooms with nylon bristles.[1] The 1996 catalog of the National Safety Equipment Outlet records the sale of water-resistant nylon tote bags, nylon handheld Safe-T Flags, and Keepr Vehicle Recovery Straps.[8] The Fall 1999 Dickies Workwear catalog shows the continued use of 100% nylon lining within their range of insulated coveralls and jackets.[13] The 2016 Hytest Safety Footwear Network catalog shows the repeated use of nylon shanks for the Wolverine brand of work boots and nylon mesh lining in the Caterpillar brand.[15]

While conducting such research, even the absence of nylon can suggest implications and raise fruitful questions. Perhaps nylon had not yet penetrated certain industries or markets. Was this absence corrected or continued at a later date? Only a comprehensive overview of catalogs across a range of dates could circumstantially prove such developments, yet that is one of the many ways in which trade catalogs can be of use for the study of materials research.

This ability of trade catalogs to be used for comparative purposes is one of their greatest benefits to researchers, yet it also points out a crucial limitation. Trade catalogs can only be compared if the library in question holds sufficient quantities of the relevant industry catalogs. Frequently, needed catalogs are available sporadically or scattered among different collecting libraries. Persistence and patience, along with a healthy dose of luck, come in handy when faced with the challenge of working with trade catalogs. Yet the benefits are real, and trade catalogs can provide information that is simply unavailable elsewhere.

References

1. Aramsco, a Herbert Abrams Company. Safety supplies for the hazardous environment. Thorofare, NJ; ca. 1989.
2. E.I. du Pont de Nemours & Company, Inc., Nylon Division. What's new in nylon: where to buy nylon products in Wilmington & Philadelphia. Wilmington, DE; 1941.
3. Hounshell DA, Smith Jr JK. Science and corporate strategy: Du Pont R&D, 1902–1980. Cambridge: Cambridge University Press; 1988.
4. J. E. Rhoads and Sons. Rhoads Texalon nylon belting: power transmission, conveying tapes. Wilmington, DE; 1957–1972.
5. Kendex Corporation. Kendex...guaranteed. Babylon, NY; 1951.
6. Kentcraft Hosiery Company. Guaranteed socks: worn by millions coast to coast. Brooklyn; 1955.
7. McKinstry ER. Trade catalogues at Winterthur: a guide to the literature of merchandising, 1750–1980. New York: Garland Publishing; 1984.
8. National Safety Equipment Outlet. National safety: America's safety source. Charlestown, NH; 1996.
9. Reliable Nylon Sling Company. Reliable Nylon Sling Co.: 225 W. Lehigh Ave., Philadelphia, Penna. . . Philadelphia; 1971.
10. Romaine LB. A guide to American trade catalogs, 1744–1900. New York: R.R. Bowker Company; 1960.
11. Smith Brothers Manufacturing Company. Price list number 435 effective May 4, 1970. Carthage, MO; 1970.
12. Walls, Nina de Angeli. Trade catalogs in the Hagley Museum and Library. Wilmington, DE: The Eleutherian Mills-Hagley Foundation; 1987.

13. Williamson-Dickie Manufacturing Company. Dickies: Fall 1999: women's wear, denim, workwear, kid's wear. Fort Worth, TX; 1999.
14. Wilson Jones Company. Wilson Jones data processing: nylon post binders, indexes and accessories. Chicago; 1970.
15. Wolverine Worldwide, Inc. Hytest Safety Footwear Network: 2016 Catalog. Rockford, MI: The Corporation; 2015.

CHAPTER 2.6

BIBLIOGRAPHIC TOOLS

DEBRA KOLAH

Many bibliographic tools that people take for granted today were not even imagined prior to the 1980s. In most cases, to access the scientific literature a researcher had to physically go to a library. Most research organizations (government and industrial laboratories as well as universities) had at least one and often several libraries. Libraries are still essential to materials research, but today many of their services are offered via a computer or phone — such that many users do not acknowledge or even realize how much they rely on the library. This entry surveys some of the most important means by which materials researchers once accessed information — tools that today are largely accessed remotely but which were invented and evolved in an era when the physical library, with its bookshelves, microfiche readers, and reference desks, was an obligatory point of passage. Professors mentored their students to research the literature. That included the books and journals located in labs, but also required students to absorb the collections physically located in the library.

This volume asks us to imagine walking into a late-twentieth-century materials research lab and looking at the tools on display. In that spirit, this entry discusses some of the tools the researchers in just such a lab would have used on their frequent trips to the library.

Chemical Abstracts

Many of the information tools that materials researchers relied on evolved over the second half of the twentieth century, transitioning gradually from a paper-only form, to paper and digital forms available in the library, to (in some cases) a paper form housed in a repository and a digital form available remotely or, more recently, to a digital-only form. One of the premier tools of materials research, *Chemical Abstracts*, offers a particularly clear example of this succession. In 1980 the CAS ONLINE was launched, making it possible for users (primarily librarians) to search the CAS REGISTRY database, which provided a centralized source of information about the structure and properties of chemical substances (Figure 2.6.1). Users with a specific model on an intelligent graphics terminal could

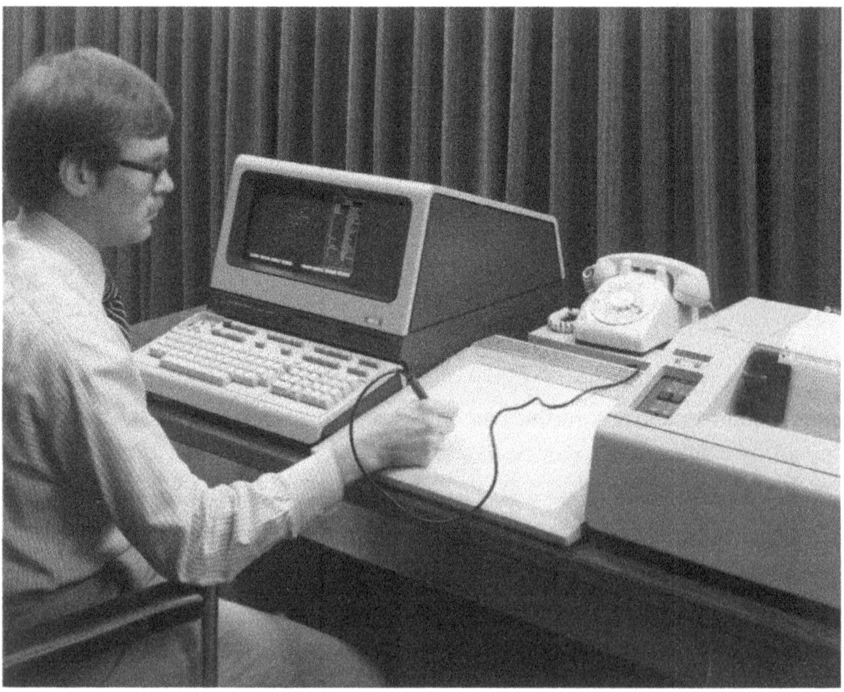

Fig. 2.6.1. Using a light pen and computer to draw a chemical structure (1976).[1] Used with permission of CAS, a division of the American Chemical Society.

select structure features from a menu, and then assemble them on the terminal monitor using a graphics tablet and stylus. These terminals could display answers to structural questions with well-drawn chemical-structure diagrams.[2]

Yet at the time, online searching was still very expensive, so most researchers relied on the print version of *Chemical Abstracts*, which listed recently published papers. The source was heavy and libraries generally only had one copy, so researchers had to wait to look at it. Thus, print and digital versions of the "same" information co-existed because each format had its advantages and disadvantages.

New Issues of Key Journals in Print

To keep current on developments in their fields (and adjoining ones), postwar materials researchers primarily relied on peer-reviewed journals. These included both the handful of established, prestigious journals that published materials-related research (such as *Physical Review Letters*) and a continually expanding array of newer, more specialized outlets such as the *Journal of Materials Research*, launched in 1986.

It is important to remember that, unlike some of the other information tools found in libraries, journals were tools that most materials researchers could both produce *and* consume. Since the 1990s (and, to a lesser extent, even before), digital tools have dramatically altered the experience of reading journal articles, but they have also altered the experience of writing for them. The number of journals has been increasing steadily for well over a century, but digital publishing has lowered the barrier to entry even further. The pace of moving an article through the publication queue was also, in general, slower: submitting a paper, getting back the referee reports, revising, and finally publishing took many months or more when manuscripts had to be typed (with a typewriter — and the "copy and paste" function involved literal paste!) and hard copies had to be mailed from author to editor to referee and back again. Paper journals also typically only allowed a maximum of three or four figures within an article, whereas today's online publishing allows the addition of supplementary figures and even videos or other media at a url for more information associated with the article.

From the late 1960s through the mid-1980s, scholars felt that they were overwhelmed with information, because of the amount of new journals being published. Although many labs and professors had access to some primary professional literature, to immerse themselves in journal articles researchers would most often come to the library to sit at large tables, skim tables of contents, and then read the most relevant papers and take notes. Individual researchers often subscribed to a few such journals, but no one had personal copies of all the journals that they needed to consult. Thus, communally owned copies — sometimes housed in a lab group office or departmental lounge but above all in a library — were indispensable. Preprints, which journals gave to authors to distribute, offered limited liberation from the tyranny of the communal copy. Eventually, photocopying would become popular for the same reason, though even in the early 1980s it was slow and expensive.

The Library

Thus, the physical library was not just a space; it was a place where researchers spent a significant amount of their working lives and where they interacted not just with books and other information tools but also with their friends and colleagues. It was, therefore, a site freighted with emotion. As Rice University materials scientist Ned Thomas remembered it:

> Researchers worked late into the night — sometimes forgetting the time. [I] actually often [got] so wound up in the reading that I forgot to eat dinner or whatever. It felt good just to have others around busily doing their work. Sometimes, by accident, I would be looking up a paper in a journal and the paper just in front or just after the one I was interested in, caught my eye. Then I would read something that I initially hadn't planned on ... sort of a gift. Sometimes, I met other researchers with similar interests, who were keen on getting me to give them the December

> issue of *Science* or *Philosophical Magazine* or *Nature* that I was reading because the librarian said that guy over there just took that issue.[5]

So what did they find in the library? Journals, of course; and increasingly, from the 1960s onward, digital tools along the lines of CAS ONLINE. But what else? Each library and researcher were different, but the following lists a few of the things that drew the latter to the former.

Table of Contents Booklet

Tables of contents (ToCs) were very popular. A researcher could scan over 100 ToCs from *Nature*, *Science*, *Journal of the American Chemical Society*, *Macromolecules*, and other journals and then go to the reading tables with the most recent journal issues and try to find the issue that contained the paper that the ToC booklet (printed semi-monthly) had pulled together.

Conference Proceedings

Conferences can be a major source of cutting edge research, particularly in science and engineering. At conferences, researchers present papers on their research and it is an opportunity to get direct feedback from colleagues in their field. The papers presented in the conference are then usually published in a volume called a conference proceeding. It may be months before such information is published as a journal article, or it may never be published.

Although now readily available online via library subscriptions or membership in a society or association, in the past these were printed volumes that were cataloged and maintained in library collections. For researchers who had not attended the conference, they provided a window into what was presented and, at the very least, an abstract of the talks.

Materials researchers who could not travel to attend conferences would use such volumes to keep up to date on the activities of the

Materials Research Society — established in 1973 by a group of scientists who "shared the belief that their professional interests were broader in scope than existing single-discipline societies and that a new interdisciplinary organization was needed"[4] — or of the American Chemical Society, formed in 1879, or of the American Physical Society, formed in 1899. Libraries collect and maintain the conference proceedings for all associations that the institution values.

Preprints

As mentioned above, preprints were copies of individual articles that journals gave to authors to distribute in order to liberate researchers from reliance on communally owned, ponderous, bound copies of journals. Most preprints were owned by individuals and often stored in filing cabinets. However, libraries also housed collections of preprints, some of which were substantial enough that rudimentary computer programs were used to create searchable databases of the preprints that were locally available. Much later, in 2001, Paul Ginsparg would win a MacArthur Fellowship for the creation of arXiv, which launched in 1991 to serve as an online repository of preprints and to eliminate the need to store preprints in emails, or on library shelves.

Handbooks

Several organizations published handbooks giving extensive reference information on various aspects of materials research. Some of these organizations were government agencies, such as the US National Bureau of Standards (now called the National Institute of Standards and Technology). Others were non-governmental standards-setting bodies such as the American Society for Testing Materials. Often, handbooks were started by an individual or organization for a specific purpose, which they then long outlived, such as *Beilstein's Handbook of Organic Chemistry* or the Chemical Rubber Company's *CRC Handbook of Chemistry and Physics*. Many libraries held multiple copies of such handbooks, which were heavily used for looking up chemical structures and physical data (melting points, for instance).

In the past, print versions of such handbooks could be found in the reference areas of libraries. Today, some handbooks are still available in print, though others — such as Beilstein — have migrated entirely online.

Browsing Books on the Shelves

Researchers were physically able to encounter the collection of books libraries held on a particular topic. A researcher looking for information on a particular topic would often find a large collection of relevant volumes located on the library's shelves — and without endless searching in databases, or exploring the online holdings of other library collections. Serendipity was the term used to describe a library user's almost magnetic affinity for books that were somehow there right at the time they needed to find them. Of course, librarians worked endlessly to ensure that collections mirrored the needs of researchers — scouring the research interests of professors, and sometimes influencing publishers to contract more books in a certain subject area.

Interlibrary Loan (ILL)

Researchers had different experiences of interlibrary loan. In geographic areas where multiple university libraries collaborated, getting a copy of a book that was not in the researcher's library was easy and fairly quick. However, it could be slow and tedious as well — and expensive for institutions. Librarians in the 1960s and 1970s used the National Union Catalog to identify ownership of materials, and books were sent from one library to another via the United States Postal Service. Additionally, printed sources such as Lee Ash's "Subject collections: a guide to special book collections and subject emphases as reported by university, college, public, and special libraries and museums in the United States and Canada," helped librarians understand what was in other libraries' collections. Because ILL was so laborious, it was usually restricted to faculty and graduate students by protocols set up by the American Library Association. Eventually, ILL became automated in 1967, with the creation of OCLC, the Ohio

College Library Center, a network of Ohio university libraries. Over the next few years, Ohio libraries came online with cataloging and sharing of resources. By 1979, Rice University had OCLC terminals in the cataloging department. There were, of course, no personal computers in the library. Librarians would use the OCLC terminal to find books that researchers needed, but it would be much later before the whole workflow would be created that allows for the easy request and delivery services that researchers take for granted today.

Librarians

Librarians were — and are — knowledgeable sources of information, and aware of the types of challenges that researchers might encounter. For example, when a researcher could not find a reference, a librarian might recommend flipping the sequence of numbers, knowing that there was often a typo in the literature. So *JACS*, vol. 24, page 18 (1981) might really be *JACS*, vol. 18, page 24 (1981). It often worked. Interestingly, these things still happen today; citation errors persist. The broad goals of librarianship have always been acquisition, organization, interpretation, and service in the name of connecting researchers with information. The basic responsibilities of classification, cataloging, and indexing are essential to librarians. No collection is accessible until it has been organized. Weeding, the elimination of the outdated or damaged print material, kept the collection in shape.

Reference desks, the physical presence of an educated librarian, sitting and waiting for the questions that researchers might have, served as a conduit to get the student or faculty member to translate thoughts to actions. The librarian, as the keeper of the books, holds vast knowledge of what the library contains, having spent time and effort in shaping the collection, working with vendors and publishers to obtain the information. More importantly, reference librarians became skilled in the art of the reference interview, often getting out of the library user unarticulated questions that led them to just the item they needed for their research. If libraries are the refuge of the scholar, the librarian often serves as the dedicated guide.

As research disciplines continued to merge, the librarian was at the forefront of interdisciplinary studies, having always had to carry the broader vision of the expansion and physical representation of human knowledge. The best librarians created networks of colleagues at other libraries whom they could call upon for support. They did not limit themselves to what was within their own walls, but sought in tenacious ways to find what the researcher needed, even if a specific item seemed unobtainable.

The teaching role of librarians was often hard-fought, with classroom time being as limited in the past as it is now, but often faculty would be taught one-on-one in the use of the card catalog, then the online system, and how to use the CD-ROMs that provided the first access to indexed journals. This skill set would then be passed on to graduate students via the faculty. In decades past, there was certainly a more widely held belief that the librarians possessed specialized skills in research, finding obscure items, and acquiring difficult to find materials.

Information Tools outside the Library

Much of the activity of research involves producing and consuming information, processing it, and moving it around. In postwar materials research, much of that activity took place in, or at least intersected with, the physical library — not all of it, though. We end with a few examples of the most important extra-library information tools.

Mailing Lists and Newsletters

The boom associated with high T_c superconductivity was communicated through early forms of mailing lists and newsletters. An archive of these exists at http://www2.ulg.ac.be/supras/activites/newshtcu.html. A couple of years later, electronic bulletin boards in the fast-moving era of cold fusion controversy affected the speed at which researchers learned about new ideas and communicated their scientific thoughts and emotions. Mailing lists and electronic newsletters "accelerated the speed at which individuals had to respond to new ideas and new information, and they affected who had access to what information at what time."[3] These informal discussions provided a venue that was untethered to the more authoritative peer-reviewed journals lining the shelves of the library.

Personal Communications

If books and journal articles are the most formal mode of scientific communication, then the most informal are conversations, letters, and telephone calls. Materials researchers rely heavily on these informal modes for at least two reasons. One is that not everything worth knowing is worth publishing. To successfully perform many experimental procedures, one needs to know the tips and tricks that the procedure's inventor did not include in the "methods" section of their journal articles. But informal communications are also crucial for research because researchers are social creatures: they cultivate networks of colleagues and friends with whom they trade favors and resources.

Thus, many journal articles well into the 1980s included citations to "personal communication," that is, to a note or conversation between the author and a colleague — personal notes back from another researcher or conversations at a conference or on the phone. Curiously, "personal communication" is not cited as much anymore, perhaps because researchers do not talk as much about how to do things — hands-on skills. Personal communications do still play an important role in science, though their form is evolving quickly. Physical letters are much less common than they once were and phone calls seem to be decreasing as well. Email and other digital communication tools, meanwhile, take up an increasing proportion of researchers' time. Notably, libraries will need to contend with this shift: in the past, a scientist's letters often wound up in the archival collections of his or her alma mater or home organization, but today those personal communications are conveyed via emails, which may or may not be preserved when the researcher dies or retires. The emerging archival field of digital forensics is busy migrating the contents of 3×5 floppy disks, and other media from the 1970s, to digitized, online databases that give ready access to the scholarly record.

Interpersonal Loan

When researchers could not find the materials they wanted in their local library, they often resorted to the formal interlibrary loan system. However, the time lag made the process tedious, and so it was common for a researcher to pick up the telephone and call a colleague to get a copy of a publication, or to send a postcard to a researcher to ask for a preprint. Such "interpersonal" loans encompassed a more heterogeneous range of items than the formal ILL system: experimental materials, blueprints, computer code, and other information.

Summary

Libraries as physical spaces, and sources of scholarly material, served a critical role to researchers. Even the mere presence of a librarian lent impetus to a researcher to get to work, and get things done. As new methods of scientific communication developed — electronic databases, email, electronic newsletters — librarians developed new skills alongside the researcher to continue to build collections, ensure access, and enable scholarly reflection in a time when physically going to the library was a requirement.

References

1. Association for Information Science and Technology. Abstracts, reviews, compilations, and indexes stored and retrieved using electronic computers. Available from: https://www.asist.org/about/history-of-information-science/chronology. Accessed June 23, 2017.
2. Chemical Abstracts Service. About CAS. Available from: http://www.cas.org/about-cas/cas-history. Accessed June 23, 2017.
3. Lewenstein BV. (1995). Do public electronic bulletin boards help create scientific knowledge?: The cold fusion case. Science, Technology, & Human Values. 1995;20(2):123–49.
4. Materials Research Society. The society. Available from: https://www.mrs.org/the-society. Accessed April 5, 2018.
5. Thomas N. Interview with Kolah D. Rice University. October 20, 2017.

CHAPTER 2.7

INTERNATIONAL STANDARDS

SHARON TSAAI-HSUAN KU

The Power of Mundaneness

Since the US Department of Defense (DoD) created Materials Science and Engineering as a discipline in the 1960s, searching[3 p223] for novel materials has constituted the main mission of the discipline.[7] The rapid growth since the 1980s of instrumentation, computational modeling, and interest in commercializing academic science has further made modern materials science laboratories places of precision measurement and control. The National Nanotechnology Initiative, the Material Genome Initiative, and the recently announced Advanced Functional Fabrics of America (part of the National Network for Manufacturing Innovation) all exemplify the materials innovation paradigm. Promoted by the White House Office of Science and Technology Policy, these national science and technology programs share the same goal of facilitating knowledge transfer from discovery to commercial manufacture by introducing finer and faster tools for design, measurement, and control. It may appear that the engine of innovation lies in the precision measurements and novel material discoveries.

However, increasing concerns about the lack of standards in scaling up benchtop practice to manufacturing for the market have hinted that novelty may not be the core of innovation after all. Instead, materials innovation must be "disciplined" by a set of mundane objects and practices, such as sample preparation, sterilization, instrument calibration, and material storage, in order to overcome the challenges of stabilizing knowledge production and transfer. For example, in the field of nanotechnology, the National Cancer Institute (NCI) has been advocating the development and use of "standard characterization assays" for the nanomedicine community, to justify the translational outcome of nanotechnology-enabled cancer research.[14]

Such mundaneness, however, contributes to the aura of self-evidence that surrounds standards, which enables them to be obeyed without proof: we follow the time given by the clock; we accept the length measured by a meter stick; we determine weight by putting things on a scale. We would never bother to question the validity of these standard objects because they define the quantities we look for. In other words, appearing self-evident makes standards powerful-but-silent dictators in materials characterization. To understand how standards function within and influence materials research thus requires explaining how they acquire the status of being taken for granted.

This chapter provides a social history of the standards that materials scientists rely on. The metric system is the best-known of these standards, but my discussion will focus instead on the production and operation of "reference materials," standards with physical embodiments produced by accredited laboratories. Reference materials play a critical role in modern materials industrialization. They have been heavily used since the early twentieth century in industrial and military materials analyses for consistent accuracy and quality assurance. In today's materials research, conducted within large scale cross-institutional collaborations, multi-center trials, and global innovation chains, producing standard reference materials is viewed as a necessary step for global manufacturing, trade, and regulatory harmony. For instance, in the 2016 US-EU nanoEHS Community of Research, producing standard reference materials for nanoscale physio-chemical and biological characterization was considered the first priority by the EU commission in order to produce scientific data with consistent quality for nanomaterials regulation.

Diverse interests and expectations have made the production of standard reference materials not just a process of making scientific knowledge, but also a process explicitly involving broader social, economic, and political calculations that reflect complicated considerations in trade and diplomacy. Tracing the production of reference materials illustrates how these different forms of calculations mingle. It argues that the production and application of standard reference materials not only turns the world into a giant laboratory wherein the power of science and technology resides; it also integrates materials research into modern bureaucratic administration and organizational management.

The Emergence of Standards

Historians of science have revealed the social roots of standardization by examining the concepts and practices of authority,[14 p18; 20 p157–159] quantification,[23 p390; 24 p73] objectivity,[31 p58; 22 p19; 32 p48; 7 p121] uniformity and universality,[20 p154; 15 p6–8; 32 p274] truth,[16 p594–596] reproducibility,[5 p29] and precision.[10 p32–35; 21 p119–123; 35 p7] They argue that standards' robust material properties are not derived purely from science, but rather that standards represent socio-political outcomes entangled with national, cultural, and economic interests: the standard ohm is English science bound tightly with the cultural values of the Victorian era;[29 p23] the meter standard is a political achievement underwritten by the belief that "rational measures would engender rational "citizenry" in the context of the French Revolution.[1 p137]

The Industrial Revolution opened a new chapter in the history of material specifications and standardization. In the United States, along with the development of new materials such as steel or cement for civil constructions, manufacturers encountered numerous quality problems in end products because of unknown material properties. The railroad system had numerous accidents, costing many lives, due to material failures. However, laboratories at the foundries that supplied materials to the railroads could not analyze the materials with sufficiently consistent accuracy to provide ongoing quality assurance. The federal government thus sensed an urgent need for standardization to ensure the quality of materials.

The near-simultaneous establishment of the National Bureau of Standards (NBS) and non-governmental standards development organizations such as the American Society for Testing Materials (ASTM) in the beginning of twentieth century was therefore not coincidental.[28 p249] Due to the high demands of standard material testing, one key mission of these standard institutes was to produce trustworthy reference materials to serve as accuracy benchmarks and as calibrators for quality assurance. NBS's Standard Sample Program was established in this context. The first project was completed in 1906 with four materials labeled "Standardizing Iron Sample A-D" to improve the material characterization of cast iron rail carwheels.[26 p4]

Early work on standard samples set a precedent for the agency's heavy investment in creating an Office of Standard Reference Materials (SRM) Program in 1964, with a profound impact on cooperative efforts between NBS and industries that continue to this day. Instead of viewing standards as a national endeavor controlled and mandated by the government, the NBS is operated under the Department of Commerce, which views standards as a "voluntary consensus" of industrial users.[18 p80] The industrial root of standardization can be traced back to President Herbert Hoover's "associative states" policy, which claimed that a partnership between big businesses and the federal government would benefit the public. The NBS, under Hoover's administration, grew in power and established standards for machines in factories and plants. To install the government–industry partnership in standards production, modern organizational management and consensus-building mechanisms were mobilized as the essential social infrastructures in the making of standard reference materials. Figure 2.7.1 illustrates the SRM flowchart, which indicates how the division of labor within the National Institute of Standards and Technology (NIST, the former NBS) was mobilized to materialize standards.

A standard reference material comes into being in three stages — pre-production, production, and post-production. The pre-production phase incorporates industrial partners through market assessment, funding preparation, and technical survey, thereby placing the SRM development in a relevant social setting. Each of the

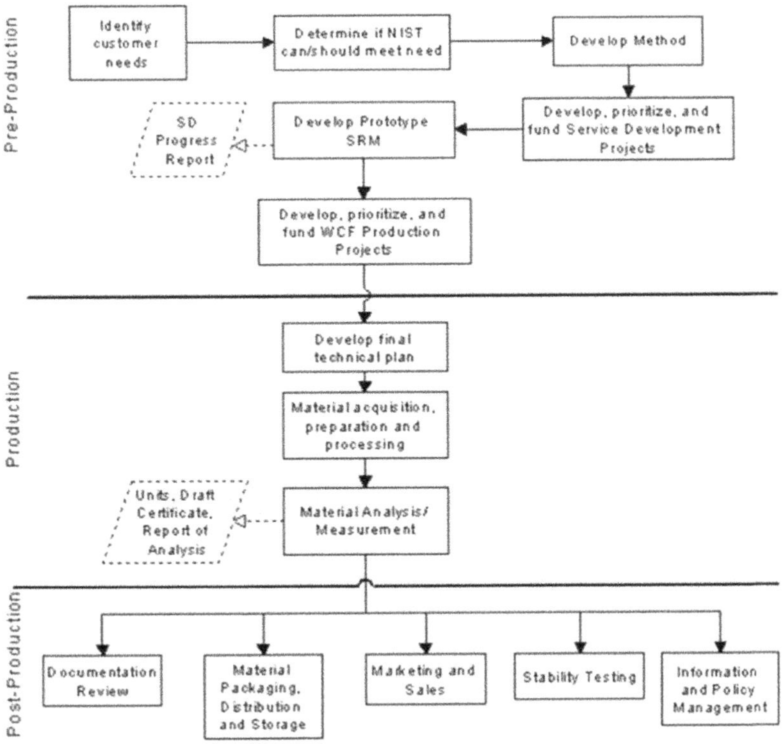

Fig. 2.7.1. The flowchart of NIST Standard Reference Materials. *Source*: Guide to NIST SRM Development and Production, SRM Office at National Institute of Standards and Technology, courtesy of Robert L. Watters, SRM Office Director.

pre-production tasks is conducted by a particular bureaucratic division. A NIST Measurement Laboratory is responsible for conducting the technical survey and deciding on the degree of accuracy and precision needed for the determination of each measurement, thereby defining the measurement procedure to be used. The Budget Division is in charge of production funds and the Statistical Engineering Division (SED) designs sampling and measurement strategies.

After consultation with SED about the initial preparation, the responsible laboratory conducts the production phase, which involves a specific number of measurements according to the advice given by SED. These data are then submitted to SED for statistical analyses.

By applying a proper statistical model consistent with NIST policy, the mean value and stated level of uncertainty can be obtained.

After all the numerical data are produced, the process moves to the post-production phase, which includes the tasks of documentation review, packaging and labeling, pricing, and marketing by the Measurement Service Division (MSD). The organizational processes of SRM production clearly shows that to produce a standard object, a quantitatively characterized substance with officially claimed certified values is beyond what any single laboratory or isolated individual could achieve. Instead, the performance, quality, and credibility of such measurements reflect the workings of the NIST bureaucracy, where coordination and collective actions among different offices take place.

The Natural and the Social Order

The bureaucratic procedure involved in developing standard reference material, as Figure 2.7.1 illustrates, indicates that material standardization relies on the organizational stability and operation of the standards institute. The material formats of standards — what to measure and how to measure in collectively agreed-upon manners — cannot be generated without the support of a well-ordered bureaucratic institution that orchestrates the diverse interests and manages the division of labor required in the process of standardization.

The process of institutionalizing the development of standards' materiality constitutes another key principle in modern standardization — traceability. The term "traceability" itself became a serious matter in the US DoD's science and technology programs in the 1960s, as the agency relied heavily on government contractors to run materials manufacturing and testing for military purposes. To ensure the quality of measurement, DoD established a policy on traceability, requiring testing equipment to be calibrated by contractors using reference standards whose calibration could be "traceable to the National Bureau of Standards."[27p115] While the government advocated the need for establishing "scientifically rigorous evidence" to quantify uncertainty and assure continued accuracy along the entire process of material production, it took a long time for the government contractors to fully adopt the policy on traceability. Performing standardized measurements through the establishment of traceability was expensive and unnecessary for most material and equipment manufacturers, testing facilities operators, and calibration vendors, since it required both extra material expenses and organizational rearrangement.[25 p4]

As support for the new approach to traceability slowly grew in the United States, the international metrology community began to endorse this approach, due to the increasing demands of global production and international trade. In the 1993 edition of the International Vocabulary of Basic and General Terms in Metrology (VIM), traceability is defined as "property of a measurement result whereby the result can be related to a reference through a documented unbroken chain of calibrations, each contributing to the measurement uncertainty." Figure 2.7.2 illustrates the pyramid of "traceability," which is constituted by a series of hierarchically arranged reference materials with different levels of uncertainty. The internal

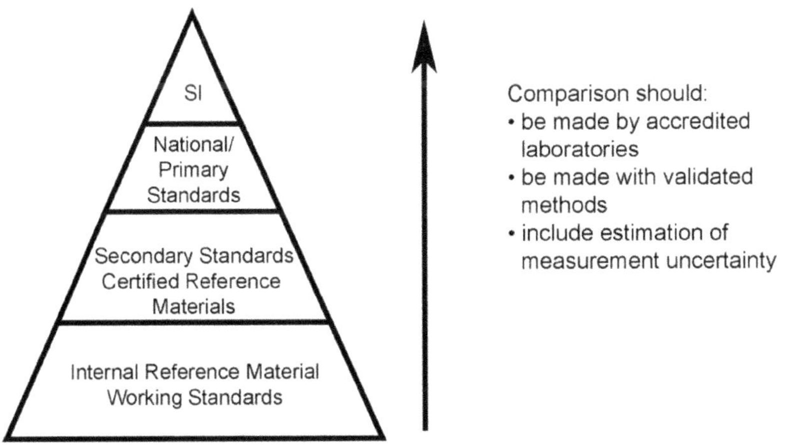

Fig. 2.7.2. The traceability diagram illustrating the unbroken chain of comparisons between laboratory reference material and SI units defined in the VIM. *Source*: Laboratory Compliance Tutorial on ISO/IEC 17025, General Requirements for the Competence of Testing and Calibration Laboratories, 2005. https://www.labcompliance.com/tutorial/iso17025/.

reference materials at the bottom are the working standards for in-house calibration, but may not be recognized beyond the local use due to large measurement uncertainty. The second-level standards are certified reference materials characterized by a metrologically valid procedure for one or more specified properties, accompanied by an RM certificate that provides the value of the specified property, its associated uncertainty, and a statement of metrological traceability. The national primary standards are those maintained by national measurement facilities with well-defined traceability and rigorously controlled uncertainty. The highest-level standards refer to the fundamental measurement derived from agreed-upon physical objects such as meter or kilogram standards maintained by the International Bureau of Weights and Measures (BIPM). A traceable measurement result depends on establishing an "unbroken chain of calibration" by systematically citing these different levels of reference materials. In addition, to ensure that the traceability chain is not broken, the laboratories undertaking the testing and calibrations should be audited by an independent accreditation agency such as the National Voluntary Laboratory Accreditation Program in the United States.

The traceability diagram demonstrates the delicately constructed procedures involved in declaring standards and accuracy. First, it demonstrates that the robustness of a standard measurement relies on a social infrastructure in which the choice of accredited laboratories and acceptable references must be decided before the measurement: the longer the chain of unbroken calibration, the higher the level of traceability, and the larger the organizational coordination and administration required to establish such a chain. Second, the practice with regard to traceability shows that accuracy is a social institution that cannot be obtained by an isolated, single measurement; rather, it is a process of comparison with existing references and can only be achieved within a seamlessly constructed social network where an "unbroken chain of calibrations" is established. This socio-material scaffolding is best illustrated by the comment made by Paul De Bièvre, the editor-in-chief of Accreditation and Quality Assurance: "Traceability is about establishing a 'trace' of the quantity value of a measurement result to a 'stated reference,' such as the result obtained by using a reference measurement procedure. It is established, not measured."[8 p497]

Traceability makes advanced materials innovation a thoroughly social endeavor, best demonstrated in the case of nanomaterials standardization. In 2004, two US federal agencies, NIST and NCI, started a collaboration for the production of a standard reference material to resolve the troublesome issue that permeated the entire nanotechnology community — the lack of reproducibility and persistent accuracy in nanosize measurement. One might find an irony in this case as the sales pitch for nanotechnology was based on the atomic precision obtained through instrumentation such as the atomic force microscope; yet it fell short of guaranteeing nanoscale accuracy that was ten times larger! The paradox can only be understood from the perspective of traceability: the atomically precise measurement might be obtained by individual labs using a single atomic imaging tool, but claiming nanoscale accuracy requires more than individual efforts, that is, the establishment of an unbroken chain of calibrations by systematically citing adequate references and building coordinative social networks. In fact, it took NIST scientists more than three years, mobilizing the entire bureaucratic infrastructure, to arrange human resources and technical information in an orderly way to produce this "gold Standard" — a reference material made by the gold nanoparticle which can only be traced to the secondary level due to the limited time, cost, and intended use.[12] Viewing materials innovation from the window of standardization thus reveals the significant and sometimes unexpected social costs underneath the linear innovation model — including the establishment of new government offices, labs, staff, and so on — to tame measurement uncertainty and construct material accuracy.[11]

The Internalization of Formal Standards

The growth of international trade has unsurprisingly resulted in an emphasis on international standards. In advanced materials research, which emphasizes global innovation and the potential of commercialization, developing internationally agreed-upon measurements and manufacturing standards to coordinate cross-state industrial complexity, technical progress, and risk assessment becomes a critical mission for contemporary material scientists and engineers. Take nanotechnology as an example: numerous technical committees and activities operated by international standard developing bodies such as ISO, ASTM, IEC, and IEEE, were formed in the past decade to address nanomaterials standardization in terms of terminologies, instrumentation, material specification, and environmental health and safety (EHS) protocols.

Defining international standards is a laborious process that relies on careful organizational administration and political mobilization. Take the mechanism used in ASTM International as an example: in order to achieve international agreement in a given time table, the institute has introduced a variety of social practices, including Robert's rules of order, inter-laboratory studies, and balloting procedures, into the process of standardization, in order to set the threshold of "consensus," as well as for members to express comments and preference for certain standards proposals. These social mechanisms for defining and achieving a consensus have had profound impacts on the materiality and consequences of standards, further raising the question of global justice. A key criticism comes from the problematic use of "international," when the actual standard processes are dominated and controlled by developed countries. For example, those involved in the SI unit standards are mainly Western countries from the North: only seventeen states from North America and Europe attended the diplomatic Conference of the Meter for setting the International Bureau of Weights and Measures (BIPM) in 1875. Even as of December 31, 2005, only fifty-one states were members of the Metre Convention, which gives them a more active role in (re)defining and maintaining the metric system.[33] In environmental standardization, the power asymmetry between the North and the South is a critical concern.[4] The international aspect of standardization has made contemporary materials research a socio-ethical-political complex at the interface of technology, trade, and diplomacy.

Conclusion: Rethinking Material Innovation through the Lens of Standardization

The increasing calls for standards and standardization in today's advanced materials research has initiated a debate on whether standardization delays or facilitates innovation. This chapter provides a brief socio-historical account of standardization in materials innovation. Using standard reference materials as a case study, the paper demonstrates the socio-technical scaffolding of the measurement traceability as a critical path of standardizing modern material science. The social history of standard reference materials further highlights standardization as a historical practice and a historiographical category in establishing modern materials science. Examining the epistemic and pragmatic validity of standards serves as a critical lens to rethink the values of precision and the profound social cost paid for pursuing the materials innovation.

References

1. Alder K. The measure of all things: the seven-year odyssey and hidden error that transformed the world. New York: The Free Press; 2002.
2. Belanger BC et al. Traceability: an evolving concept. In A century of excellence in measurements, standards, and technology: a chronicle of selected NBS/NIST publications, 1901–2000. Special NIST Publication 958; 2001.
3. Bensaude-Vincent B. The construction of a discipline: materials science in the United States. Historical Studies in the Physical and Biological Sciences. 2001;31(2):223–48.
4. Clapp J. The privatization of global environmental governance: ISO 14000 and the developing world. In Levy D, Newell P, editors. The business of global environmental governance. Cambridge, MA: MIT Press; 2004.
5. Collins HM. Changing order: replication and induction in scientific practice. Chicago: University of Chicago Press; 1985.
6. Daston, Lorraine, editor. Biographies of scientific objects. Chicago: University of Chicago Press; 2000.
7. Daston L, Galison, P. Objectivity. New York: Zone Books; 2007.
8. De Bièvre P. Traceability is not meant to reduce uncertainty. Accreditation and Quality Assurance 2003;8(11):497.
9. Doiron T. The history and current status of traceability. American Society for Precision Engineering; 2004.

10. Gooday G. Precision measurement and the genesis of physics teaching laboratories in Victorian Britain. British Journal for the History of Science. 1990;23(1):25–51.
11. Ku S. Forming interdisciplinary expertise: one organization's journey on the road to translational nanomedicine. Wiley Interdisciplinary Reviews: Nanomedicine and Nanotechnology. 2012;4(4):366–77.
12. Ku S, Klaessig F. A matter of size does not matter: material agency and the state authority in nanotechnology standardization. In Schlaudt O, Huber L, editors. Standardization in measurement: philosophical and sociological issues. London: Pickering and Chatto; 2015.
13. Kula W. Measures and men. Szreter R, translator. Princeton: Princeton University Press; 1986.
14. Latour B. Pandora's hope: essays on the reality of science studies. Cambridge: Harvard University Press; 1999.
15. National Cancer Institute. Process Overview. Accessed 22 December 2018. Available from: https://ncl.cancer.gov/working-ncl/process-overview#a.
16. Mallard A. Compare, standardize and settle agreement: on some usual metrological problems. Social Studies of Science. 1998;28(4):571–601.
17. Marks RB, Hebner RE. Government/Industry interactions in the global standards system. In Bolin S, editor. The standards edge: dynamic tension. Ann Arbor, MI: Sheridan Books; 2004.
18. NBS Special Publication 552: Research and Innovation in the Building Regulatory Process.
19. National Institute of Standards and Technology. Guide to NIST SRM development and production. 2004. Available from: http://tsi.nist.gov/ts/htdocs/230/232/ABOUT/SRMGuide1.pdf (limited access, only within NIST Campus).
20. O'Connell J. Metrology: The creation of universality by the circulation of particulars. Social Studies of Science. 1993;23(1):129–73.
21. Olesko KM. The meaning of precision: the exact sensibility in early nineteenth-century Germany. In Wise MN, editor. The values of precision. Princeton: Princeton University Press; 1995.
22. Porter TM. Objectivity as standardization: the rhetoric of impersonality in measurement, statistics, and cost-benefit analysis. In Megill A, editor. Rethinking Objectivity. Durham: Duke University Press; 1992.
23. Porter TM. Making things quantitative. Science in Context. 1994;7(3): 389–407.
24. Porter TM. Trust in numbers: the pursuit of objectivity in science and public life. Princeton: Princeton University Press; 1995.
25. Ramesh B, Harrington G, Rondeau K, Edwards M. A model of requirements traceability to support systems development. Technical Report NPSSM, Naval Surface Warfare Center, Dahlgren Division; 1991.
26. Rasberry SD. Standard reference materials — the first century. NIST Special Publication 260-150; 2003.

27. Riordan JJ, Brzezinski C. Current DoD policies and problems in calibration. In Proceedings of the 1966 Standards Laboratory Conference. National Bureau of Standards Miscellaneous Publication 291, 115–18; 1967.
28. Russell AL. Standardization in history: a review essay with an eye to the future. In Bolin S, editor. The standards edge: future generations, 247–60. MI: Sheridan Press; 2005.
29. Schaffer S. Late Victorian metrology and its instrumentation: a manufactory of ohms. In Bud R, Cozzens S, editors. Invisible connections: instruments, institutions and science, 23–56. Bellingham, WA: SPIE Optical Engineering Press; 1992.
30. Schaffer S. Accurate measurement is an English science. In Wise MN. editor. Values of precision. Princeton: Princeton University Press; 1995.
31. Shapin S, Schaffer S. Leviathan and the air-pump: Hobbes, Boyle, and the experimental life. Princeton: Princeton University Press; 1985.
32. Slaton A. Reinforced concrete and the modernization of American building, 1900–1930. Baltimore: Johns Hopkins University Press; 2001.
33. Taylor B. The International System of Units (SI). NIST Special Publication 330; 2008.
34. Timmerman S, Berg M. Standardization in action: achieving local universality through medical protocols. Social Studies of Science. 1997;27(2):273–305.
35. Wise N. The values of precision. Princeton: Princeton University Press; 1997.

CHAPTER 2.8

COMPUTER SIMULATIONS

JULIA R. S. BURSTEN

The relationship between computer simulation and materials science is complex, intimate, and symbiotic. One of the most important fields of study within materials science has been the study of semiconducting materials, which are precisely those materials used to build computer processors. Likewise, many early developments in computer simulation techniques arose initially as solutions to problems of modeling the physical and electronic behaviors of materials. This entry chronicles the joint evolution of computational techniques and theories of materials from the mid-twentieth century through the early twenty-first century.

The study of the properties of solids and fluids coalesced into the modern field of materials science around the middle of the twentieth century, contemporaneous with the first wave of digital computing machines introduced into academic, government, and industrial laboratories. At this time, one of the central challenges to modeling the behavior of materials was computational: the rise of quantum theory and its new mathematical tools, as well as new wartime needs for more precise models of material behavior in non-equilibrium settings (namely, in nuclear explosions), together demanded computationally intensive solutions to physical problems. These demands strained the limits of pen-and-paper calculations. Digital computer programs promised the ability to map the properties of materials in more detail than ever before, and in so doing aided the search for solutions to the material challenges of improving early computers. In the ensuing decades, computer simulations of materials have often grown in sophistication through a feedback loop of increased computational power generated by, and in turn generating, more precisely developed construction materials for computers.

Most of the earliest computations on electronic computers in the physical sciences were carried out under the auspices of the Manhattan Project and allied laboratories. These computations predicted and described the behavior and impact of nuclear particles during the explosion of atomic bombs. These computations can be understood as falling into two categories: (1) critical-mass computations and (2) the computation of the movement of matter affected by the bomb's explosive force, that is, the hydrodynamics of implosions and explosions.

The story of computer simulations in materials sciences begins with these latter computations. Although critical-mass computations relied on the computing power of human, punched-card, and eventually electronic computers, these computations were not simulations of material behavior in any meaningful sense. On the other hand, the techniques used to compute the hydrodynamics of implosions and explosions involved mathematically describing a hypothetical explosion as a flowing fluid and then running the clock forward, so to speak. They were among the earliest computer simulations, and they simulated a material behavior (the explosion of atomic bombs) that was too complex and too dangerous to be studied solely by repeated field testing.

These early simulations were dynamical simulations of fluid behavior, modeled on numerical solutions to nonlinear dynamical problems in fluid dynamics. A central challenge they faced was developing a method for solving the Navier–Stokes equations, which describe viscous liquids such as the matter immediately surrounding atomic implosions and explosions. The Navier–Stokes equations are nonlinear partial differential equations and so, except in very limited cases, they defy analytic solutions. Instead, Navier–Stokes problems are typically solved by numerical methods, a technique that relies on brute-force computational power to generate increasingly accurate approximations of an analytic solution. Punched-card analog computers were introduced into wartime physics through the Manhattan Project in order to solve these problems.[21] Later, in 1945, physicist and mathematician John von Neumann collaborated with physicists Stan Frankel and Nicholas Metropolis to run simulations of thermonuclear implosions on ENIAC, one of the first digital computers.[21] These early applications of digital computation to obtain numerical solutions for dynamical problems in materials science laid the groundwork for applying numerical principles to the investigation of material properties in solids and liquids.

The success of this approach led researchers on the Manhattan Project to apply digital computing methods to other problems in the behavior of materials. A group of physicists that included Metropolis, Edward Teller, and Marshall Rosenbluth, as well as Arianna Rosenbluth and Augusta Teller, sought generalized numerical methods for obtaining equations of state for solids and fluids of increasing complexity. They developed the application of a statistical technique now known as the Monte Carlo Method (MCM) to solve problems in solids and fluids. Metropolis and the mathematician Stanislaw Ulam initially proposed MCM in an earlier paper as a technique for solving computational problems in particle physics.[23] The Manhattan Project group's application of MCM to materials problems — that is, computations on particles considered as members of solids or fluids, rather than on particles considered as objects with a certain likelihood of producing a nuclear chain reaction — was a landmark improvement over straightforward numerical methods. The central innovation of MCM is to perform computations on a random sample of a set of data rather than the full set. The Manhattan Project group's first paper used MCM to generate equations of state for classical systems of atoms in two dimensions, computed much more efficiently than brute numerical methods could have.[22] Shortly thereafter, members of the group reported MCM results for equations of state for three-dimensional classical systems and for two-dimensional arrays of atoms with non-bonding electronic interactions, also knowns as Lennard-Jones potentials.[28] These simulations were crucial early steps in the use of digital computers to solve many-body problems, and MCM is still widely used as a modeling tool across the quantitative sciences. The early MCM papers from the Manhattan Project, accompanied by a series of results from cognate research endeavors at Lawrence Livermore National Laboratory and physicist John Kirkwood's laboratory at the California Institute of Technology, form the theoretical basis for much of the simulation modeling in condensed matter physics throughout the 1950s, 1960s, and into the 1970s.

The group at Lawrence Livermore National Laboratory, led by Berni Alder and Thomas Wainwright, developed an alternate approach to simulating the behavior of many-body systems.[2–6] Their method, molecular dynamics, aimed to solve many of the same sorts of problems as MCM, but with a dynamical rather than a statistical approach, so that non-equilibrium systems could also be simulated. Molecular dynamics treats the atoms in a solid or a fluid as classical (non-quantum) spheres and computes the systems of forces acting on one, some, or all of the atoms in a many-body system. Like MCM, it is a computationally efficient and predictively powerful set of simulation techniques that is still in regular use today.

The progression of computational techniques from numerical methods to MCM and molecular dynamics forms the early trajectory of computer simulation in the modeling of physical systems. Until the development of the microprocessor, discussed below, most of the energy devoted to simulation in physical modeling was expended testing, refining, and extending these techniques. The situation was somewhat different for chemistry, the other key theoretical basis of materials science. There, scientists addressed a problem that, while computationally similar, originated from a very different set of theoretical challenges.

Quantum mechanics posed a special problem for chemistry, namely how to harness the theory's physical insights about the behavior of electrons to explain molecular and crystalline behavior, and how to use this new understanding of electronic behavior to improve upon experimental techniques and expand the capabilities of chemical synthesis. The overarching goal was a quantum theory of matter, which quantitatively described the movement of electrons not only in atoms, but in molecules and solids as well. The problem was the inordinate complexity of solving quantum mechanics' dynamical equations for a system more complex than a pair of electrons. Early computational schemata in quantum chemistry sought to reduce the complexity of quantum mechanics' dynamical equations conceptually, by finding and eliminating terms in the mathematics of quantum theory that did not contribute to the prediction of electronic behavior in molecules and crystals. These early quantum theories of the chemical bond — the valence bond theory of Walter Heitler, Fritz London, and Linus Pauling,[12; 26] and the molecular orbital theory of Robert Mulliken, Freidrich Hund, Erich Hückel, John Clarke Slater, and Sir John Lennard-Jones[13; 14; 20; 24; 29; 30] — were, like the numerical methods solutions to Navier–Stokes, computation-driven workarounds developed to address the impossibility of analytic solutions to pressing problems.

However, unlike the dynamical problems of flight and nuclear reactions, refining theories of electronic behavior in molecules was not essential to the military efforts that dominated research funding and dictated who got the expensive computing machines in the first decade of digital computing. Through the mid-1950s, computations in quantum chemistry were carried out largely by hand, often with the assistance of hired human computers, or with analog punched-card computers such as the IBM 604 (for example, as in [1]). The earliest digital computer simulations in chemistry were carried out in the mid-1950s, after the early fluid-dynamical simulations, and largely after the development of first-generation commercial digital computers. A notable exception was physicist Alvin Meckler's use of the MIT Whirlwind I, a digital vacuum-tube computer commissioned by the military, to perform computations using Slater's approach to electronic structure computation in the early 1950s.[31]

The computational targets of mid-century quantum chemistry were models of the electronic structure of molecules and crystals. One of the touchstone approaches to these targets was the Hartree–Fock method, also known as the self-consistent field method, for approximating electronic wavefunctions, which was developed from molecular orbital theory. The Hartree–Fock method is an iterative method for generating an approximate solution to the wavefunction of a many-electron atom or molecule. The basic architecture of recursion built into early digital computers helped to advance the Hartree–Fock method and its successors, such as linear combination of atomic orbitals, density functional theory, and configuration interaction, which are still in use today. Density functional theory, in particular, forms much of the theoretical basis for modern quantum chemistry.[34]

Early computer simulations in both physics and chemistry were severely limited by costs in money and time. The IBM 704, an early mass-produced computer that debuted in 1954 and which was used in simulations such as those in [33], could perform 4,000 arithmetic operations per second (4 kHz) — significantly faster than human computers, but slow enough that disputes over computing time were common, as were days or weeks of lost effort when a computational or programming error was discovered. The machines were large, expensive, and typically shared among a variety of laboratories with different computational projects, further limiting the rate at which computations could be performed.

Speeding up computing rates would become the singular focus of developments in computer hardware. From the late 1950s onward, those efforts revolved around the construction of integrated circuits containing miniaturized components. Integrated circuits anchored the electronic circuitry necessary for the operation of a computer on a single "chip" of semiconducting material, typically silicon, germanium, or a compound such as gallium arsenide. By the early 1970s, integrated circuits had become complex enough that most of a computer's processing power could be contained on a single chip or "microprocessor." To achieve that degree of circuit integration and miniaturization, engineers required a more detailed quantitative picture of the properties of semiconducting materials. Computing the electronic structure of materials was now not just performed on computers; it was carried out in order to enhance computer performance.

To better understand electronic structure, chemists extended and quantified conceptual models of the distribution of electrons in solids such as the jellium (uniform electron gas) model and the muffin-tin model. These conceptual models of electron distribution relied essentially on simplifying assumptions about the behavior of electrons in solids, in a fashion similar to the valence-bond and molecular-orbital theory models of electron behavior in molecules discussed above. Rather than applying numerical or statistical methods directly to quantum equations of state, as the MCM models had, the conceptual models of solids and of molecules aimed to make the behavior of electrons more understandable or tractable to the human minds interpreting the models. These models achieved this goal by introducing explicitly unrealistic assumptions about electronic behavior, such as the jellium model's assumption that electrons are uniformly distributed in a solid, in order to make effective predictions both about the behavior of individual solids and about patterns of behavior in similarly structured groups of solids.

The power and speed of 1970s microprocessing offered researchers an early opportunity to unite solid-state conceptual models with molecular conceptual models, thereby contributing to a more robust theory of electronic behavior. This theory, in turn, was used to model, understand, and improve solid-state electronics — in short, to build better computer hardware. The wider availability, portability, and computing power of integrated circuits and microprocessors allowed theorists to apply new computational tools to problems in fluid mechanics, molecular structure, and to the very solid state physics that governed the behavior of the materials on which those circuits were built. At the same time, software developments, particularly the invention of the Fortran and Lisp computer languages, made computation more portable both within and between research programs, and easier to test for errors. This widening of communication channels between researchers improved the rate at which the electronic properties of semiconductors, as well as of other molecules and materials, could be calculated.

In molecular structure, researchers in the 1970s pursued less simplified models of the behavior of electrons in atoms, molecules, and condensed matter. The increased computational power of microprocessors made it possible for researchers to perform *ab initio* calculations of electronic distribution in increasingly complex molecules. *Ab initio*, or "from the first," calculations made predictions about electronic structure directly from numerical solutions to Schrödinger's equation. This method stands in contrast to the approaches of valence-bond theory and molecular-orbital theory described above, which made predictions based on simplified, conceptual models of solutions to Schrödinger's equations. In 1975, James B. Anderson published a landmark *ab initio* paper using what would come to be known as the Quantum Monte Carlo Method (QMCM) for computing Schrödinger's equation for a system of three protons and two electrons.[8] QMCM is significantly more computationally intensive than the Hartree-Fock–based methods, but it bypasses a stage of approximation by incorporating electron correlation, and thus offers a more precise and detailed electronic configuration for some systems, including both molecules and solids. As the size of microprocessors continued to shrink, engineers required these more precise and detailed pictures of electronic structure in order to build usable integrated circuits at a smaller scale. Quantum electronic behaviors such as long-range electron correlation and even quantum tunneling are much more likely to impede the performance of a microprocessor on a one-millimeter chip than on a three-centimeter chip. The shrinking size of microprocessors continues to demand increased precision in calculations of electronic structure today.

In fluid mechanics and condensed matter physics, increased computational power meant the possibility of simulating the behavior of macroscopic objects and flows by treating not just statistical samples but exact behaviors in fluids and materials. These models range widely, including large-scale molecular-dynamics models of stresses and strains in homogeneous and heterogeneous solids; modified electron-gas and lattice-dynamics models of crystals; finite-elements models of thermal and mechanical properties in solids; semi-empirical quantum models of solids and fluids; and numerical simulations of boundary-layer, turbulent, and supersonic flows. Some of these models are atomistic, in contrast to earlier simulations' continuum treatments of solids and fluids, and through the 1980s and into the 1990s the trajectory of computer simulation in materials modeling appeared as if it would be a simple matter of obtaining sufficient computer performance to obtain exact, atomistic simulations of all material behavior. This was an extension of the intuition shared by many pioneers of quantum theory and expressed succinctly by Dirac in 1929: "The underlying physical laws necessary for the mathematical theory of a large part of physics and the whole of chemistry are thus completely known, and the difficulty is only that the exact application of these laws leads to equations much too complicated to be soluble."[10 p1]

Atomistic and *ab initio* computer simulations in materials science have come a long way, and it is possible to model material behavior more quantitatively and precisely because of these models. However, these models have turned out to have limitations that are no longer thought to be merely due to the limits of a given year's or decade's computational power. Despite technological advances in parallel computing, the prospect of exactly solving the Schrödinger equation for, say, one of the suspension cables in the Golden Gate Bridge, remains elusive at best and counterproductive at worst. Instead, computer simulations of materials in the twenty-first century have shifted toward improving multi-scale or hybrid modeling techniques, which combine multiple component models to generate simulations that describe a variety of types of material behavior within a single simulation. Hybrid models combine at least two, and frequently three or more, theoretical frameworks for simulation. A unique challenge of these methods has been developing methods to stitch together the component simulations into a multi-scale model, which has become the subject of scientific as well as philosophical discussion.[9; 27; 32]

Not only does developing methods of connecting component simulations from different theoretical frameworks resolve a central computational challenge of modeling materials in the twenty-first century, but it also parallels broader developments in twenty-first century materials science. Designing materials with tunable properties for application in a wide variety of computer-driven electronic systems is no longer a matter of understanding merely solid state physics, or merely solid state physics and the electronic structure of molecules. New materials, such as compounds with nanoscale architectures and organic semiconductors, comprise the forefront of materials science research. Many of these materials are being developed as solutions to problems posed by hardware limits on computer miniaturization, rigidity of displays, and the impacts of widely available computing devices on the natural and human environments in which these devices reside. In order to develop usable simulation models of organic and nanoscale semiconductors, researchers have had to broaden the range of theories they use to inform their understanding of material behavior. Theoretical insights from organic chemistry and biology in particular now make regular appearances in computer simulations of modern materials. Developing simulations flexible and powerful enough to incorporate these insights will be a central focus of research in computer simulations of materials in the coming decades, and the emphasis on connecting component simulations from multiscale methods holds promise as a way forward.

This overview has not addressed the role of graphical representations of simulations in materials science in generating understanding of material properties, nor has it addressed the impact of the internet on the development and dissemination of simulation methods. Both of these features of computer simulation in materials science impact the conceptual and computational insights that have driven research in materials simulation, and they merit further consideration at another time. As computer simulation moves through the twenty-first century with materials science, the complex feedback between developments in computing technology and human understanding of natural and synthetic materials appears to continue. Historian Ann Johnson devoted much of her too-short career to careful scholarship of these relationships, and readers interested in pursuing the subject further would do well to survey her work.[16–19]

References

1. Alder B, Frankel S, Lewinson V. Radial distribution function calculated by the Monte-Carlo method for a hard sphere fluid. The Journal of Chemical Physics. 1955;23(3):417–19.
2. Alder B, Gass D, Wainwright T. Studies in molecular dynamics. VIII. The transport coefficients for a hard-sphere fluid. The Journal of Chemical Physics. 1970;53(10):3813–26.

3. Alder B, Hoover W, Young D. Studies in molecular dynamics. V. High-density equation of state and entropy for hard disks and spheres. The Journal of Chemical Physics. 1968;49(8):3688–96.
4. Alder B, Wainwright T. Phase transition for a hard sphere system. The Journal of Chemical Physics. 1957;27(5):1208–9.
5. Alder B, Wainwright T. Studies in molecular dynamics. I. General method. The Journal of Chemical Physics. 1959;31(2):459–66.
6. Alder B, Wainwright T. Studies in molecular dynamics. II. Behavior of a small number of elastic spheres. The Journal of Chemical Physics. 1960;33(5):1439–51.
7. Allen M, Tildesley D. Computer simulation of liquids. Oxford: Oxford University Press; 1989.
8. Anderson J. A random-walk simulation of the Schrödinger equation: H_3^+. The Journal of Chemical Physics. 1975;63(4):1499–503.
9. Batterman R. The tyranny of scales. In Batterman R, editor. Oxford handbook of the philosophy of physics. Oxford: Oxford University Press; 2013:256–86.
10. Dirac P. Quantum mechanics of many-electron systems. In Proceedings of the Royal Society of London A: Mathematical, Physical and Engineering Sciences. 1929;123(792):714–33.
11. Dove M. An introduction to atomistic simulation methods. Seminarios de la SEM. 2008;4:7–37.
12. Heitler W, London F. Wechselwirkung neutraler Atome und homöopolare Bindung nach der Quantenmechanik. Zeitschrift für Physik A Hadrons and Nuclei. 1927;44(6):455–72.
13. Hückel E. Quanstentheoretische beiträge zum benzolproblem. Zeitschrift für Physik A Hadrons and Nuclei. 1931;72(5):310–37.
14. Hund F. Zur frage der chemischen bindung. Zeitschrift für Physik A Hadrons and Nuclei. 1932;73(1):1–30.
15. Jiles D. Introduction to the electronic properties of materials, 2nd edition. Cheltenham: Nelson Thornes Ltd.; 2001.
16. Johnson A. The shape of molecules to come. In Lenhard J, Küppers G, Shinn T, editors. Simulation: pragmatic constructions of reality. Dordrecht: Springer Netherlands; 2007:25–39.
17. Johnson A. Modeling molecules: computational nanotechnology as a knowledge community. Perspectives on Science. 2009;17(2):144–73.
18. Johnson A. From Berkeley to Boeing: civil engineers, the cold war, and the origins of finite element analysis. In Wise M, editor. Growing explanations: historical perspectives on the sciences of complexity. Durham, NC: Duke University Press; 2004: 133–58.
19. Johnson A. The virtual tool: social and epistemological issues in computer-aided chemical process design. In Baird D, McIntyre L, Scerri E, editors. The philosophy of chemistry. Boston Studies in the Philosophy and History of Science. Dordrecht: Springer; 2005: 273–91.

20. Lennard-Jones J. The molecular orbital theory of chemical valency. I. The determination of molecular orbitals. Proceedings of the Royal Society of London A: Mathematical, Physical and Engineering Sciences. 1949; 198(1052):1–13.
21. Metropolis N, Nelson E. Early computing at Los Alamos. IEEE Annals of the History of Computing. 1982;4(4):348–57.
22. Metropolis N, Rosenbluth A, Rosenbluth M, Teller A, Teller E. Equation of state calculations by fast computing machines. The Journal of Chemical Physics. 1953;21(6):1087–92.
23. Metropolis N, Ulam S. The monte carlo method. Journal of the American Statistical Association. 1949;44(247):335–41.
24. Mulliken R. Electronic structures of polyatomic molecules and valence VI. On the method of molecular orbitals. The Journal of Chemical Physics. 1935;3(7):375–78.
25. Nielaba P, Mareschal M, Ciccotti G, editors. Bridging time scales: molecular simulations for the next decade. Lecture notes in physics series 605. Berlin: Springer-Verlag; 2002.
26. Pauling L. The nature of the chemical bond. Application of results obtained from the quantum mechanics and from a theory of paramagnetic susceptibility to the structure of molecules. Journal of the American Chemical Society. 1931;53(4):1367–400.
27. Phillips R. Crystals, defects and microstructures: modeling across scales. Cambridge: Cambridge University Press; 2001.
28. Rosenbluth M, Rosenbluth A. Further results on Monte Carlo equations of state. The Journal of Chemical Physics. 1954;22(5):881–84.
29. Slater J. Molecular energy levels and valence bonds. Physical Review. 1931;38(6):1109–45.
30. Slater J. Quantum theory of matter. New York: McGraw-Hill Book Company, Inc.; 1951.
31. Slater J. Solid state and molecular theory: a scientific biography. New York: John Wiley & Sons, Inc.; 1975.
32. Winsberg E. Handshaking your way to the top: simulation at the nanoscale. Philosophy of Science. 2006;73(5):582–94.
33. Wood W, Jacobson J. Preliminary results from a recalculation of the Monte Carlo equation of state of hard spheres. The Journal of Chemical Physics. 1957;27(5):1207–8.
34. Zangwill, A. A half century of density functional theory. Physics Today. 2015;68(7):34–39.

CHAPTER 2.9

INFORMATION DISPLAY AND RECORDING

CYRILLE FOASSO

Information display and recording instruments are seldom mentioned in history of science scholarship. They nevertheless play a key role in the scientific research process since they are the conduit through which the material world, via measurement instrumentation, speaks to scientists. They serve as a visual interface, but also as a memory of the measured phenomena. Recorders even enabled the discovery of new phenomena that could not otherwise be observed directly (see, for instance, the nuclear reaction chart in Figure 2.9.1b). Like every other link in the chain of measurement, they must correspond to the phenomena; that is, their rapidity, response time, accuracy, range of values, and other features must be appropriate for the system under investigation. The experimenter must have the requisite talent and experience to design or choose suitable instruments and to know how to interpret the results they display. This critical sense of the experimenter has become more important, as, over the course of time, self- and shop-built instruments have given way to black-boxed, off-the-shelf devices.

The history of information display dates to classical measurement instruments, which gave scientists (or natural philosophers) discrete data points. Users of such instruments had to make a great number of measurements and write out point-by-point readings on a graph by

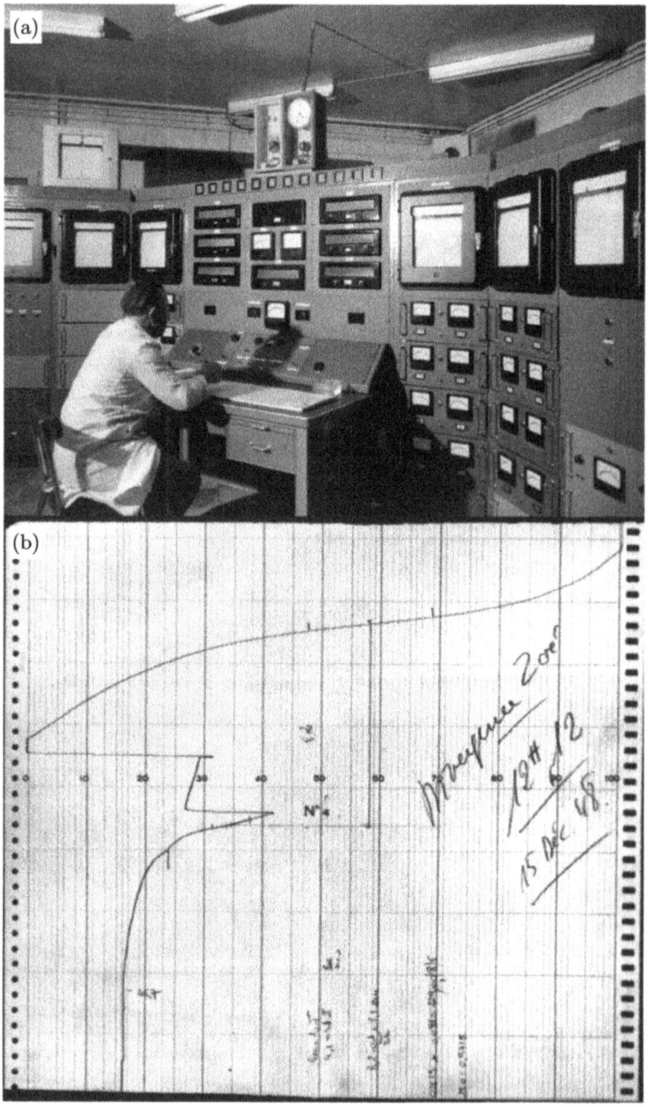

Fig. 2.9.1. a) Photograph of a typical 1950s installation: the ZOE atomic pile control room. Title: Service de la pile de Châtillon, tableau de commande de la pile EL1 (alias ZOE). (Châtillon, octobre 1956). Archives historiques CEA @ CEA/Pierre JAHAN – J 55. b) Nuclear materials: The graph is the proof that the first chain reaction in the ZOE atomic pile had occurred. Title: Diagramme d'enregistrement de la première divergence du réacteur ZOE le 15 décembre 1948 à 12 heures 12 (Châtillon). *Credit*: Archives historiques CEA @ CEA-FAR_SA_N_00317.

hand. They then had to link the points together, usually by finding a mathematical formula for this purpose. Subsequent developments have sought to automate or speed up various aspects of this essential procedure.

From the First Display and Automatic Recording Instruments to the Chart Recorder

By the eighteenth century, the first automatic recorders had emerged, such as Pajot d'Ons en Bray's anemometer in France. Invented around 1734, it could trace the speed and the direction of the wind on a roll of paper driven by a clockwork mechanism. Another example is James Watt's famous graphic pressure indicator, which appeared at the end of the century.[9]

By the nineteenth century, instruments that produced graphs had worked their way into many scientific and technical fields. Many inventions during that period prefigured aspects of, and eventually contributed to, the archetypical data-inscription instrument of the postwar era, which still exists today: the chart recorder.[7 p260–264] This recording instrument, based on a clockwork mechanism, drives a strip of paper on which an input signal is recorded as an ink trace showing variations with time in the magnitude of the measured signal. The input may be mechanical or electrical. Where the input is electrical, the chart recorder relies on a mechanical analogy to the electrical signal, communicated via a component such as a galvanometer. Galvanometric coils were often used to drive the marking devices of early chart recorders.

However, because analog chart recorders using galvanometric movement to drive a pen had limited sensitivity, a first improvement led to the potentiometric recorder, in which a servomechanism and a sensor drove the marking pen instead, reducing inaccuracy by a factor of ten. Today, all potentiometric chart recorders for laboratories and industry contain a microprocessor controller. Yet analog recorders also persist, despite their long-predicted extinction. Some industries require them, for example where safety rules require continuous recording and retention of certain measured parameters.

The Electronic Revolution: The Cathode-Ray Oscilloscope

In parallel with electromechanical recording and display methods, another technique developed from research on cathode rays at the end of the nineteenth century: the analog oscilloscope or cathode-ray oscilloscope.[8] The central part of such an instrument is a cathode-ray tube, in which the electron beam issued from a cathode is deflected by electrically driven electrodes on a phosphor-coated screen. An oscilloscope draws a graph of the variation with time of the magnitude of a measured electric signal. The cathode-ray tube sounded the death knell for other readout instruments developed at the beginning of the 1900s, such as the electromechanical oscillographs invented by André Blondel, William Duddell, Édouard Hospitalier, and others.

By the 1930s, progress on vacuum tubes, cold cathodes, and increasing voltages led to the first oscilloscope manufactured for work outside the laboratory. But it was the progress in electronics during World War II that made the oscilloscope the most versatile and useful instrument available for signal measurement and display after the war.[7 p251] Ribet-Desjardins, the great French producer of oscilloscopes in the 1950s and 1960s, even published a brochure relating the history of measurement, which claimed that the oscilloscope was the essential measurement instrument scientists had always dreamed of and that without it science and technology could never have reached the level achieved in the postwar era.[6 p54]

Postwar electronic instruments firms became famous for the production of oscilloscopes. Three world class manufacturers were Du Mont (Allen B. Du Mont Laboratories) of New Jersey (Figure 2.9.2), and Heathkit from Michigan, and Tektronix, a firm in Oregon that focused entirely on oscilloscope design and production after World War II and remained the most popular brand until the end of the 1980s.[4] Among practicing researchers, a "Tek" was almost synonymous with an oscilloscope. The firm's inventions included the 1963 Direct View Bistable Storage Tube that enabled display not only of repetitive wave signals but also single shot events. The Tektronix model 7104 incorporating this feature became the most advanced analog oscilloscope ever.

Fig. 2.9.2. Dual beam cathode ray oscillograph, 322 A type, Allen B. Du Mont Laboratories, ca. 1955. Inv. 40330. *Credit*: © Musée des arts et métiers-Cnam. Photo by Denis Pruvrel.

From Analog to Digital: Computer and Data Acquisition

In the 1980s, oscilloscope technology was revolutionized by the introduction of digital storage techniques using fast analog-to-digital converters and memory chips. By the mid-2000s, most new oscilloscopes were digital. Ultimately, though, the move from analog to digital contributed to the eclipse of oscilloscopes in favor of computerized data acquisition, recording, and display. Since the 2000s, the instrumental chain has largely been digitized and now relies on software that runs on a personal computer. Software packages allowing computer-based data acquisition, analysis, and display were available to laboratories in the mid-1980s, such as LabVIEW, which was introduced by National Instruments in 1986. Measurements of physical systems taken by sensors (essentially analog in nature) are converted into digital records and digital representations that are processed by the computer. The advantage over traditional oscilloscopes is the ease of data backup and transcription in conventional data processing and display formats on personal computers. This kind of software is used in laboratories where technicians still build the experiments themselves.

But more and more often, scientists do not have time to build specific experiments, nor do laboratories have the skilled technical staff for such work. In most cases, commercial instrument manufacturers now offer measurement apparatus with specific interfaces that give the user the means to drive the experiments, acquire, process, and display the data, and even write the experiment report. The instrument has become a black box for the user. To understand the trajectory from home-built data recording technology to black-boxed instruments, let us examine some case studies of instruments used in the metallurgy section of the French Conservatoire National des Arts et Métiers over the course of the twentieth century.

Metallurgy in the Conservatoire des Arts et Métiers: Some Illustrative Instruments

The chair of metallurgy of the CNAM was created in 1890. It was to be headed by great French scientific figures who had a long-lasting influence on metallurgy in France. The first chairman was Urbain Le Verrier (1848–1907), engineer of the Ecole Polytechnique and the Corps des mines. He was to devote his efforts to bringing industry and teaching closer together. According to Le Verrier, industry should develop laboratories because scientific research was the only way for metallurgy to grow into something more than purely empirical practice. Léon Guillet (1873–1946), his successor, continued this influence through the Revue de Metallurgie that he led for twenty-five years in the footsteps of its founder, Henry Le Chatelier. Jean Cournot (1895–1993), a graduate from the Ecole Polytechnique, followed him, serving from 1942 to 1966.

During the first part of the twentieth century, the galvanometer reigned as the premier measuring and display instrument,[3] as illustrated by the Ferry telescope (or optic pyrometer) used in the chair of metallurgy at CNAM (Figure 2.9.3). In metallurgical research, or during iron and steel production processes, one essential parameter

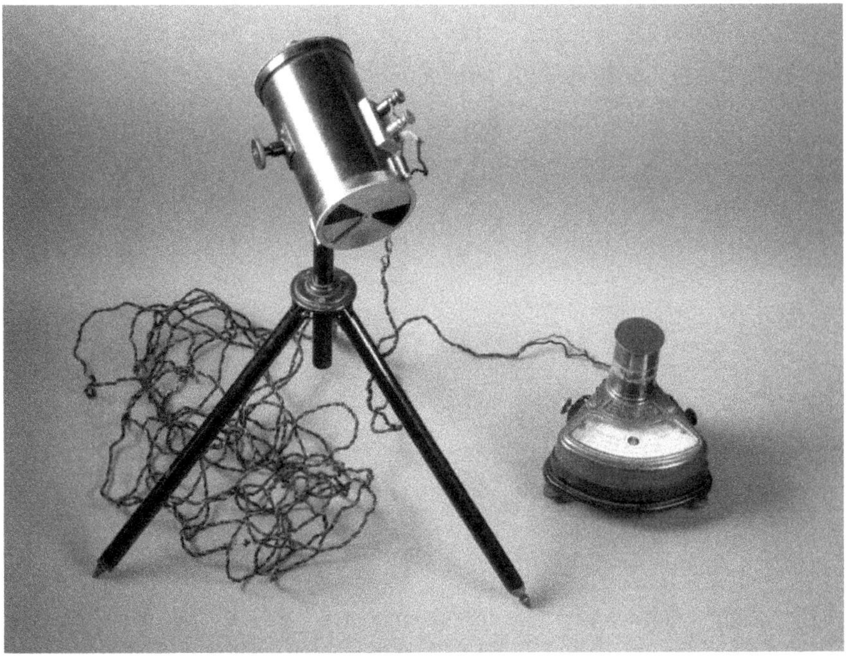

Fig. 2.9.3. Ferry telescope or optical pyrometer, Compagnie pour la fabrication des compteurs, 1908. Inv. 14179. *Credit:* © Olivier Delarozière 2017. The instruments shown here now belong to the scientific instruments collection of the Musée des arts et métiers in Paris.

is the ability to measure the temperature of the metal. During the 1920s, the optical pyrometer invented by the French scientist and instrument maker Charles Ferry (1865–1935) was in common use. Thanks to the telescope, it was possible to target the part of the metal where one wanted to measure the temperature. A mirror then concentrated the radiation onto a thermoelectric couple linked to a galvanometer.[2 p40]

For phenomena that are transient and/or evolve over time (like time–temperature curves), the Le Chatelier–Broniewsky galvanometer (Figure 2.9.4) made it possible in the 1920s to record the deviations of a double galvanometer on a photographic plate. With the double mirror of a galvanometer users could study functions with two variables, thanks to the Lissajoux prism. The prism transformed the

Fig. 2.9.4. Galvanometric photographic recorder of the Le Chatelier–Broniewsky type, ca. 1930. Inv. 40017 *Credit*: © Musée des arts et métiers-Cnam. Photo by Denis Pruvrel.

horizontal deviations into vertical deviations: the photographic plate recorded the variations given by the first mirror on the ordinate, and those given by the second on the abscissa.[5 p133]

After World War II, oscilloscope-type instruments with cathode ray tube screens were the key instruments, as we may see in the two following examples.

The Ferrosigmatest (Figure 2.9.5) was a common device in the 1950s for the control and non-destructive sorting of metals. The figure observed on the screen showed the differences in the magnetic characteristics of the steel. It allowed, for example, the control of thermal treatments and control of the smallest steel parts. It was also suitable for the control of non-ferromagnetic metals (austenitic steels, brass, aluminum, bronze, brass, and others). The instrument's display uses a galvanometer and a scope.

Fig. 2.9.5. Non destructive testing instrument, "Ferrosigmatest," LEGPA (Laboratoires de Physique Appliquée), model F-108, 1950-1960. Inv. 45334. *Credit*: © Musée des arts et métiers-Cnam. Photo by Denis Pruvrel.

Another device representative of work in metallurgical research is an apparatus for non-destructive testing of defects in metals using an ultrasonic beam (Figure 2.9.6). It was developed according to an idea suggested in 1928 by the Soviet engineer Sergei I. Sokolov: an ultrasonic wave is sent to the part to be checked, where it is reflected, passes over the faults, and transmits the signal to a receiver, which can then give information on the location and shape of the faults. After the Germans, the British (Donald Sproule and Floyd Firestone) were particularly active in research on this technique during World War II: Kelvin & Hughes was the industrial partner of their works and began producing the first commercial machines in 1952. This type of ultrasonic detector has been adapted and has been successful in the medical field in obstetrics, where it is better known as an ultrasound machine. Typical of the 1950s, the display uses a cathode ray tube with a phosphoric screen.

Fig. 2.9.6. Ultrasonic detector, Kelvin & Hugues Ltd, Model MK5, ca. 1955. Inv. 45335. *Credit*: © Musée des arts et métiers-Cnam. Photo by Denis Pruvrel.

Our last instrument is a scanning electron microscope (Figure 2.9.7) that was in use from the 1970s onward in the metallurgical chair at the CNAM. The display and recording instruments are a chart recorder, a photographic camera mounted on the screen of an oscilloscope, and a television coupled with a camera. No computer is to be seen, whereas today the computer would be the only means of data acquisition, processing, recording, and display in the modern version of this microscope.

The leap to computer control and data recording, and the loss of generations of analog data recording technologies, should make us pay heed. Beware the dangers of the black box! In the past, scientists learned to interpret the indications and the precision or accuracy given by a needle in front of a scale or by a pen trace on paper. Today, the whole measuring chain in the black box is designed by others; computers can give results of experiments with n digits

Fig. 2.9.7. Scanning Electron Microscope, Cameca MEB 07, 1970. Inv. 44711
Credit: © Olivier Delarozière 2017.

after the decimal point where just one digit is significant.[1] They can (re)create colors for phenomena, and so on. New instruments require new critical sense.

References

1. Chevalier JP [last chairman of the department of Metallurgy, EPN Ingénierie mécanique et matériaux]. Interview with the author. CNAM. April 12, 2017.
2. Grard CAM, Cournot J. Métaux et alliages. Paris: Berger Levrault; 1930.
3. Hackmann WD. Galvanometer. In Bud R, Warner DJ, editors. Instruments of science. an historical encyclopedia. New York: Garland; 1998: 257–59.
4. Hiscocks, PD. Oscilloscope development, 1943–1957. March 15, 2011. Consulted on www.ee.ryerson.ca on February 7, 2017.
5. Le Chatelier H, Broniewski W. Sur un enregistreur photographique. Revue de métallurgie. 1912;9:133. Or: Guilet L. Les méthodes d'étude des alliages métalliques. Paris: Dunod; 1923: 53.
6. Malifaud P. De la mesure des premiers âges à l'ère électronique. Paris: ed. Ribet-Desjardins; 1962.
7. Morris AS, Langari R. Measurement and instrumentation: theory and application, 2nd edition. Academic Press; 2016.
8. Phillips JV. Oscilloscope. In Bud R, Warner DJ, editors. Instruments of science. An historical encyclopedia. New York: Garland; 1998: 430–31.
9. Zelbstein U. Histoire des enregistreurs. Brève réflexion sur l'évolution des techniques d'enregistrement vue à travers trois exemples de réalisation du 18e et du 19e siècles. Revue de métrologie. 1987;65(9):931–37.

CHAPTER 2.10

MECHANICAL TESTING

JUSTIN CARONE

Mechanical testing refers to the process by which manufacturers and researchers determine the physical characteristics of a material, such as strength and hardness, and therefore its suitability for a particular application. This process is driven by a desire to predict how materials will behave when they are pressed into service. The aspiration to predict the behavior of materials across various contexts of production and use underlies many technological advancements, from the construction of railroads to the invention of nuclear reactors.[6] This predictive impulse is not limited to testing materials; it also shapes how institutions are organized and how individuals conceive of their roles within them. Mechanical testing, then, aims both to control materials in a way that ensures a profitable manufacturing operation and to control the relations among manufacturers, management, and laborers within a testing laboratory in a way that is consistent with the impulse to predict. Laborers and the tools of their trade, management, production operations, and the nature of raw and finished materials intermingle within organizations that predictably support industrial profit, and that is the context in which mechanical testing procedures were established. Mechanical testing is therefore both about understanding materials and about preserving technical practices and labor relationships that have historically made industrial production profitable.[7]

Mechanical testing, as it is understood today, emerged in the Anglo-American world during the mid-nineteenth century.[4] As the complexity of manufacturing tasks increased, industry researchers developed new testing methods to ensure that materials would behave as desired. When expanding railroad infrastructure, for example, companies needed to confirm that the rail materials would not fail when subjected to the stress of regular use.[6] To provide this assurance, researchers devised some of the first mechanical tests, including impact testing, which is intended to determine the energy required to cause complete failure in railroad steel. Before selling their products to railroad companies, steel manufacturers routinely performed mechanical tests to prove the effectiveness of their rail materials.[1]

Early forms of mechanical testing in the steel industry employed post-production materials — that is, materials that had already been fashioned into something resembling their final shape, as in Figure 2.10.1. Manufacturers encountered two significant challenges in this approach.[6] First, upon being dropped from specified heights, railroad steels would often bend rather than break, making it difficult to calculate the energy that they could absorb in practice before failing. Second, the use of post-production materials meant that manufacturers would need to produce finished batches prior to learning if they could withstand the strains of mechanical testing and eventual use, constituting a major detriment to profits. To address these challenges, manufacturers sought more predictable, transferable, and quantifiable data points, prompting researchers to devise one of the first machine tests, the pendulum impact test (PIT), and standards to describe its optimal use and findings, toward the end of the nineteenth century.

Following the development of the PIT allows us to identify the ways in which standards reassure their users of the certainty of predictions. The PIT involves a hammer that, when attached to an arm, acts as a pendulum.[6] When released from a raised position, the hammer swings a pre-calculated distance before coming into contact with a test material. The test material is meant to approximate, or in early tests was itself, the post-production material as it would exist in the context of use. In order to reproduce the contexts of use,

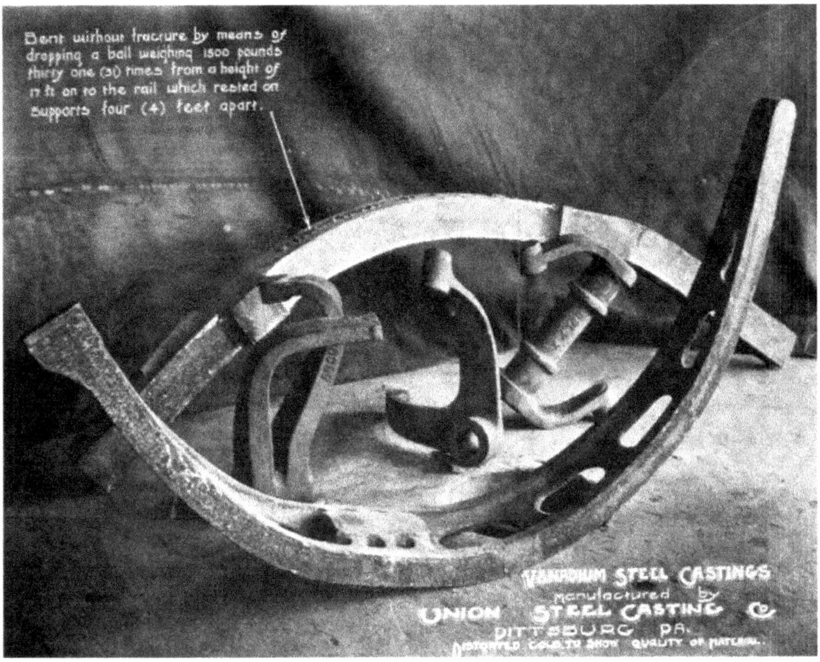

Fig. 2.10.1. The results of mechanical tests performed at the Union Steel Casting Company, Pittsburgh, PA.

practitioners ensure the material is tested at the range of temperatures, and often beyond, that the post-production material would be subject to in use. When struck, the test material breaks, with the distance that the hammer continues to travel after impact then being measured. If the material presents no resistance, then no energy is lost and the reading is zero. As the resistance grows, a more significant amount of energy must be expended to break the material, resulting in a higher reading. Increasingly over the twentieth century, machine tests like PIT used specially produced test specimens with more controlled behavior than the raw or post-production materials used up until this point.[6] These specimens, although similar in composition to post-production materials, cost manufacturers a fraction of the time and resources to produce. Because of their standardized nature, these specimens also helped to establish a new metric for steel fitness, the Ductile-Brittle Transition Temperature (DBTT).

Fig. 2.10.2. A sample of steel following a pendulum impact test.

DBTT is the temperature at which a material, when struck with a specified amount of energy, does not simply break, but shears in a particular fashion.[6] A material that does so is significantly more likely to fail at that temperature than is a material that retains ductility, or the ability to be deformed prior to breaking (Figure 2.10.2).[1] In early impact tests, however, researchers could neither identify the DBTT as a vital characteristic in predicting a material's use nor accurately measure such a characteristic, given a lack of standardized procedures with which to compare materials and produce consistent results.[6] Beyond standardized materials, then, researchers required institutions that could ensure predictability across various contexts through the generation and distribution of "best practices."

Standards organizations performed this function for manufacturers and researchers involved in mechanical testing. The International Organization for Standardization (ISO) and the American Society for Testing and Materials (ASTM), for example, were established at the start of the twentieth century, in part to serve as intermediaries between companies and governments (see also Sharon Ku's contribution to this volume).[5] They developed specifications for the design of PIT machines and the production of test specimens, which then governed the testing-machine manufacturing operations of companies like Tinius Olsen in the United States.[8] Furthermore, the standards that these organizations set, and the design of the testing machines themselves, regulated the practices through which predictable results were achieved via mechanical testing, including the production of notches in test specimens, the frequency with which the machines were to be calibrated, and the proper ways in which workers were to interpret information.[5]

Industrial management's regulation of the practices of individual workers was formalized at the turn of twentieth century in what is today referred to as rationalized labor. Rationalized labor was put into practice as industries were pressured to produce the greatest number of goods in the least amount of time, to become more efficient, in order to remain competitive in an increasingly global economic landscape. To achieve this goal, two key claims about the nature of work were made. The first claim was that in order to increase efficiency, tasks must be divided into two categories: conception and execution.[3] This division assumed that for best results, one group should plan out in detail the precise movements and the optimal pace of the work, and another group should perform that work. The second claim was that certain individuals have a natural propensity for one of these two types of work — that there is a "right man for the job." The idea that certain individuals have the natural capacity for abstract, conceptual thinking, while others do not, and are instead most well suited for physical labor, continues to operate today across a range of industries. The separation of conception and execution, and the assumption that such a division reflects the natural capacities of individuals, made the hierarchical organization of labor seem not only more efficient, but a natural result of human difference. The work of mechanical testing can reveal that while these divisions are still in place, the natural categories they assume do not accurately reflect the skills and capacities of today's workers.

Because an industrial labor force was required in order to carry out standardized mechanical tests, the predictive impulse displayed by engineers and executives in materials industries also had consequences for individual workers. Like the time-motion studies and practices of scientific management, which re-organized American industrial production at the start of the twentieth century,[8] the exacting standards of mechanical testing shaped the identities of the workers who performed the tests. Above all, standardized mechanical testing encouraged managers to view workers only in terms of their skills and capacities — that is, the number of tasks that they could perform to specification in a narrow timeframe — which could boost productive potential and profit-making.[11] Some labor historians have suggested that standardized rules reduced workers to mere automata, yet such an account fails to capture the complex interplay of laborers' self-characterizations and those used by managers to reproduce inequalities of opportunity.[3]

Historians of industrial labor have focused on the alienating effects rationalized labor regimes have on workers;[10] however, this focus neglects the skills and experiences that individual workers bring to their labor. Although rationalized labor regimes can produce rigid and narrow notions of workers' competencies, they still rely on the individual expertise of those workers, even if in a limited capacity.[11] The mechanical tests that workers perform to determine DBTT, for example, are set to exacting standards.[2] Yet, contrary to the industry's expressed interest in producing predictable, transferable, and quantitative data points, the actual practice of impact testing hinges on a qualitative evaluation.[5] Determining DBTT requires workers to employ perceptual and interpretive capacities that they have developed on the job. Even as mechanical testing continues to move further toward automation, with digital readouts that obviate the need for hand-drawn graphs and plot diagrams, the ability of individual workers to identify shear patterns and inspect failed specimens remains invaluable.[5]

In addition to reflecting social relations consistent with the predictive impulse, mechanical testing devalues workers' unique knowledge, which they gain through their experience. Workers require this knowledge to carry out the work of mechanical testing, but, because they are known largely by the end results of their labor, the number of tasks they can perform, and the amount of profit they can help produce, their knowledge is excluded from directing decision making processes at the companies for which they work. If the principal motivation of mechanical testing is to predict the behavior of materials across various contexts and this motivation, in turn, demands institutional and individual behaviors that are equally predictable, then opportunities for workers to direct institutional change are constrained. The institutionalization of the predictive impulse, primarily via standards-setting, prevents workers from introducing novel processes or changing the conditions that surround mechanical testing. Most crucially, challenges to the hierarchical organization of labor become difficult to imagine when workers' contributions are expressed only in standards-compatible ways, that is, through single data points concerning the ductility of some materials. Opportunities for the redistribution of authority also become fewer, thereby locking workers into forms of labor in which their skills and experiences can only be expressed through the standardized language that manufacturers, researchers, and standards organizations have laid out in advance. Thus, the predictive impulse that underlies mechanical testing suggests a future in which both institutional forms and individual identities are predetermined.

In sum, the practices and standards of mechanical testing have developed alongside an impulse to predict the behavior of materials across various contexts and their suitability for industrial, profit-making applications. Because the predictive impulse guides empirical inquiry into material properties, however, it is also taken up in institutional and individual commitments to predictability. In this way, pre- and post-production materials, institutional arrangements, and the identities of individual workers are mutually formed in a self-perpetuating cycle aimed at predicting future states with certainty. That is to say, materials, institutions, and individuals are produced, organized, and defined alongside one another in a continuous process that preserves the social relations and organizational structures required to sustain the productive regimes of industrial capitalism.

References

1. American Vanadium Facts: A Journal of the Industrial Application of Vanadium, Through the Use of Ferro-vanadium ... American Vanadium Company; 1914.
2. ASTM E23-16b: Standard test methods for notched bar impact testing of metallic materials. ASTM Int [Internet]. 2016; Available from: www.astm.org.
3. Braverman H. Labor and monopoly capital: the degradation of work in the twentieth century, 25th anniversary ed. New York: Monthly Review Press; 1998.
4. Chinn R. Hardness, bearings, and the rockwells. Advanced Materials and Processes. 2009;167(10):29–31.
5. E. Director of Technology, Tinius Olsen. Personal communication. 2017.
6. Siewert TA, Manahan MP, editors. Pendulum impact testing: a century of progress. W. Conshohocken, PA: ASTM; 2000.
7. Slaton AE. Reinforced concrete and the modernization of American building, 1900–1930. Baltimore: Johns Hopkins University Press; 2001.
8. Tait C. President, Tinius Olsen. Personal communication.
9. Taylor FW. The principles of scientific management. New York: Harper; 1911.
10. Thompson EP. The making of the English working class, new edition. London: Penguin Books; 2013.
11. Townley B. Reframing human resource management. New York: Sage Publications Ltd.; 1994.

CHAPTER 2.11

CLIMATE TEST CHAMBERS AND EXPOSURE TESTING

ANDREAS T. HAKA

Alarming reports of major pipeline and cable breakdowns due to corrosion — above all for commercial suppliers of gas and water, telegraph services, and electricity, as well as for railroad companies in transport — led to the search for solutions not only in the United States but also in Great Britain, Germany, and Japan at the end of the nineteenth century. According to the West of Scotland Iron and Steel Institute, the amount of metal lost due to corrosion between 1890 and 1923 was approximately 41 percent of all the metal produced during that time. So, a world metal production of 1,760 million tons in the mentioned period meant a loss of 718 million tons.[30]

The most serious manifestations were, to use modern definitions, *surface corrosion* and *contact corrosion* (*bimetal corrosion*). However, what exactly corrosion is was still being debated at the time and no clear distinctions were yet being made between the various kinds of corrosion.[14]

In the United States, the Bureau of Standards opened the discussion about corrosion in 1900 and commissioned the present-day California Institute of Technology, then still a small polytechnic in Pasadena, to conduct the first extensive scientific analyses.[3; 20] In Great Britain, the Board of Trade adopted the topic of corrosion and commissioned the Royal School of Mines in London early on with the scientific investigation of this topic. Special interests mostly guided the examination of corrosion problems in the United States and Great Britain during this period, and separate solutions were sought for concrete topics from industry.

In Germany, however, scientific corrosion research developed differently. A research alliance was formed in 1907 between various German agencies that had been working individually since the turn of the century (the Commissions of Gas and Waterworks Specialists, the Association of Electrical Engineers, and the Association of German Streetcar and Minor Railway Managers) to conduct systematic research on corrosion, under the scientific leadership of Fritz Haber (1868–1934), director of the Institute for Electrochemistry at the Karlsruhe Polytechnic.[13] The commission's initial findings were compactly presented for the first time in 1910 in Berlin at the International Exhibition "On the Protection of Gas and Water Piping — Telluric Currents along Electrical Pathways" (*Zum Schutz der Gas- und Wasserröhren — Erdströme elektrischer Bahnen*).[3]

Extensive academic investigations on topics related to corrosion that continued in the United States, Great Britain, and Germany were primarily based on work by Wyndham R. Dunstan, Hooper A. D. Jowett, and Ernest Goulding (1905) and by Emil Heyn and Oswald Bauer (1910).[7; 17] Industrial and academic research in the aforementioned countries employed outdoor exposure testing stands to simulate realistic corrosive wear of metallic materials. But from the start, the long testing duration of these analyses was regarded as disadvantageous. The search for forms of artificial weathering thus began almost immediately, leading to the development of climate test-chamber systems.

Outdoor Exposure Testing

Isolated natural weathering experiments began in the United States, Great Britain, and Germany in the 1890s. Industrial firms first conducted their own outdoor exposure tests in order to examine their products for corrodible weak spots and, where needed, to make improvements to their designs.[21] Industrial firms involved in the working and processing of materials were also preoccupied with warranty issues in connection with component-part failure due to corrosion. Scientifically defined outdoor exposure tests then started to move into academic research at the beginning of the twentieth century. The emergence of the chemical industry was the trigger, as academic researchers began examining new types of commercially manufactured substances, particularly protective coating products for metallic materials, such as varnishes, oil paints, lacquers, and varnish paints.[4; 24]

The first instance of the institutionalization of corrosion research occurred in Britain with the founding of the Corrosion Research Committee of the Institute of Metals in 1910.[12] The outdoor exposure test was specifically indicated as one tool of their corrosion analysis. The first refinement in outdoor exposure testing, which at this time was still being referred to as a "natural trial" or as "atmospheric testing," also occurred within this context. With industrial experimental data serving as a basis, the first multiannual outdoor exposure tests were performed, already subdivided by county, city, and industry. Separate outdoor exposure tests under marine weathering conditions were carried out mainly with the support of the shipbuilding industry.[16] Academic researchers perceived these testing procedures as independent because their materials samples were subjected not only to natural weathering but also to partial submersion in dune sand or to alternating tidal exposure to sea air and seawater.[18] It was of critical importance for the academic research that this involved several distinct tests that all reproduced fully natural processes without any human influence. Starting in 1923, a larger testing station in southern England at Birnbeck Pier was operated for almost a full decade by the Corrosion Research Committee, and from 1924 to 1931 by the Department of Scientific and Industrial Research.[12]

Outdoor exposure testing stands were initially built out of wood, irrespective of the local climate requirements (Figures 2.11.1 and 2.11.2). Metallic materials, and later also samples made from glass and plastic, were initially simply laid onto a testing stand and secured with horsehair. Later models had wooden rails affixed to the weathering surfaces, from which the samples could be suspended or

Fig. 2.11.1. Outdoor Exposure Testing built out of wood, Zoellner-Werke AG, Berlin 1909.

Fig. 2.11.2. Outdoor Exposure Testing built out of wood, Deutsche Versuchsanstalt für Luftfahrt, Berlin 1923.

into which they could be slotted.[28] The exposure surface bearing the samples was usually set at an angle of about 15 to 40 degrees to the weathering. The exposure slope was initially in emulation of regular rooftops. It was first scientifically defined in the 1920s.

In 1913 the American Sheet and Tin Plate Company in Pittsburgh conducted a four-year-long series of tests on iron sheets, the very first trial of its kind in which metallic materials from different producers were compared in outdoor exposure testing. Just a year later, the German association of ironworkers (the Verein Deutscher Eisenhüttenleute) conducted equally elaborate outdoor exposure tests with samples provided by the ironworks Hoesch in Dortmund, Bremerhütte in Siegerland, and Bismarckhütte in Upper Silesia.[1] Outdoor exposure testing stands were set up on the roof of the materials testing agency Berliner Materialprüfungsamt, as well as in Hörum on the Island of Sylt, and on the shop rooftop of the Hoesch iron and steelworks in Dortmund for comparative analysis of samples exposed to air in different climates.

However, the outbreak of World War I led to the abandonment of large-scale experiment, both in British universities and German civilian industry, owing to material shortages on both sides.[8] The Royal Prussian Ministry of War took up the subject of corrosion on behalf of Germany's military, delegating it to its Artillery Testing Commission, and commissioning a "rust protection panel" with which the academic research institutions Kaiser-Wilhelm-Institut für kriegstechnische Wissenschaft, the Kaiser-Wilhelm-Institut für Chemie, and the Berliner Materialprüfungsamt were associated.[19] This panel was supposed to consider corrosive damage to ammunition and technical weaponry. In 1920 the panel was converted into the civilian Reichsausschuss für Metallschutz and issued the world's first scientific journal dedicated to the topic in 1925, entitled *Korrosion und Metallschutz*.

One of the central topics brought up for international discussion in this journal was a corrosion testing method, which was still awaiting exact definition. Until that time, corrosion testing had been performed to the best of the tester's abilities, on the basis of available data. This also applied to the performance of outdoor exposure trials and the associated evaluation of weathered samples. Comparability figured increasingly as an essential aspect of international research, though, and was finally regulated for the first time by the association of German engineers (Verein Deutscher Ingenieure) in 1930 within the context of the establishment of German industrial standards (the Deutsche Industrie Normen) as DIN-Normen sheet no. 4850. This is a technical specification that was mandated by law for all German companies.

Comparability was also the underlying topic of the largest outdoor exposure test in sea air or seawater ever to be conducted during the first half of the twentieth century. The shipping concern Germanische Lloyd in Hamburg initiated this trial in 1921. The weathering surfaces employed were onboard 102 vessels of various sizes traveling along various maritime routes around the world and owned by seven different shipping companies. After four years of exposure, a total of 4,010 iron and steel pipes affixed to the ships on places ranging from the keel to the deck were subjected to corrosion analyses.[2] These tests, some of which were partly continued on different ships, yielded decisive insights into corrosive processes under dynamic load conditions, which are now considered the basis of stress corrosion cracking. The extensive use of microscopes in this corrosion analysis was equally decisive for the further distinction between types of corrosion.

The emergence of light metals as structural materials, for example, aluminum and magnesium alloys, from the 1920s on, motivated the chemical industry, and particularly the aviation industry, to consider outdoor exposure testing as a realistic testing procedure in materials research for this horizon of applications.[23; 26] In Germany, the aviation testing institute Deutsche Versuchsanstalt für Luftfahrt operated, in addition to its own weathering stands on its premises in Berlin, a separate outdoor exposure testing stand in the Baltic near Warnemünde in order to be able to test aluminum alloys under marine climate conditions as well.[27] Outdoor exposure testing stands continue to be an indispensable testing instrument not only in examining the corrosiveness of metallic materials but also in analyzing and testing other working materials under trying climatic conditions.

Climate Chamber Testing

Artificial weathering of material specimens first occurred in the laboratory in a test tube. The first documented trials took place at the material-testing institute of the Technische Hochschule Berlin in 1883.[22 p37]

The materials were boiled, brought into contact with plain water and chemical substances, and the corrosive effect on metallic materials was analyzed. Outdoor exposure trials during the first decade of the twentieth century set off the compilation of such data, mainly with regard to influential factors to which materials are subjected upon outdoor exposure in a city environment, in the countryside, in industrial regions, and at sea.

However, neither the limited test-tube trial nor time-consuming outdoor exposure testing could satisfy the need by technical production for short developmental periods. Technical development and production got faster and faster, so the time-consuming tests increasingly failed to keep pace. Neither could they provide an exact solution to the problem of how much the materials used in the manufacture of components or groups of components were susceptible to corrosion. For that reason, the search for a time-saving testing method with variable test parameters started just a few years after outdoor exposure testing stands became the accepted testing procedure in corrosion research. One obvious starting point for artificial outdoor exposure testing was the hothouse, as an established space with an artificial climate. Initially only temperature, air humidity, and sunlight were used as adjustable conditions of the environment. Miniaturization and the integration of further climate variables, such as, for instance, a lamp for artificial sunlight and water vapor was a subsequent step. Sodium chloride was mixed into the water vapor soon afterwards, thereby introducing an essential variable exposure condition of marine climates.

Besides the introduction of additional environmental variables, another central aspect of this developmental phase was the reduction in time needed for the testing procedure and a more rapid selection of the data. A whole range of specific improvements to the detection of corrosive phenomena in laboratory testing were made in this context in the United States and Great Britain as well as in Germany. An early version of a climate chamber (according to the definitions used in modern environmental testing) was the "rust apparatus"

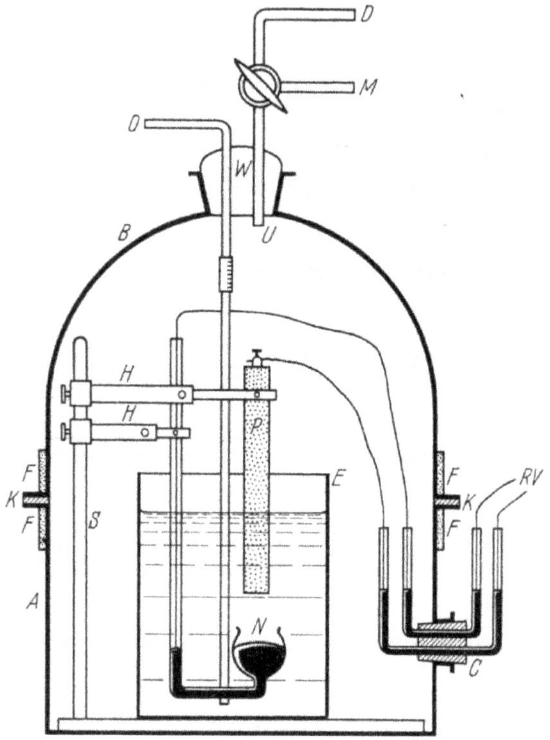

Fig. 2.11.3. Diagram of Viktor Duffek's *Rost-Apparat.*

(Figure 2.11.3), invented by Viktor Duffek, head of the corrosion department of an Austrian crucible steel works, Firma Gebrüder Böhler & Co AG Gußstahlwerke) located in Kapfenberg.[5] Duffek developed his "*Rost-Apparat*" on the basis of research from 1907 by William Walker, Anna Cederholm, and Leavitt Bent in the technical chemistry laboratory at the Massachusetts Institute of Technology (MIT) and work from 1913 by Fritz Wüst (1860–1938), the founding director of the Kaiser Wilhelm Institut für Eisenforschung, as well as by Ulick Evans (1926).[9; 10; 29]

This "rust apparatus" ascertained the corrosion rate of steel by measuring the strength of current. A bell jar (Figure 2.11.2), into which oxygen was introduced under pressure, held a vessel containing the electrolyte in which the specimen and a mercury electrode were submerged. The strength of current was recorded by a measurement instrument upon the application of high-impedance resistance.

A resistance that generated very low current densities in the specimen was chosen. The onset of rusting produced a kink in the current-strength curve.[6; 11] The testing of a sample took up to five days and was supposed to simulate the course of rusting over a period of one year. The subsequent analysis examined the material for weight reduction, bud formation, pitting, cracking, and alterations to its solidity and structure. Duffek, who had been experimentally improving the apparatus repeatedly since the beginning of the 1920s, finally presented it at the exhibition Berliner Werkstoffschau in 1927 after it had undergone a three-year testing stage.

The Duffek apparatus was an excellent device for measuring the rate of corrosion. As such, it was precursor to many devices in the field of climate chamber testing and had a lasting effect as regards design, measuring technique, and handling. The Deutsche Forschungsanstalt für Luftfahrt in Berlin, the Scientific Institute for Aircraft Materials in Moscow, and the National Bureau of Standards in Washington introduced, at about the same time, innovative testing chamber models that differed almost exclusively in their testing regimes rather than in design.[18] Most of these testing chambers were constructed from transparent insulated glass and resembled small enterable hothouses measuring a few square meters, in which metallic specimens were hung grid-like in a system of shelves positioned at defined distances. The cut surfaces of the metallic test objects were sealed in rubber. In most cases these climate chambers were equipped with air-humidity controls for the injection of water vapor that could optionally be combined with a sodium-chloride solution. In addition, they contained a sprinkler unit for plain water and seawater, an adjustable temperature control ranging between 5 and 30 degrees Celsius, and the possibility to switch on artificial sunlight.

In 1934 the firm Siemens & Halske developed the first alternating-climate testing chambers for manufacturers in the field of electrical engineering to test the suitability of their products for export to South America and Africa.[25] The alternating climates covered dry and hot environments at 60 degrees Celsius and 4 percent relative air humidity, as well as humid and wet at 40 degrees Celsius and 90 percent relative humidity. The first alternating-climate testing chambers capable of simulating below-zero temperatures were developed at the behest of the German military during World War II to test the corrosiveness and functionality of technical weaponry.[15] Ingenieurbüro Karl Weiss, a Thuringian firm in Greiz, offered the first alternating climate chamber in serial production conceived for civilian use in 1955.

Further rapid developments in climate chamber engineering have now made it feasible to perform complete mechanical and medial testing of any material. Appropriate testing chambers make it possible to analyze them under mechanical stress at conditions prevailing anywhere on Earth and in space.

References

1. Bauer O. Rostversuche mit kupferhaltigen Eisenblechen. Mitteilungen aus dem Materialprüfungsamt zu Berlin-Lichterfelde West. 1920;38:85.
2. Beitrag zur Erforschung der Korrosionsursachen an Kondensatorrohren; Mittel zur Verhütung der Anfressungen. Korrosion und Metallschutz. 1925;1:146.
3. Besig F. Aus der Praxis der Bekämpfung der Rohrkorrosion. Korrosion und Metallschutz. 1929;5:99–102.
4. Dubois NA. The protection of iron and steel by paint films. Industrial & Engineering Chemistry. 1913;5(12):968–71.
5. Duffek V. Ermittlung der Rostgeschwindigkeit von Qualitätsstählen. Korrosion und Metallschutz. 1926;2(7–8):183–87.
6. Duffek V. Untersuchungsmethoden über Rostgeschwindigkeit und -neigung von Qualitätsstählen mit besonderer Berücksichtigung der Korrosionsquellen an rostsicheren Chromstählen. Korrosion und Metallschutz. 1927;3(3):49–53.
7. Dunstan WR, Jowett HA, Goulding E. The rusting of iron. Journal of the Chemical Society. 1905;87:1548–74.
8. Emil Maass. Zum 60. Geburtstag. Korrosion und Metallschutz. 1929;5:1.
9. Evans UR. The fundamental principles of corrosion. Journal of the Society of Chemical Industry. 1926;45(30):504–16.
10. Evans UR. Metals and Metallic Compounds. London: Edward Arnold; 1923.
11. Fry A. Korrosionsprüfung metallischer Werkstoffe. In Siebel E, editor. Die Prüfung der metallischen Werkstoffe. Berlin: Springer; 1939.
12. Haas M. Korrosionsforschung in England. Korrosion und Metallschutz. 1926;2:133–34.
13. Haehnel O. Was versteht man unter Korrosion? Aus dem Telegraphischen Reichsamt. Korrosion und Metallschutz. 1926;2:42–43.

14. Haher F, Goldschmidt F. Der anodische Angriff des Eisens durch vagabundierende Ströme im Erdreich und die Passivität des Eisens. Zeitschrift für Elektrochemie. 1906;12(4):49–74.
15. Haka A. Soziale Netzwerke im Maschinenbau an deutschen Hochschul- und außeruniversitären Forschungseinrichtungen 1920–1970. Berlin: Logos; 2014.
16. Hayßen L. Rostschutz von Eisenkonstruktionen im Seewasser. Korrosion und Metallschutz. 1928;4:182.
17. Heyn E, Bauer O. Metallographie. Kurze gemeinfaßliche Darstellung der Lehre von Metallen und ihren Legierungen, unter besonderer Berücksichtigung der Metallographie. Berlin: W. de Gruyter; 1909.
18. Kroenig WO. Korrosionsprüfungen. Korrosion und Metallschutz. 1938;14:53.
19. Maass E. Der Reichsausschuss für Metallschutz. Korrosion und Metallschutz. 1925;1:1–3.
20. Maass E. Die wirtschaftliche Bedeutung der Aufgaben des "Reichsausschusses für Metallschutz." Korrosion und Metallschutz. 1926;2:1–2.
21. Maass E, Liebreich E. Zur Frage der Korrosion von Metallen. Angewandte Chemie. 1924;37(46):897–99.
22. Martnes G, Guth M. Das königliche Materialprüfungsamt der Technischen Hochschule Berlin: Springer; 1904.
23. Meißner K. Der Einfluß der Alterungsbehandlung auf die Korrosionsfestigkeit von Aluminiumlegierungen. Korrosion und Metallschutz. 1925;1:206–8.
24. Normung. Einheitsverfahren für die einfache Prüfung von Farben und Lacken. Korrosion und Metallschutz. 1925;1:141.
25. Prüfraum mit künstlichem Tropenklima. *Zeitschrift des Vereins deutscher Ingenieure*. 1934;78(47):1387.
26. Rackwitz E. Korrosionserscheinungen und Korrosionsversuche an Leichtmetallen für den Flugbetrieb. Korrosion und Metallschutz. 1927;3:171–77.
27. Rackwitz E, Schmidt E. Prüfung von Leichtmetallen auf Korrosion. Korrosion und Metallschutz. 1927;3:58.
28. Rackwitz E, Schmidt E. Prüfung zur Beurteilung der Korrosionsbeständigkeit von Metallen gegen Witterung und Seewasser. Korrosion und Metallschutz. 1929;5:9–12.
29. Walker W, Cederholm AM, Bent LN. The corrosion of iron and steel. Journal of the American Chemical Society. 1907;29(9):1251–64.
30. West of Scotland Iron and Steel Institute report. Korrosion und Metallschutz. 1925;5:180.

PART III

TOOLS FOR MAKING MATERIALS

TOOLS FOR MAKING MATERIALS: INTRODUCTION

CYRUS C. M. MODY

One of the ironies of "materials research" is that there was no such thing until just the moment when the materials themselves became dispensable to research on materials. Until the mid-twentieth century, you needed to have some material stuff in order to do chemistry, the physics of solids and liquids, metallurgy, mineralogy, and similar investigations. Everything else in this volume you could do without — you could do quite good mineralogy, for instance, with the naked eye. But by the 1950s, the tables had turned. To do materials research you now needed expensive equipment, elaborate institutions, specially constructed buildings — but materials themselves were no longer absolutely necessary for materials research.

Indeed, the concepts of "materials science" or "materials research" were coined with the express aim of coming up with materials on paper, rather than making or synthesizing them from raw materials provided by nature. Arthur von Hippel — for whom the Materials Research Society's highest award is named — articulate this vision like so: "instead of taking prefabricated materials and trying to devise engineering applications consistent with their macroscopic properties, one [someday soon] builds materials from their atoms and molecules for the purpose at hand.... [The engineer] can play chess with elementary particles according to prescribed rules ... foreseeing inherent limitations of materials."[10 p315] That is, von Hippel argued that in the past, engineers imagined and then built a design that would work with existing materials; in the future they would imagine and then make materials that would work with an existing design.

As the entries in this section of the volume show, of course, materials were still very much at the heart of postwar materials research. The toolkit for making and manipulating materials grew and diversified enormously in the second half of the twentieth century. But so did the toolkit for describing the properties of materials that did not (yet) exist. When Mendeleev proposed his periodic table in 1869, he predicted some of the properties of several as-yet-undiscovered elements, but only their properties as elements — he could say very little about their properties as materials such as melting point or conductivity.[5 p33–43] By the end of the nineteenth century, organic chemists could do a bit better. Instead of stumbling into synthetic dyes as William Perkin had done in 1856, by 1877 they could apply Griess's method to predictably generate an almost unlimited series of synthetic dyes.[3] Moving from "chemicals" to something more like "materials," the phase diagram became an important tool for visualizing the properties of (abstract) materials in the early twentieth century in metallurgy and chemical engineering. By the end of the twentieth century, though, chemists and materials scientists could model materials *in silica* in great detail, without ever having to make the modeled material.

Thus, what set the materials of postwar materials research apart was not that materials could be made that did not exist in nature. For that, the tools in part I were more than sufficient. Rather, what (roughly) distinguished the tools in this section is that they could be used to make materials that had never before existed, yet the properties of which were already somewhat known. This is only a rough criterion, of course, but one that conveys the atmosphere of late-twentieth-century materials research quite effectively.

Buckminsterfullerene as an Accidental but Pre-Known Material

It is useful to follow an extended example to see how this distinction worked in practice. One of the most iconic classes of new materials of the late twentieth century are the fullerenes — closed cages of pure carbon. The first of these, C_{60} or buckminsterfullerene, was discovered accidentally in 1985, using an almost unique and yet thoroughly typical postwar materials research experimental setup. And yet, it turned out that buckminsterfullerene was already well known — on paper. And it was only through the interplay of experimental carbon and theorized carbon that fullerenes became the Nobel Prize–winning and possibly world-changing substances they are today.

The story begins with the AP2, a monstrous, nearly room-size machine — "AP" short for "apparatus" — built at Rice University by a chemist, Rick Smalley, and his graduate students in the early 1980s. Smalley had built an earlier version while a postdoc at the University of Chicago, but by the 1980s there were only two such instruments in the world, the one at Rice and a copy that Smalley had built for colleagues at Exxon (the AP3). Neither the AP2 nor the AP3 was originally intended to make *new* materials so much as extremely small particles of *existing* materials, what today we would call nanoparticles. They did so by zapping a disc of a given substance — initially Smalley looked at metals, later semiconductors such as silicon and gallium arsenide — with a laser, then instantaneously blowing the remnants of the laser blast into a supersonic stream and across a second laser beam to cool them down to near absolute zero before they could recombine.

Many postwar materials researchers worked in this way — they would build a one-of-a-kind machine for making one-of-a-kind materials, then exercise their monopoly over that substance by writing article after article on their properties. Most such machines stayed one-of-a-kind; only a few proved useful enough that other researchers built their own versions and/or manufacturers marketed commercial versions. Informal markets — such as Exxon's paying Smalley to build the AP3 for them — were more common than you might think, but rarely accounted for more than a handful of copies. The entries in this section of the volume almost all describe tools that "made it" and transitioned from unique, homemade, kludgy set-ups to polished commercial tools. But readers should keep in mind that even very successful experiments such as the AP2 sometimes stayed unique (or nearly so).

One way for unique tools to achieve success is to be essentially rented out. That is, their inventors can find collaborators who want to draw on the tool's unique capabilities without having to build one themselves. An opportunity to do this with the AP2 presented itself when Smalley's Rice colleague, Bob Curl, mentioned the apparatus to a British chemist, Harold Kroto, at a conference. Kroto immediately saw that the AP2 offered a way to simulate the conditions under which the interstellar medium is created: intense heat (in a star) followed by ejection into intense cold at extremely low pressure (in deep space). Kroto had a pet theory that carbon compounds were an important component of the interstellar medium, and so he asked Smalley to place a carbon disc in the AP2.[1;2] For several months Smalley refused — partly because the Exxon team had already looked at carbon and found nothing, and partly (I believe) because he thought the carbon experiment might interfere with his ongoing research on semiconductor clusters.

In the late summer of 1985, though, the semiconductor research came to a natural pause and Smalley relented. Kroto flew to Houston, Smalley's graduate students ran the carbon experiment several times, and over the next few weeks, Smalley, Kroto, and Curl wrestled with the unexpected results. For it turned out that with a little coaxing, the AP2 — quite anomalously — would spit out clusters of exactly 60 carbon atoms much more often than clusters of any other size. Add in heated discussions over barbecue dinners, some chest-thumping (both Smalley and Kroto were quite competitive and strong-willed), and a little model-building with gumdrops and toothpicks — and *voilà*, the group came to the conclusion that they had created a closed cage molecule made up of carbon atoms arranged in the shape of a classic soccer ball.

This was an astounding result which eventually led to Smalley, Curl, and Kroto winning the 1996 Nobel Prize in Chemistry. And yet, their "discovery" was also very old news. Many people had proposed such a molecule over the years. David Jones had speculated about closed carbon cages in his humorous "Daedalus" column in *New Scientist* as early as 1966. Somewhat more serious articles appeared in the United States in 1965 and again in 1985, Japan in 1970, the Soviet Union in 1973, the United Kingdom in 1981, and Spain in 1983. The Spanish team even worked out "soccerene's" official IUPAC designation. These were all theoretical studies, but at least one chemist — Orville Chapman, a National Academy of Science member at UCLA — had initiated a program to synthesize C_{60} in 1981.

Even more curiously, closed-cage carbon molecules had almost certainly been glimpsed well before the AP2 experiments. Since the 1950s, electron microscopists had occasionally reported small regular structures in micrographs of otherwise disordered soot, which only after 1985 were recognized as spherical fullerenes or their cousins, carbon nanotubes. Before 1985, apparently, none of the theorists who had proposed fullerene were aware of each other, nor of electron microscopists' experimental observations (and vice versa), and the Smalley-Curl-Kroto team was ignorant of all of them.[7]

The fullerene story contains practically all the routes to novel materials that one could imagine: hypothesizing a material and predicting its properties before it exists; trying to synthesize one from scratch; deliberately looking for a material conjectured to exist in nature; and stumbling across a material while looking for something else. Before 1985, these different approaches to fullerenes were all conducted in isolation; in the late 1980s they gradually coalesced; in the 1990s they completely fused around C_{60}'s elongated cousin, the carbon nanotube (CNT). CNTs were (re)discovered by an electron microscopist, Sumio Iijima, who remembered those old anomalous structures in soot and went back to try and find them again. At about the same time, they were hypothesized by theorists such as the "queen of carbon" Mildred Dresselhaus. Soon, they were being modeled at places like NASA, where computers could do chemistry with them *in silica* even when no one could make sufficient quantities for laboratory experiments.[6] Archaeologists and astronomers also set out to track down ancient nanotubes in meteorites and medieval swords made of Damascus steel.

And finally, Smalley — adapting techniques from the petrochemical industry — figured out how to synthesize gram-quantities in his Rice laboratory and started distributing them to colleagues via a quasi-company called Tubes@Rice. That, then, turned out to be a stepping stone toward the formation of a firm, Carbon Nanotechnologies, Inc., which planned to make tons of tubes as additives in high-performance plastics, paints, glass, and other common commercial substances. Again, that transition — from a laboratory experiment capable of making quantities in the nanogram-to-kilogram range to an industrial concern capable of making tons of a material — is typical of the tools surveyed in this section of the volume and in postwar materials research more generally.

Bombs, Rockets, Chips, Oil...

Fullerenes are a rich if quirky example of how postwar materials researchers made and/or found new materials. Much more representative examples — including many of the entries in this section — could be drawn from the four technological domains in my heading: bombs, rockets, chips, and oil. Or, put more formally: nuclear weapons and energy; the space and missile programs; microelectronics; petroleum and its derivatives. These four sectors exerted wide-ranging influence over many science and engineering disciplines, particularly the fields involved in materials research.

Indeed, it is telling that all four of these sectors were entangled with fullerene research. Smalley was a former Shell chemist, and later a business partner with petrochemical executives who brought their industry's expertise to the synthesis of nanotubes. In addition, the only copy of the AP2 was owned by an oil company — though, curiously, Exxon seems to have originally wanted the AP3 to aid in their nuclear energy program. Hence, the samples they put in it were things like uranium and graphite (the constituents of the fuel and moderator rods in nuclear reactors). As for "chips," the samples Smalley was putting in the AP2 just before the buckyball discovery were semiconductors (provided by Texas Instruments), and when C_{60} research lost steam in the late 1980s he briefly went back to semiconductors. His later interest in nanotubes was sparked in large part by hopes that they could be used as transistors or other integrated-circuit components. Finally, the other visionary application of nanotubes that he often touted was for the cables in a "space elevator" reaching from the earth to geosynchronous orbit. Not to mention the astrochemistry motivation for doing the buckyball experiment in the first place.

Oil, nuclear energy, microelectronics, space — these sectors also shadow most of the entries in this section of the volume. The editors have chosen not to feature most of the canonical synthesis tools used in petrochemistry — such as distillation apparatus — in this section, since they are often elaborate versions of technologies that have existed since the nineteenth century if not much much longer. However, this section does include chromatography, a technique which could only be commercialized and achieve widespread use in materials research thanks to lucrative markets in the petrochemical industry. Ultracentrifuges and high-pressure apparatus also have at least tangential connections to oil, although they are much more closely related to isotope separation for nuclear weapons/energy. Like chromatography, both have also diffused across all fields of materials research.

That diffusion is worth a special note. Historians of science have written much about so-called "big science" — giant technoscientific programs such as the Manhattan Project that so dominated mid-twentieth century research. Indeed, big science programs were important for many of the tools surveyed here, as the entry on nuclear reactors attests. Yet one feature of big science that historians often overlook is how such programs take techniques from and provide them to "little science" laboratories scattered all over the world and all over the disciplinary landscape. The entry on complex heating techniques, for instance, shows how big science is always mutually implicated in little science — microelectronics giants such as Bell Labs borrowed techniques such as zone refining from less influential corners of the innovation system and in turn developed other techniques, which spread outward from the microelectronics industry into many fields of materials research. Moreover, big science interacts with little science to a surprising degree — as shown by the interplay between microelectronics and space exploration in the entries on ion implantation and electrochemical cells.

Of these four domains, nuclear research is probably the most evocative with respect to materials. The race to build an atomic bomb depended on the creation of many novel, high-performance materials. It even led to the creation of new elements with magical properties, such as plutonium: as Manhattan Project veteran Leona Marshall noted, "when you hold a lump of it in your hand, it feels warm, like a live rabbit" due to the continual nuclear decays going on inside.[8 p660] Yet it was the space and microelectronics domains that were foremost in the early articulations of "materials science." Arthur von Hippel, for instance, followed the lines quoted above with an extended thought experiment on picking the right materials for a supersonic jet. The frames and engines of jets, missiles, rockets, and spaceships experience unusual combinations of stress, high and low temperature, and high and low pressure, and are also subject to idiosyncratic military requirements, such as a low radar profile. Jets, missiles, rockets, and spaceships also depend on fast, reliable electronics, meaning that new materials were needed that could both power fast circuits and survive the demanding aerospace environment. Indeed, the integrated circuits in your computer and phone were originally invented for use in missiles.

Only equally unusual materials could meet those requirements, and by the late 1950s many science policymakers in the United States believed that such materials were not forthcoming — that they represented a "materials bottleneck" slowing efforts to win the Cold War (and prepare for the next hot war). Departments and laboratories dedicated to "materials science" — as opposed to traditional chemistry, physics, or metallurgy — resulted from the effort to overcome that materials bottleneck. Nuclear technology was not absent from panic over the materials bottleneck, though when it did enter the conversation it often piggybacked on aerospace and microelectronics.

For instance, one of the Air Force's great dreams when it backed von Hippel's "molecular engineering" concept was of a nuclear-powered airplane. Similarly, the Air Force later plowed millions of dollars into research on silicon-on-sapphire and other exotic electronic materials that could survive a nuclear burst.

... And Life

Microelectronics and aerospace applications were, then, the bread-and-butter of early materials science. Those were the domains where visionaries such as von Hippel foresaw, and funders such as the Air Force demanded, near-term applications. And, indeed, we now have unbelievably sophisticated materials for those purposes. Since the late 1980s, for instance, high-performance jet-engine turbine blades have been grown as giant single crystals, despite also featuring intricately complicated structures and containing intentional inhomogeneities in the concentration of alloys. Similarly, in microelectronics we have now reached the point where more than half of the periodic table can be found in a cellular telephone. Neither of those achievements would have been possible without the wholesale reorganization of parts of physics, metallurgy, chemistry, mechanical engineering, and other specialties as "materials science."

Missing from the original disciplinary constellation surrounding materials science were the life sciences. Yet if you browse a materials research journal today or attend a meeting of the Materials Research Society, biomaterials topics are ubiquitous. Aerospace and microelectronics applications have not exactly gone away, but the past thirty years has seen a steady growth in materials research that draws on biologists' expertise. "Life" now ranks as a co-equal domain of materials research alongside rockets, bombs, chips, and oil.

In some ways this is not new. Chemists have always looked to life for inspiration, and arguably the first truly modern material — celluloid — was both derived from a natural product (camphor) and intended as a substitute for another (ivory). Nevertheless, the trend of the past thirty years is distinctive in that it has involved disciplines other than synthetic chemistry in research on both the production of novel materials *by* organic systems and the production of novel materials *for* organic systems. An example of the latter is the long search for materials for implants — artificial hips, pacemakers, and others — that can survive the unusual conditions inside a human body without harming that body. An example of the latter is the attempt to use specially engineered viruses to make the intricate, tiny patterns in integrated circuits.

Particularly since the turn of the millennium — and the contemporaneous creation of national nanotechnology initiatives in many countries — there has been a third type of biomaterials research that is more hybrid and harder to categorize. A good example is Nadrian Seeman's work using DNA as a construction material with which he and his colleagues can form almost any shape imaginable.[9] This work does not take advantage of DNA's properties as "life's instruction set" or even its more prosaic biological use as a template for constructing proteins. But Seeman's work does take full advantage of everything biologists have learned about DNA over the years — such as how to construct precise extended sequences of base pairs, what DNA's mechanical properties are, and what the van der Waals and other short-range forces do to the physical form of a DNA strand.

The biological turn in materials synthesis has led to some unusual career trajectories among the current generation of practitioners. One member of that generation with whom I am acquainted is Henry Hess, currently a professor in the Biomedical Engineering department in Columbia University's School of Engineering and Applied Science (both "biomedical engineering" and "school of engineering and applied science" are themselves signs of the times). Yet Hess's PhD training was in physics, not biology or engineering. He then took a postdoc in the bioengineering department at the University of Washington, working for Viola Vogel, a prominent figure in the ascendance of biomaterials research. After that, he became an assistant professor in a materials science and engineering department before taking his current post at Columbia.

And what kind of research could qualify one for jobs in all these different disciplines? Hess is best known for taking the molecules involved in motion in living cells — for example, muscle cells or bacteria — and adapting them as stand-alone "machines." For example, he can take the protein kinesin and "train" it to walk in a circle around a surface. One could imagine applications in microelectronics (using the kinesin or something like it either to construct circuit components or even to carry information itself), in pharmaceuticals (the kinesin could carry a drug molecule to where it was needed), or prosthetics (large arrays of kinesin or something similar could convert chemical energy into motion in a manner analogous to natural muscle). What is perhaps most intriguing for the future of the concept of "materials" is the ambiguity inherent in the idea of "large arrays" of engineered biochemical systems. An individual kinesin walking around is not a material; but a large array might be. That, at least, is a likely path for the discipline. In the meantime, the same research will also lead to more "traditional" biomaterials such as advanced polymer coatings to allow implants to survive longer inside a patient's body.

Heating, Beating, Zapping, Spraying, Growing

So far we have considered the why and how of modern materials synthesis. Now, finally, let us look at the what. What is it that materials researchers do to materials to prepare them for materials research? A complete answer to that question is impossible — there are so many techniques and they evolve so rapidly that one could never list them all. But it is possible to list some broad classes of things that can be done to materials.

Some of the oldest materials synthesis "techniques" are still among the most important: varying temperature and pressure or, as many young materials scientists learn to say, "heating and beating." Heating puts energy into a system; sometimes it is just as important to take energy out. Thus, this section includes entries on both complex heating and refrigeration. Some of the reasons for heating a system have been well understood for a long time — see the entry in part 1 on distillation. The reasons why *mechanically* manipulating a material would change its qualities were more mysterious until the twentieth century, when phase diagrams entered practice as a way to picture the behavior of alloys and other mixtures at different temperatures and pressures. It was not until mid-century, though, that metallurgists had a good sense of the role of dislocations — caused by mechanical manipulation — in the properties of metals. Indeed, the introduction of the *concept* of dislocations into metallurgists' toolkit was an important event in the incorporation of metallurgy and solid state physics into materials science — even if blacksmiths have been intentionally introducing *actual* dislocations into metals for millennia. That said, the village smithy did not have access to the technologies profiled in this section, such as ultracentrifuges and high-pressure apparatus.

If "heating and beating" have been around forever, the most distinctive materials preparation techniques of the twentieth century could be labeled "zapping and spraying." In the late nineteenth century scientists learned how to "zap" materials with electromagnetic radiation and other beams intended to, again, put energy into a physical system. Slightly later they learned how to "spray" materials with molecular and atomic beams intended to add matter to a physical system. "Zapping" and "spraying" are obviously lay terms, but materials scientists generally recognize them as as related to each other, and as distinctive from other preparation techniques. For instance, in the late 1960s researchers interested in engineering micron and submicron structures — particularly for the microelectronics industry — founded a conference series nicknamed the "three beams" meeting, to bring together research on various zapping and spraying techniques. Confusingly, as Figure 3.0.1 shows, the "three beams" really involved four separate-but-connected constituents: optical photons, X-rays, electrons, and ions. Photons and electrons count as zapping in my terminology: researchers would shine these beams at a stencil (called a mask) and where the beam went through the mask and hit the sample it would change the sample's chemistry. Ion beams were more like spraying; they changed the sample's chemistry by adding to its make-up.

The types of people who employed the three beams also tended to talk about "growing" layers of materials, in preparation for then shining their beams at those layers. For most of the twentieth century,

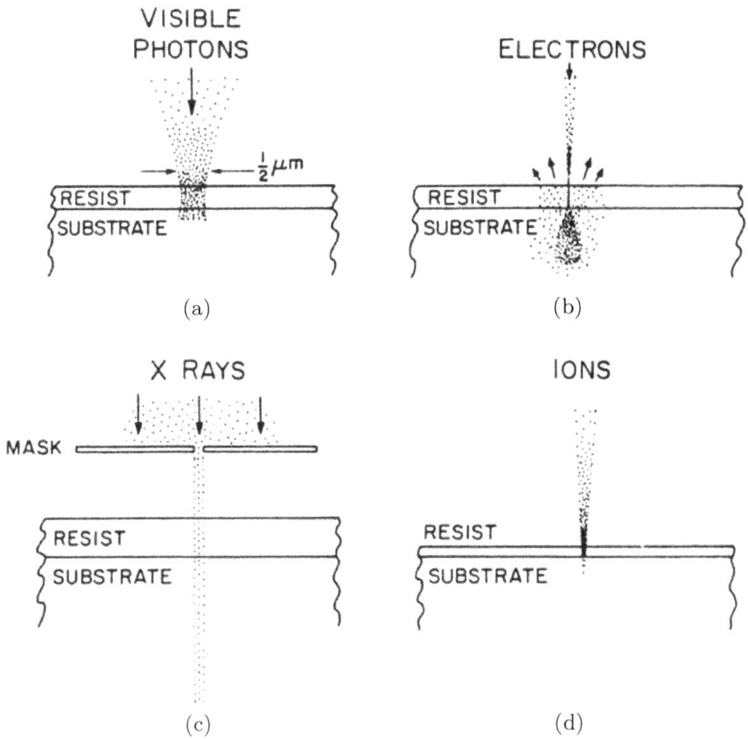

Fig. 3.0.1. The "three beams:" photon (a and c), electron (b), and ion (d). Reprinted from [11]. *Credit*: Copyright 1981, with permission from Elsevier.

growing in this sense was a metaphor — though I suspect it has its roots in medieval and early modern ideas about crystals as living entities which could grow in a non-metaphorical way. By the mid-twentieth century, though, materials scientists talked of growing layers of a material onto a substrate in (usually) one of two ways: either by evaporating a material and exposing it to a colder surface onto which it would condense; or by exposing a surface to a gas under conditions in which constituents in the gas would adhere to or chemically combine with constituents in the surface material. One of the most important examples of the latter is the growth of silicon dioxide on a silicon surface exposed to oxygen. The fact that silicon so easily forms an insulating dioxide layer is the reason it is the dominant material in modern microelectronics.[4 p89] Santa Clara County should really have been nicknamed Silicon Dioxide Valley!

Over time, more and more sophisticated means of zapping, spraying, and growing were invented, many of which are surveyed in this section: vapor deposition, molecular beam epitaxy, ion implantation. Many of these techniques were used both for making and characterizing materials: lasers, for instance, are profiled in part 4, but could as easily have been placed in this section. And as they were invented, these techniques found a place in the "three beams" community, such that the nickname came to refer to an imagined past, rather than reflecting the diversity of the present. Tellingly, the three beams community has taken the same biological turn as the rest of materials science. I met Henry Hess, for instance, when he was a plenary speaker at the Gordon Conference on the Chemistry and Physics of Nanostructure Fabrication, a kind of sister meeting to the three beams conference. When it first began in 1976, this Gordon Conference was almost exclusively focused on the traditional three beams as applied to microelectronics. By the early 1990s, though, that format had become stale and the conference organizers began to welcome practitioners of other sample preparation and characterization techniques. By the mid-1990s, they were eagerly looking for the early pioneers of bio-based materials science to show the rest of the community the way forward. And by the time I attended in 2006, people like Hess were almost as common as those who began their presentations with gestures to Moore's Law. One could say that the metaphor of growing materials is no longer a metaphor: it is increasingly an accurate description of how materials are made. To reflect that change, the editors have included a few entries in this section which give some historical depth to the emergence of biomaterials techniques: particularly chromatography and bioorganic synthesis.

Conclusion

The gradual shift toward biomaterials adds a new technological domain for materials research to orient to: biomedicine. That shift adds new patrons: in the United States, the National Institutes of Health is now almost as likely to fund materials research projects as the Pentagon, the Department of Energy, NASA, or Intel. It also adds new stakeholders. The end users of mid-century materials knowledge were typically members of the military. That started to change as transistor radios and later personal computers brought advanced microelectronics to consumers. But the end user of a hip replacement or artificial muscle is not just a consumer: they are a person in whom advanced materials are, literally, incorporated.

But the shift toward biomaterials brings continuity as well as change. I began this essay with the claim that what marks modern materials research from its antecedents is the dispensability of the materials themselves. Or, more accurately, that researchers now make novel materials with a great deal of foresight as to the characteristics of the materials they are about to make. Such foresight was indispensable to the marquee technological domains of mid-century. The Cold War, after all, only remained cold because neither side used its nuclear weapons *and* because both sides were fully confident that the materials in those weapons would behave as expected if they were used. That same foresight applies just as much to biomedicine. Ever since the thalidomide controversy of the 1950s, regulators of pharmaceuticals and biomedical technologies have insisted that their effects on human bodies be known *before* large numbers of humans consume them. Biochemists have possessed sufficient predictive power to meet that criterion for individual molecules for several decades now. But it is only in the past twenty to thirty years that computing power and theoretical materials science have come close to that point for complex materials. As foresight into the behavior of novel biomaterials advances, we should expect a proliferation of new means for *making* biomaterials along the same lines as the proliferation of means for making nuclear, aerospace, electronic, and petromaterials that occurred after World War II. Thus, we hope that the history of that earlier proliferation, as covered in this section of the volume, may clarify the present-day development of biomaterials techniques.

References

1. Aldersey-Williams H. The most beautiful molecule. London: Aurum; 1995.
2. Baggott J. Perfect symmetry: the accidental discovery of buckminsterfullerene. New York: Oxford University Press; 1994.
3. Belt Hvd, Rip A. The Nelson-Winter-Dosi model and synthetic dye chemistry. In Bijker WE, Hughes TP, Pinch T, editors. The social construction of technological systems. Cambridge: MIT Press; 1987.
4. Brock DC, editor. Understanding Moore's Law: four decades of innovation. Philadelphia: Chemical Heritage Foundation; 2006.
5. Gordin MD. A well-ordered thing: Dmitrii Mendeleev and the shadow of the periodic table. New York: Basic Books; 2004.
6. Johnson A. Modeling molecules: computational nanotechnology as a knowledge community. Perspectives on Science. 2009;17(2):144–73.
7. Loeve S. Point and line to plane: the ontography of carbon nanomaterials. In Teissier P, Mody CCM, van Tiggelen B, editors. From bench to brand and back: the co-shaping of materials and chemists in the twentieth century. Nantes: Cahier François Viète; 2017.
8. Rhodes R. The making of the atomic bomb. New York: Simon & Schuster; 1986.
9. Tyrrell B. From theoretical biology to 'where's the biology?' a history of structural DNA nanotechnology. Unpublished manuscript.
10. von Hippel A. Molecular engineering. Science. 1956;123(3191):315–17.
11. Wolf ED, Ballantyne JM. Research and resource at the National Submicron Facility. In NG Einspruch, editor. VLSI electronics, vol. 1. Cambridge, MA: Academic Press; 1981.

CHAPTER 3.1

MODERN HEATING TECHNOLOGIES

PIERRE TEISSIER

Fire, as Gaston Bachelard observed, haunts the psychology of modern societies.[2] Its psychological power can be traced deep into ancient mythologies. The Titan Prometheus, who stole fire from Olympus and gave it to humankind, is a popular hero of European memory. A broad, comparative anthropological study could link blacksmith gods from Africa to Asia to Europe and demonstrate their resonance with the social practices suggested by archaeological artifacts such as weapons and potsherds. For the European tradition, Hephaestus (or Vulcan), the Greek (or Roman) god of metalworking, fire, and crafts, is an important figure in collective cultural memory. Stories of his subterranean forge depict furnaces as frightening and powerful tools that transform raw, unformed materials into artificial, carefully shaped artifacts.

The romanticized public image of furnaces provided by classical mythology was reframed by their increasing use from the Middle Age onward, including in alchemy and industrialization. Medieval and early modern alchemists developed high-temperature techniques to produce and purify metals, glasses, and ceramics. Literary fiction and visual art provided long-lasting stereotypes of alchemical practices.[19] Such representations often drew a close association between intellectual labor and material transformation, as we can see in a seventeenth century oil painting by David Ryckaert III (Figure 3.1.1).

Fig. 3.1.1. David Ryckaert III, *An alchemist seated at a furnace, turning away in thought*. *Credit*: London: Wellcome Trust Library, Creative Commons, https://wellcomeimages.org/indexplus/image/V0017664.html.

Between 1750 and 1850, the European iron industry was dramatically transformed thanks to the adoption of several new making processes.[14 vol. 3 p585–614] By replacing charcoal with coke in blast furnaces, British craftsmen and engineers increased heating temperatures and produced better cast iron with less fuel. This process, known as "puddling," was gradually implemented in Europe during eighteenth century, which contributed to improve the quality and the quantity of manufactured engines that supported the "First Industrial Revolution" (1750–1850), based on textiles and steam engines.

Thinking about how different cultures have represented heating technologies across millennia indicates that the history of the furnace, and the way it has been used, can reveal a great deal about scientific and technological imagery, the intellectual and manual labor that goes into manipulating materials, and the connection between practical know-how and theoretical knowledge. Yet, in spite of their scientific and economic importance in the twentieth century, furnaces have received little attention from science and technology studies (STS) scholars. At least five features of furnaces can be put forward to explain their relative invisibility: 1) their physical appearance — simple black boxes made of ceramic bricks heated by electric resistance; 2) their low economic value — with an average price of around $10,000 for one unit; 3) their interstitial position — between science, industry, and state; 4) their low symbolic capital — due to their links with empirical operations by technicians and routine syntheses by chemists, usually considered as "dull" and "dirty" practices;[18; 22; 37; 41 p71–87; 36; 32] and 5) their long-lasting tradition — since medieval alchemy and early modern industry — and, thus, slow apparent transformation. Furnaces thus appear as *invisible tools* in the collective memory of materials. Yet, they are precisely the type of "things that talk" according to Lorraine Daston's assertion "that things are simultaneously material and meaningful."[13 p17] To give voice to high-temperature devices, let us follow briefly, first, some of their scientific and industrial developments during the twentieth century and, then, the role they played in the making of the US microelectronics industry in the 1950s.

Historical Trends in Heating Technology

To gain a historical perspective on heating technologies in materials research, it is worth looking back to the "Second Industrial Revolution" (ca. 1870–1920). This context was characterized by economic expansion in the chemical, electrical, and mechanical industries, driven by new practices of (and findings from) industrial research and development (R&D). Heating technology advanced through the industrial production of alloys and metals, ceramics, and glasses, as well as mineral products.[14 vol.4 p493–542] Last but not least was the introduction of electric furnaces in the chemical and metallurgical industries during the last decades of the nineteenth century. Although the Siemens company had first used an electrical device in 1879 for

Fig. 3.1.2. Unknown author, Henri Moissan using a Moissan-Chaplet electric arc furnace. *Credit*: Wikimedia Commons.

the industrial melting of iron, Henri Moissan designed a more efficient "electric arc furnace" in his laboratory at the College of Pharmacy in Paris (Figure 3.1.2). He aimed to synthesize refractory materials, including artificial diamond (in vain) and calcium carbide (with success). His laboratory prototype was quickly transferred to industry and scaled up for the production of increasingly large quantities of brand materials in the new industry of electrometallurgy.[49]

After Moissan's achievements, the industrial use of electric arc furnaces increased for the making and study of various solid-state materials such as alloys, cements, ceramics, crystals, and glasses. Indeed, electric furnaces allowed a more controlled and broader range of temperature (up to 3,000°C) than existing gas devices like Bunsen burners (up to a few hundreds of degrees) and welding torches designed by Henri Sainte-Claire Deville (up to 1,000°C). From the late nineteenth century to the 1920s, coke blast furnaces (Bessemer,

Martin-Siemens) had been replaced by electric arc furnaces for the industrial production of steel and other materials like ceramics and glasses.[8 p293] This industrial trend, known as electrothermics, which has received scant historical attention, has the potential to refine the classical picture of the "Second Industrial Revolution," characterized by the tripod of organic chemistry, electricity, and the internal combustion engine.[9 p29]

In the early 1920s, researchers and industrialists perceived high-temperature devices as "modern tools." Development of these tools was nurtured by several institutions around the world: the Faraday Society (London), the Franklin Institute (Philadelphia), the International Union of Pure and Applied Chemistry (Zurich), the Electrochemical Society (Pennington, NJ), and others.[48] At a Franklin Institute meeting in March 1923, Edwin F. Northrup, a former Princeton University physics professor, surveyed six categories of furnaces for high-temperature research: "Apparatus which focuses radiation from the sun, hot flames, the electric arc, and the electric resistance furnace, vacuum bulbs which heat a surface by electron impact, and the high-frequency induction furnace."[38 p676] The four last categories were "electric furnaces" characterized by different heating means: arc, resistance, electrons, and induction. Some alternative classifications could be found around this time. A collective book by French physicists and chemists identified one additional category — that of spark and gas electric furnace — without considering the electronic furnace.[27] In addition, their categories were refined according to components (metallic or carbon resistance), conditions of uses (high or low frequency), and gaseous atmospheres (vacuum or high pressure). But Northrup's categorization was nonetheless influential.

According to Northrup, the "high-frequency induction furnace" (Figure 3.1.3) was "the final and ideal method" for research since it provided a uniform volume of heating with no contamination from wire coils (no contact) and a continuous range of temperature that was only limited by the temperature of fusion of the walls. Besides the method of crystallization in a crucible, high-temperature physicists and chemists were excited by two novel techniques for

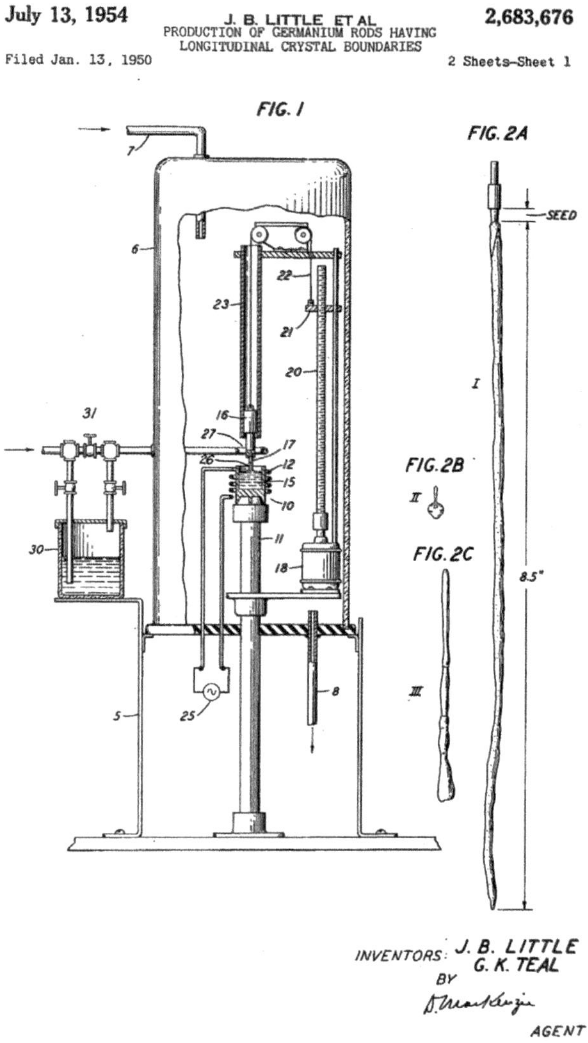

Fig. 3.1.3. A convenient form of high-frequency induction furnace.[39 p682]

growing crystals from melt: the flame-fusion "Verneuil process" (ca. 1891) to synthesize artificial ruby and the Czochralski method for pulling metallic single crystals (ca. 1916).[37] Strong financial interest drove the industrial production of artificial gems at high temperature and pressure.[23] Most industrial domains sought to benefit from the increasing thermal and material efficiency of furnaces. The energy intensity, which is the energy required to make a given mass of materials, was actually divided by twofold between 1880 and 1940 for the production of pig iron in the United States.[18 p6]

Interwar France offers an interesting entry point for exhibiting the multiple actors involved in high-temperature R&D across scientific, industrial, and political spheres. In the context of early national industry policy, the minister of trade and industry (1915–1919), Étienne Clémentel, promoted a four-tier organization of industry going from iron at the bottom to cast iron and steel, mechanics and engines, and, finally, textiles and chemistry at the top.[28 p189] Both coke blast furnaces and electric furnaces were identified as key actors for the production of cast iron and steel. This national policy echoed the establishment in 1921 of a private Foundation for the Development of Scientific Research by Edmond de Rothschild. The Rothschild Foundation had the dual goal of providing grants for "an elite of scholars" and orienting scientists toward "the applications of science" so as to favor "the economic forces of the nation."[29] The first committee to be set up, that same year, was for high-temperature research, under the leadership of an heir of Moissan, Paul Lebeau. Its editorial activity and research aroused the interest of metallurgists like Henry Le Chatelier, Georges Charpy, and Léon Guillet, who had managed the *Revue de métallurgie* since its foundation in 1902.[11; 31] Both engineering schools (Centrale, CNAM, Mines) and universities were involved in the dynamics of high-temperature R&D. In 1926, the Gas Company of Paris funded a chair at the University of Paris entitled "fuel chemistry" that would become that of "high temperature" in 1932.[10] In addition, the French Ministry of Aeronautics funded PhD grants in several laboratories to develop light materials with good mechanical properties. All this disciplinary research, infused by private and public funds, contributed to a blossoming French community of chemists, engineers, metallurgists, and physicists involved in the production and measurement of high temperatures both for research and applications for gases and solids. This was typical of an "instrumental research community" shaped by the material nexus of furnaces under the umbrella of technoscience.[24 p1–13]

This nationwide focus had international ramifications. The French community contributed to the field of high-temperature research that was structured after World War II by the context of Cold War. From 1956 onward, three successive international symposia on high-temperature technology were organized by the Stanford Research Institute in California.[44–46] Discussions were divided into methods, materials, and processes. They gathered between five- and eight-hundred participants from around the Western (that is, non-communist) bloc of the Cold War: academic and industrial researchers with administrative experts and financial entrepreneurs. The main sponsors of the events were US military, nuclear, and space agencies. By this time, refractory compounds with good mechanical properties had become strategic materials for the design of reactors, rockets, and shields. A large consensus supported the craft design of new apparatus to produce still higher temperatures and develop still cleaner processes.[3] Different techniques were supposed to find their places in different spheres: "plasma furnaces" as promising industrial devices because of their high electricity–heat conversion rate; "image furnaces" as high-technology research tools; and "solar furnaces" as cheap tools for developing countries.[16] The influential culture of microelectronics R&D pushed a general trend toward the ultra-purification of materials, including metals.[1]

The 1960s can be understood as a golden age for high-temperature R&D. Since then, several trends related to changing research on materials have modified the landscape: 1) Standardization of commercial furnaces (as opposed to tailor-made apparatus built in-house), driven by an industrial demand for control,[9 p27–28] has transformed most high-temperature apparatus into humdrum black-boxes;[35] 2) The relative dominance of high-temperature "dry routes" was gradually challenged by alternative methods: epitaxy, electrochemical processes (deposition, intercalation), hydrothermal syntheses (wet chemistry), and sol-gel (soft chemistry);[47 p271–287] 3) The increasing precision of characterization instruments and the diversification of synthetic methods moved interest from 3D crystals to a larger range of morphologies — which would lead to nanoscience and technology in the 2000s — used in new commercial compounds — advanced glasses, gels, films, liquid crystals, and others.[5; 33]

Telling Quotations
Comparing European and American chemical skills, a famous chemist from Bell Labs, Francis J. Di Salvo wrote: "In the United States, synthesis of solid-state compounds has been considered out of date and a little dull, and few academic departments have even one professor involved in synthesis of new solid-state compounds."[15 p164–165] Fifteen years later, in another context, a materials scientist from Oxford University confirmed the low symbolic capital of metallurgical furnaces: "The image of materials for computers, aeroplanes and cars is more exciting than that of dirty blast furnaces in the steel industry."[20]

Heating Technology in 1950s US Microelectronics

A survey of materials research and industry in the United States shows that furnaces were routine devices in the two decades after World War II. They paved the way for the dominant crystal synthesis technique, known as "dry chemistry" (as opposed to "wet chemistry" in liquid solutions). High-temperature synthesis had two stages: first, inorganic powders were melted at high temperature in a crucible, then the liquid crystallized as it gradually cooled. Since crystal growth is essential to understanding the structural and physical properties of solids, many solid state physicists could hardly work without a furnace in the early 1950s.[22] As the physicist John B. Goodenough put it: "Physicists wanted single crystals.... The ceramics laboratory that I inherited at Lincoln Laboratory had high-temperature furnaces, a hand press, a hood, a powder x-ray diffractometer, and chemical benches.... At Oxford, there was little money for equipment in the Inorganic Chemistry Laboratory.... I acquired furnaces for my own synthesis needs."[17]

Furnaces were also used to purify materials, to modify their morphology (defects, grains, interfaces, and other features), to test their thermal resistance, and to observe their behavior in various thermal conditions. This is, for example, the case of "solid oxide fuel cells" (SOFC), the behavior of which was (and is still) analyzed at high temperature and high pressure in a laboratory furnace to simulate their conditions of operation.[25] Characterization of materials was another common application. It became usual in the second half of the twentieth century to analyze structural and physical properties of crystals at high temperatures with standardized furnaces coupled with X-ray diffraction machines or microscopes. Other high-temperature techniques, such as thermal differential analysis (TDA), were used to measure glass transition temperatures of amorphous materials (glasses and polymers). Their multiplicity of uses turned furnaces into a nexus of interaction among chemists, engineers, metallurgists, physicists, and technicians. In the late 1950s, US policy-makers considered these cheap tools to be essential characterization instruments for the formation of materials science and engineering (MSE). A memo prepared by the staff of the US President's Science Advisory Committee in March 1958 stated: "In various universities one finds the faculties engaged in planning interdepartmental efforts to establish a new materials science and engineering. One of their problems is the lack of buildings and equipment. There are many research tools such as electron microscopes, electron diffraction units, high-temperature furnaces, and x-ray diffraction units, each of which represents a cost of approximately $20,000."[40 p23–24]

The ability of furnaces to bring about new connections among researchers was amplified by architectural strategies in the design of research buildings, as shown by the materials research laboratory of the University of Pennsylvania.[12] Indeed, the building, which dated from about 1960, was designed around ten "central facilities" — well-equipped, specialized experimental rooms — to which all researchers and students of the laboratory had access. One can imagine that many casual discussions started in front of the closed doors of the operating furnaces of the "materials processing facility." (See Brit Shields's contribution to this volume.)

High-temperature tools and processes also helped to transfer research materials into commercial devices. This was the case for

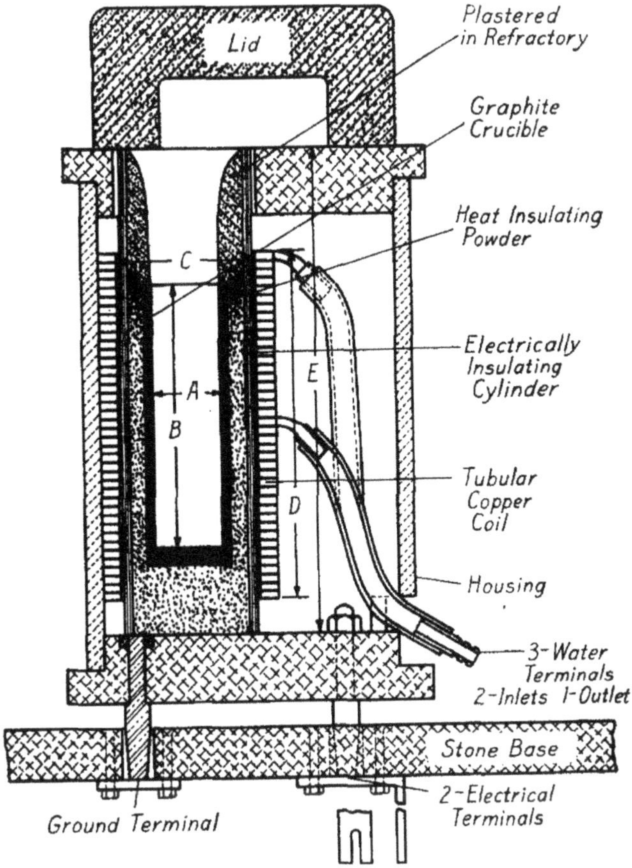

Fig. 3.1.4. Little John B, Teal Gordon K, Production of germanium rods having longitudinal crystal boundaries. *Source*: US Patent 2683676 A, Submitted 13 January 1950, Published 13 July 1954, Assignee Bell Telephone Labor Inc.

the early development of the microelectronics industry, as shown by the examination of several "institutional locations."[30] At AT&T's Bell Laboratories, ca. 1948–1952, the chemist Gordon Teal and the engineer John Little adapted earlier metallurgical methods developed by Jan Czochralski to design a "crystal puller," patented in 1950 (Figure 3.1.4), for growing, from melt, single crystals of germanium, and then silicon. As stressed by historian of electronics David Brock, this "was a key milestone on the road to a silicon junction transistor."[7 p12] The chemist William Pfann, from the Metallurgy Research Department of Bell Labs, carried on their effort by inventing "zone melting," a high-temperature process to purify solid-state compounds, including semiconductor materials.[39] High-temperature purification and crystal growth of semiconductor materials enabled solid state physicists to make and understand junction transistors.[27 p26–30] Interdisciplinary R&D has been characteristic of Bell Labs since the 1930s.[21]

Then, around 1952–1954, Texas Instruments, under Teal's leadership, scaled up Bell Labs' high-temperature synthetic methods to become "the first firm to commercialize the silicon transistor."[30 p310] Around 1958–1960, at Fairchild Semiconductor, Sheldon Roberts and his co-workers designed robust, simple, practical heating apparatus and crystal pullers that "enabled the firm to be the first to develop and market two major electronic devices — the double diffused planar transistor and the integrated circuit."[30 p314] Thus, a quick look at materials R&D in the United States during the 1950s and 1960s shows that furnaces and high-temperature techniques were essential for making advanced materials and microelectronics devices through the networked interaction of laboratories, factories, and markets. This contributed to "the expansion of both military technology and consumer culture across the Cold War."[30 p316] Investigating the role of furnaces in the US semiconductor industry therefore permits us to endorse the dominant historiography of "research on materials" focused on MSE and the electronics industry in the United States during the Cold War.[4]

Conclusion

A more careful study of heating technology during the twentieth century may give several fresh perspectives to science and technology studies as well as history of science and technology. At least four types of questions could be investigated. First, the topic would provide an original probe to compare national specificities. For example, although France and Germany were comparable in solid-state research as a whole, their high-temperature science and technology were quite different: the French interwar community was reinforced by nuclear and space policy after 1945, whereas the lack of a nuclear or space industry in Germany hindered high-temperature research, both in industry itself and in academic institutions.[24 p272; 6 p32]

Second, the field of high temperature can be considered as a case of "American hegemony," in which the United States used its larger volume of science and technology to implement policies regarding US allies during the Cold War.[26] The long-term history of high-temperature technologies would strengthen the Cold War model of American hegemony and show how technological devices could be shaped to become political tools. A comparison with other models of techno-scientific politics could thus be fruitful.[43]

Third, high-temperature R&D constitutes a fresh ground to test different models of the interaction among science, industry, military, and state. It could lead to a reappraisal of the role of heavy industry with regards to aerospace, microelectronics, and nuclear industry in the making of military-industrial complexes in the post–World War II context.

Last, but not least, the focus on customers and communities, companies and institutions, instruments and practices, or systems and territories, involved in the blossoming field of high temperature would provide stimulating and surprising "historical alternatives" to the mainstream narrative on research materials largely molded by the US Cold War lens of MSE and the microelectronics industry.[42]

References

1. American Society for Metals. Ultra-high-purity metals. Papers presented at a seminar of the American Society for Metals, October 21 and 22, 1961. Metals Park, Ohio; 1962.
2. Bachelard G. The psychoanalysis of fire. Beacon Press; 1987. 1st ed. in French 1938.
3. Bensaude-Vincent B, Teissier P. Robert Collongues (1924–1998). In Koertge N, editor. New dictionary of scientific biography. Detroit: Charles Scribner's Sons; 2008;2:159–164.
4. Bertrand E, Bensaude-Vincent B. Materials research in France: a short-lived national initiative (1982–1994). Minerva. 2011;49:191–214.

5. Boilot J-P. Interview (French) with Bensaude-Vincent B. Palaiseau, Polytechnique. December 12, 2000. https://www.sho.espci.fr/spip.php?article121.
6. Brauer G, Riley RF. Handbook of preparative inorganic chemistry, vol 1. Academic Press; 1963.
7. Brock DC. The backdrop to Moore's Law: developments in semiconductor electronics to 1965. In Brock DC, editor. Understanding Moore's Law. four decades of innovation. Philadelphia, PA: Chemical Heritage Foundation; 2006. Chapter 2.
8. Cardwell D. The Fontana history of technology. London: Fontana Press; 1994.
9. Caro P. Interview (French) with Teissier P. Paris: French Academy of Science. May 3, 2005.
10. Charle C, Telkes E. Les professeurs du Collège de France. Dictionnaire biographique 1901–1939. Paris: CNRS; 1988.
11. Chezeau N. De la forge au laboratoire: naissance de la métallurgie physique (1860–1914). Rennes: Presses universitaires de Rennes; 2004.
12. Choi H, Shields B. A place for materials science: laboratory buildings and interdisciplinary research at the University of Pennsylvania. Minerva 2015;53(1):1–42.
13. Daston L, editor. Things that talk: object lessons from art and science. New York: Zone Books; 2008.
14. Daumas M, editor. Histoire générale des techniques. 5 vols. Paris: Presses universitaires de France; 1979.
15. Di Salvo F. New and artificially structured electronic and magnetic materials. In Psaras P. A., Langford H. D., editors. Advancing materials research. Washington, DC: National Academy Press; 1987: 161–75.
16. Foëx M. Compte-rendu sur le "Symposium des hautes températures" d'Asilomar (1959). Bulletin de la Société chimique de France. 1961;28:117–23.
17. Goodenough JB. Interview with Bensaude-Vincent B., Hessenbruch A. Austin: University of Texas. May 2001. https://sho.spip.espci.fr/spip.php?article28.
18. Gutowski TG, Sahni S, Allwood JM, Ashby MF, Worrell E. The energy required to produce materials: constraints on energy-intensity improvements, parameters of demand. Philosophical Transactions: Mathematical, Physical and Engineering Sciences. 2013;317(1986):1–14.
19. Haynes R. The alchemist in fiction: the master narrative. In Schummer J, Bensaude-Vincent B, Van Tiggelen B, editors. The public image of chemistry. Singapore: World Scientific; 2007.
20. Hirsch P. Interview with Bensaude-Vincent B. Oxford, Oxford University. December 12, 2002. https://sho.spip.espci.fr/spip.php?article126.
21. Hoddeson L. The roots of solid-state research at Bell Labs. Physics Today. 1977;30(3):23–30.
22. Hoddeson L, Braun E, Teichmann J, Weart S, editors. Out of the crystal maze: chapters from the history of solid-state physics. New York: Oxford University Press; 1992.

23. Jayaraman A. A brief overview of gem materials: natural and synthetic. Current Science. 2000;79(11):1555–65.
24. Joerges B, Shinn T. Instrumentation between science, state and industry. Dordrecht: Kluwer Academic Publishers; 2001.
25. Joubert O. Interview (French) with Teissier P. Nantes, University of Nantes. May 2, 2017.
26. Krige J. American hegemony and the postwar reconstruction of science in Europe. Cambridge, MA: MIT Press; 2006.
27. Lassman TC. From materials research to materials science: the industrial origins of an academic discipline. Baltimore, MD: Master of Arts; 1996.
28. Lebeau P, editor. Fours électriques et chimie. Paris: Presses universitaires de France; 1924.
29. Lebeau P, Meyer A. Cérémonie en l'honneur du XXVe anniversaire de l'Institut de Biologie physico-chimique. 2e Fondation Edmond de Rothschild pour le développement de la Recherche scientifique. Paris: Hermann; 1953.
30. Lécuyer C, Brock DC. The materiality of microelectronics. History and Technology. 2006;22(3):301–25.
31. Letté M. Henry Le Chatelier (1850–1936) ou La science appliquée à l'industrie. Rennes: Presses universitaires de Rennes; 2004.
32. Letté M. Le rapport d'Étienne Clémentel (1919). L'avènement administratif des technocrates et de la rationalisation. Documents pour l'histoire des techniques. 2011;20:167–95.
33. Livage J. Interview with Bensaude-Vincent B. Paris, University of Paris. 4 January 2001. https://www.sho.espci.fr/spip.php?article122.
34. Lochte-Holtgreven W. High temperature research activity in germany. In Stanford Research Institute, editors. Proceedings of an International Symposium on High Temperature Technology, held at Asilomar, California, October 6–9, 1959. New York: McGraw-Hill; 1960.
35. Lucas J. Interview with Teissier P. Paris, Quartier latin. May 9, 2005.
36. Mody C. A little dirt never hurt anyone: knowledge-making and contamination in materials science. Social Studies of Science. 2001;31(1):7–36.
37. Nassau K. Dr. A. V. L. Verneuil: The man and the method. Journal of Crystal Growth. 1972;13–14:12–18.
38. Northrup EF. High temperature investigation. Journal of the Franklin Institute. 1923;195(1169–46):665–86.
39. Pfann WG. How zone melting was invented. Progress in Crystal Growth and Characterization. 1983;6:iii–iv.
40. Psaras PA, Langford HD, editors. Advancing materials research. Washington, DC: National Academy Press; 1987.
41. Riordan M, Hoddeson L. Crystal fire: the birth of the information age. New York: Norton; 1997.
42. Sabel S, Zeitlin J. Historical alternatives to mass production: politics, markets and technology in nineteenth-century industrialization. Past and Present. 1985;108:133–76.

43. Sismondo S. Science and technology studies and an engaged program. In Hackett E, Amsterdamska O, Lynch M, Wajcman, editors. The handbook of science and technology studies. Cambridge, MA: MIT Press; 2008.
44. Stanford Research Institute, editor. Proceedings of the symposium on "High Temperature — a Tool for the Future," Berkeley, California, June 25–27, 1956. Menlo Park, CA: Stanford Research Institute; 1957.
45. Stanford Research Institute, editor. Proceedings of an International Symposium on High Temperature Technology, held at Asilomar, California, October 6–9, 1959. New York: McGraw-Hill; 1960.
46. Stanford Research Institute, editor. Proceedings of an International Symposium on High Temperature Technology, held at Asilomar, California, 8–11 September, 1963. London: Butterworths; 1964.
47. Teissier P. Une histoire de la chimie du solide contemporaine. Synthèses, formes, identités. Paris: Hermann; 2014.
48. Trumbore FA, Turner DR, The Electrochemical Society, 1902–2002: a centennial history. Pennington: The Electrochemical Society; 2002.
49. Viel C. Henri Moissan, pharmacien, premier Français prix Nobel de chimie, 1852–1907. Paris: Pharmathèmes; 2006.

CHAPTER 3.2

REFRIGERATORS AND VACUUM FLASKS

SIMÓN REIF-ACHERMAN

To my mother, who was present at the beginning of this work and who would have been very happy and proud to see it published.

The use of refrigeration in comfort cooling and the processing and preservation of food and drink goes far back in history. The Chinese used ice in food around 2,000 BCE and the Romans mixed snow with fruit juices to make sorbets. In the Middle Ages, the snows of Lebanon were carried by camel caravans to the palaces of the caliphs of Baghdad and Damascus and the sultans of Cairo. Until the 1850s, however, natural ice was the only available source to meet the demands for conservation and transportation of fresh foods, the manufacture and storage of some alcoholic and non-alcoholic beverages, butter, and frozen creams, and other provisions for the populations of growing cities. Ice was stored during winter months in cold, northern climes in specially constructed buildings near ponds, lakes, or rivers, from where it was delivered to regions nearer the equator with horse-drawn wagons and railroads, with a great quantity of it exported in the holds of sailing ships.

Reliance upon natural ice began to show significant drawbacks in the second half of the nineteenth century. Dependence on the seasonal conditions and consequent uncertainty of ice provision, the constantly increasing contamination of water sources in the vicinity of large cities from which ice was harvested, the significant increase in ice consumption that accompanied the dramatic growth of cities, and the emergence of industries like brewing, chemical manufacture (which used ice in processes to crystallize certain salts), intercontinental transport of frozen meat, and ice-based air cooling, to mention just a few, ensured that natural ice could not meet the demand. These conditions created an incentive to pursue mechanical systems for artificial refrigeration.

From Natural to Artificial Cooling

An intermediate stage between natural and artificial cooling during the mid-1800s was characterized by the small-scale use of so-called refrigerating mixtures. The dissolution of certain salts in water is an endothermic process, which produces a drop in temperature. Sodium nitrate and sulfate, potassium nitrate, calcium chloride, and others salts, in different combinations, therefore, allow heat transfer and subsequent temperature reduction. Based on the energetic balance involved in dissolving processes of these salts, this reduction could be up to 30°C, and reach below the freezing point of water (Figure 3.2.1).[18] The mixtures' cost and technical difficulties in recovering the dissolved salts turned out to be the most important disadvantages that prevented ice manufacture on a larger scale based on this principle.

The principle of producing cold by evaporation at pressures below one atmosphere allowed the French civil engineer Edmond Carré (1833–1894) to design simple, inexpensive refrigeration equipment on an industrial scale (Figure 3.2.2), with which he was able to obtain 0.34 kg of ice per each 0.06 kg of evaporated water.[6] The apparatus, composed of a simple, inexpensive pneumatic machine, a flask of resistant glass for the water to be frozen, and a reservoir for sulphuric acid to absorb the aqueous vapor formed, was used in restaurants of many European cities for small-scale production of cool beverages and ice in those places where an abundant winter harvest of natural ice was not possible. Complex machines based

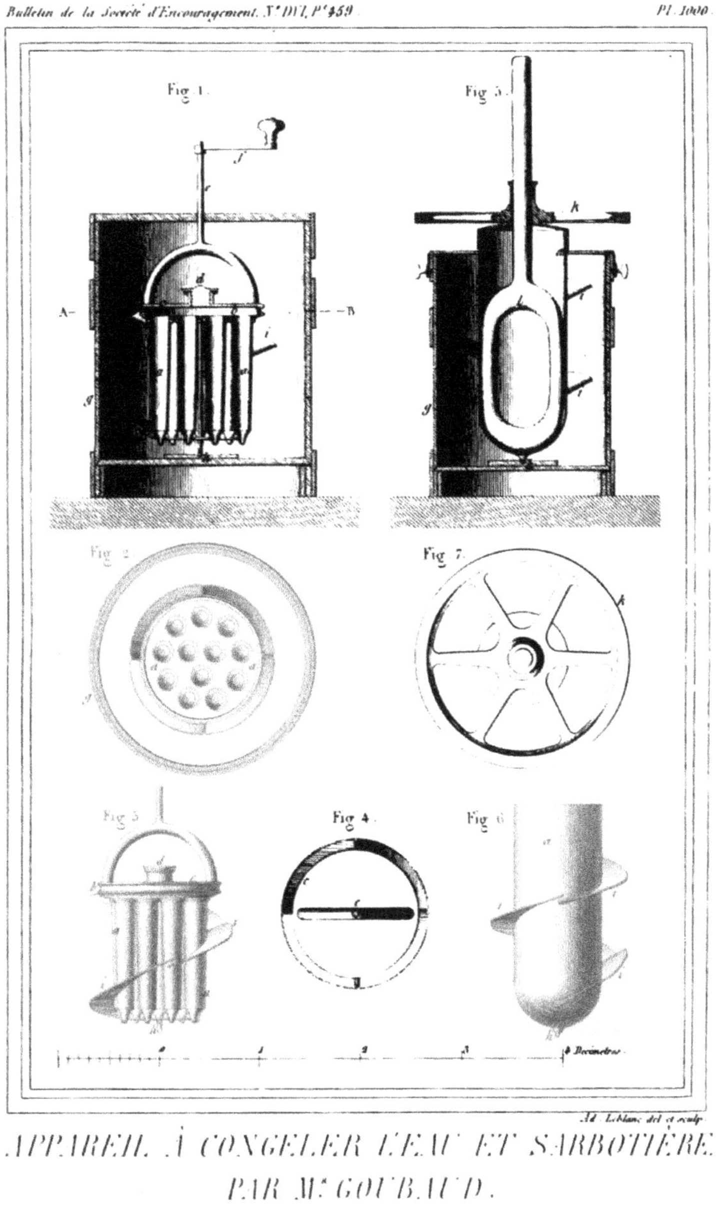

Fig. 3.2.1. Typical "glacier de famille," based on the principle of refrigerating mixtures.[7]

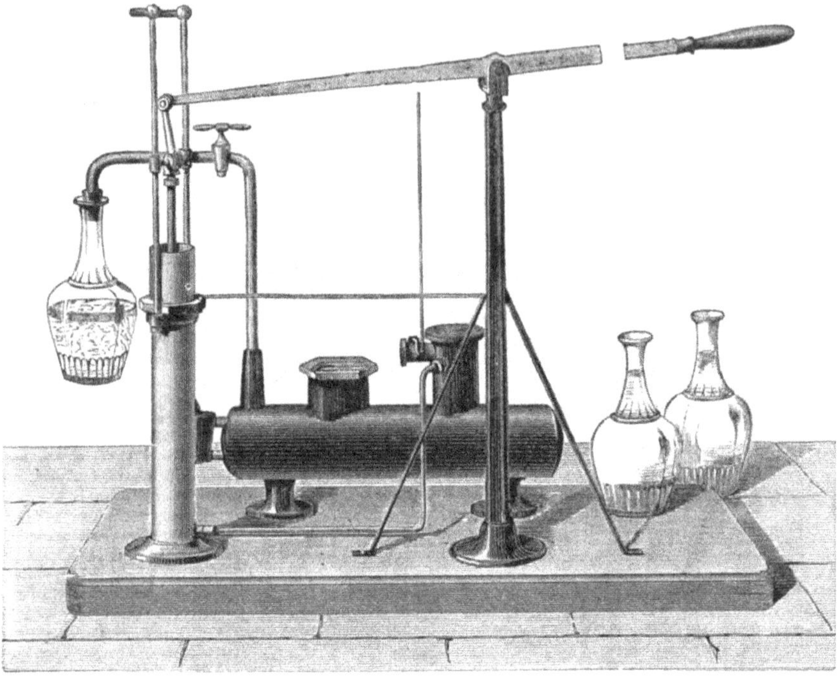

Fig. 3.2.2. Edmond Carré's apparatus.[1]

on this principle, working under vacuum conditions, achieved daily productions of 12,000–15,000 kg of ice, mainly for breweries. The main disadvantages of these machines were their lower thermal efficiency, the difficulty of handling and repairing them when necessary, and the excessive costs associated with handling and regenerating the acid.

Meanwhile, the roots of mechanical air conditioning trace back to the American physician John Gorrie (1803–1855), who built machines to produce cold air to relieve his patients suffering from yellow fever. He produced several models between 1844 and 1850 at the hospital in Apalachicola, Florida, with the cooling produced by the expansion of compressed air.[9] Gorrie's work then became the basis for the design of the first practical closed-cycle air machine by

Fig. 3.2.3. Kirk's regenerative air machine.[3]

the Scottish marine engineer Alexander Carnegie Kirk (1830–1892). Kirk produced his machine, which alternated compression and expansion processes, between 1862 and 1874. The equipment was inspired by the cycle developed and patented some decades before by the Scottish clergyman Robert Stirling (1790–1878) for improving the thermal efficiency of a variety of processes (Figure 3.2.3).[12] The new model brought temperatures of −13°C within reach in apparatus that could produce 2,000 kg of ice per day.

An open-cycle refrigerating machine, based on this same principle and constructed by the French engineer Paul Giffard (1837–1897) in 1873, used the expanded cold air as a refrigerant, or brought it into contact with water to promote its freezing. A typical vertical machine of this class was able to obtain, with an 18 horsepower motor (about 1.8 kg of coal per hour equivalent), 100 kg of ice or, equivalently, about 650 m^3 of conditioned air at 0°C per hour. Although these machines were popular in the last quarter of the nineteenth century in applications related to marine refrigeration, food preservation, and fabrication of chemical and pharmaceutical products, practically all types, using both open and closed cycles, became obsolete by the beginning of the following century and were replaced by other, more compact devices that offered better performance.

Liquid Vapor Refrigeration

Machines using liquefiable vapor compression systems were, perhaps, the more widely diffused refrigeration apparatus in this period. From a strictly theoretical viewpoint, they only differed from air machines in the working substance used and in their inability to run as open cycles. However, from a practical viewpoint, the substantially reduced costs associated with the energy required for their operation, as well as their more energy-efficient performance, distinguished them and encouraged their wide adoption. The theory underlying the system, which explained the alternation of liquefaction and vaporization processes, was the same for the several refrigerant liquids proposed.

Open versus Closed Refrigeration Cycles

In an open-cycle refrigeration process, the air or other substance to be cooled is itself used as the medium for the cooling cycle, and is renewed at the start of each cycle. In a closed-cycle process, the same substance is recycled and energy is transferred to the environment or item to be cooled by another means of thermal exchange.

Ethyl ether systems were the first to be proposed, as early as 1834, perhaps because ether was readily manufactured, already in use as a solvent, and easy to work with as a liquid at room temperature and atmospheric pressure. Jacob Perkins (1766–1849), an Anglo-American inventor, mechanical engineer, and physicist, envisioned an apparatus whose initial objective was to preserve food and constructed a machine that produced refrigeration by evaporating ether under a vacuum and then pumping the vapor into a water-cooled heat exchanger to condense it for reuse.[14]

Although the machine failed to achieve commercial success, the idea was revived by the American civil engineer and professor at Yale University, Alexander Catlin Twining (1801–1884), and the Scottish inventor and journalist James Harrison (1816–1893), the latter considered to be the first person to find a commercial application for

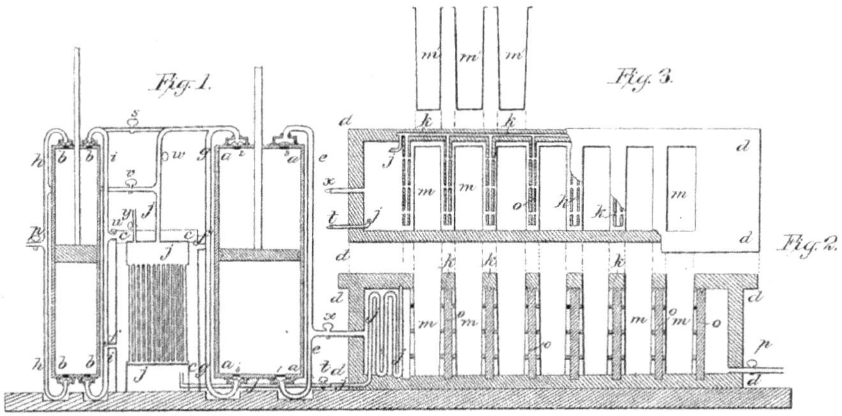

Fig. 3.2.4. Different views of Twining's compression machine.[20]

the vapor compression process. By carrying out vacuum experiments between 1848 and 1850, Twining was able to construct machines with which he could produce 1.2 kg of ice by the vaporization of 1 kg of ether at a temperature of 0°C (Figure 3.2.4). Approximately 22.3 m^2 of surface area of thin copper pipe was enough to produce 1,000 kg of ice in 24 hours, irrespective of the initial water temperature.[19] Meanwhile, in Australia, Harrison developed and built, with the collaboration of the brothers Daniel (1830–1866) and Henry Siebe (1832–1887), the first commercial ice-making machine based on this principle, which could produce 5,000 kg of ice per day with a charge of 400 liters of ether and a power of 24 horsepower (Figure 3.2.5).[10] These machines were built and sold in Great Britain, mainly to breweries and paraffin and oil works, well into the twentieth century.

If it was true that the ether compression machines offered a comparative advantage for uses in hot climates because of the low working pressure in the condenser, their fabrication nevertheless concluded in the 1930s. They were hampered by excessive costs associated with their coal and water consumption for power and condensing purposes, the large size of machines that continued to grow as new systems were merged with old, and some troublesome physical properties of the refrigerants. Ether deterioration, in terms of loss of material due

Fig. 3.2.5. Harrison's ice making machine.[2]

to the volatile nature of the compound, its repeated vaporizations and condensations, and the presence of holes or inappropriate seals in the machines themselves, were another reason for the machines' decreasing performance.

Methyl ether,[17] sulfurous dioxide,[15] carbon dioxide,[13] and ammonia,[8] were the most-used volatile substances in successful ice-making machines proposed on the basis of the principle of vapor compression in the second half of the nineteenth century. Machines using these substances were introduced, respectively, by: the French civil engineer Charles Tellier (1828–1913); the Swiss physicist Raoul Pictet (1846–1929); the American aeronaut, scientist and inventor Thaddeus Sobieski Coulincourt Lowe (1832–1913); and the Scottish mechanic David Boyle (1837–1891) and (mainly) the German engineer Carl Paul Gottfried von Linde (1842–1934).

Each substance exhibited specific advantages. The lower boiling point of methyl ether (−23.6°C at one atmosphere) compared with that of sulfuric ether (34.6°C at the same pressure), guaranteed far lower temperatures. Sulfur dioxide's perfect stability, indecomposability after continuous changes of state without any chemical action on the metals, as well as excellent self-lubrication, and the fact that it did not represent any explosion or fire danger, made this compound an attractive alternative. Carbon dioxide's non-flammable and non-toxic characteristics represented great advantages, but the high pressures necessary for its liquefaction, far higher than those used in boilers working at that time, became the main obstacle for its practical utilization. Ammonia exhibited more strengths than weaknesses. Its more convenient expansion and cooling properties, easier accessibility, and higher reported ice yields compared with other refrigerants (on average, by energy unit), allowed the construction of smaller machines.[11] No single substance, however, was universally adopted, and all served, to a greater or lesser degree, in different situations.

A different liquefiable vapor absorption system that appeared in both domestic and industrial settings used a physio-chemical process rather than a mechanical one. It was based on the great volatility of anhydrous ammonia and the great affinity between this substance and water. In the continuous cycle, which had almost all the characteristics of modern machines, pure ammonia vapors driven off from a liquid mixture with water in a generator or boiler under pressure passed directly to a condenser, where it was liquefied by indirect thermal contact with cold water. The liquid ammonia produced then passed through an expansion valve and then into an evaporating or refrigerating chamber. It was then re-absorbed afterwards by the water in which it was initially dissolved. Weak liquor in the generator was allowed to flow into the absorber in order to keep the solution in a condition to absorb ammonia.[16] The classic machine of this type (Figure 3.2.6), developed by the French engineer Ferdinand Philippe Edouard Carré (1824–1900), was a manageable and effective apparatus for its epoch, with a performance reported by the

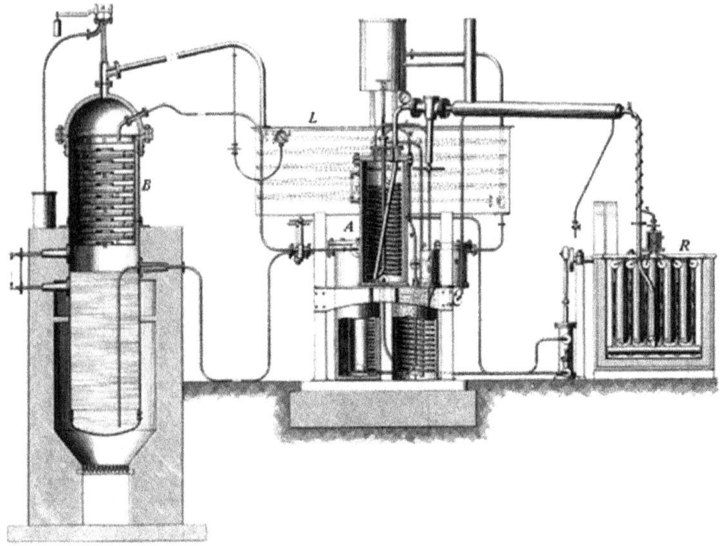

Fig. 3.2.6. F. Carré's continuous ammonia absorption ice machine.[16]

author ranging from 8 to 12 kilograms of ice per each kilogram of coal burned, depending on the dimensions of the apparatus.

From Making Cold to Maintaining Cold

The evolution of mechanical refrigeration changed significantly in 1885. The fundamentals of thermodynamics became better understood in the late-nineteenth century and new types of refrigeration attracted the attention of scientists and investors. Other markets joined beer brewing and frozen meat, which had heavily influenced the earlier evolution of mechanical refrigeration equipment. Cold produced for different food industries solved the scientific difficulties associated with bacterial action in food and subsequently created the possibility of safely storing it for longer and, more importantly, allowed the establishment of a large-scale international food market.

An outstanding development in cryogenics in the 1890s, involving the figure of James Dewar (later Sir) (1842–1923), closes the history of refrigeration in this period.[4] The Scottish chemist and physicist needed better containers to store and prevent rapid evaporation of liquid gases produced in the machinery he had built for his cryogenic experiments at the Royal Institution in London. A double-walled calorimeter he had made two decades before for the measurement of specific heats, with an inner brass chamber surrounded by an outer brass chamber and the space between them evacuated, would become the basis of the instrument. The vacuum space of the containers, now blown from glass, was lined with charcoal, the absorbing power of which in turn removed the last traces of gas left after his air pump had done its best. In addition, coating the inner glass with a thin layer of silver reduced losses by heat conduction and radiation. Dewar demonstrated in 1893 the possibility of increasing the holding time of stored liquefied gases by a factor of ten or more.

The first vessels were made in London, and afterward became available in large quantities from Germany. Oddly, Dewar never patented the design of his flask. Instead, Dewar's German glassblower and manufacturer, Reinhold Burger (1866–1954) secured the patent in his country in 1903, and four years later in the United States (Figure 3.2.7), after he realized the vacuum bottle's commercial applications and designed a smaller domestic version, which included a protective metal shell for the delicate glass walls. The fragility of the article, in spite of this improvement, was solved when the American physicist William Stanley Jr. (1858–1916) discovered that a welding process he was using could be used to insulate a vacuum bottle with steel instead of glass.

The well-known name "Thermos," from the Greek word *therm*, which means hot, was registered in 1906 and remained a patented trademark until July 1970, when, after a lengthy court battle in the US courts between rival manufacturers, it became a generic term in the English language. The flask received a tremendous amount of favorable press in a relatively short time. Half a century after its patenting, vacuum bottles were known almost universally for use with food and beverages, and proved useful in science, medicine, and industry.

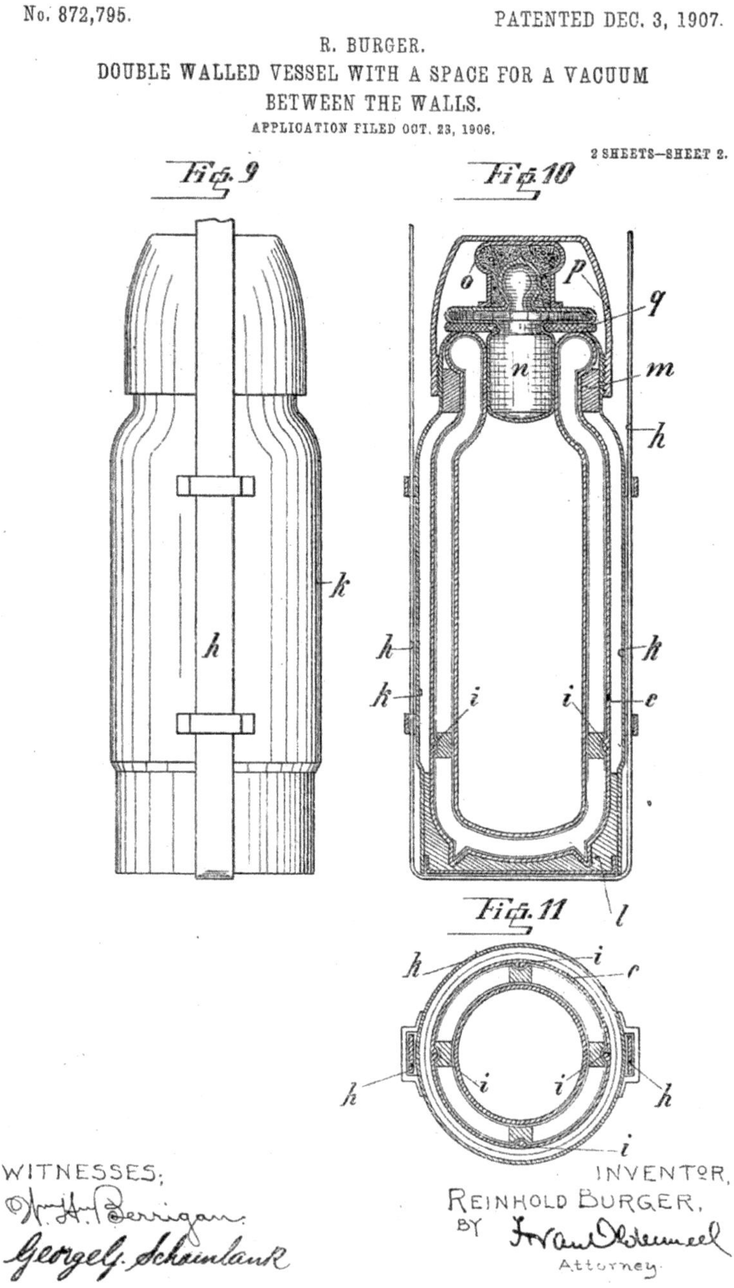

Fig. 3.2.7. Drawing accompanying the vacuum flask patent (1907).[5]

References

1. Anonymous. Description of the vacuum ice machine. Manufacturer and Builder. 1874;6:187.
2. Anonymous. Harrison's ice making machine. The Engineer. 1861;11:231.
3. Anonymous. Kirk's maschine zur kälteerzeugung und eisbereitung mittelst expansion der luft. Dingler Polytechnisch Journal. 1863;170(67):241–45.
4. Bragg W. History of the vacuum flask. Nature. 1940;145:408–10.
5. Burger R. Double walled vessel with a space for a vacuum within the walls. United States patent US872795. December 4, 1907.
6. Carré E. Sur de nouveaux générateurs de froid. Comptes rendus hebdomadaires des séances de l'Académie des Sciences. 1867;64:897–98.
7. De Silvestre A. Rapport sur un appareil à faire de la glace présenté par M. Goubaud. Bulletin de la Société d'Encouragement pour l'Industrie Nationale. 1846;45(505):401–3.
8. Dienel HL. Linde: History of a technology corporation, 1879–2004. New York: Palgrave MacMillan; 2004.
9. Gladstone J. John Gorrie, the visionary. American Society of Heating, Refrigerating and Air-Conditioning Engineers Journal. 1998;40(12):29–35.
10. Gosney WB. James Harrison and the birth of mechanical refrigeration. Proceedings of the Institute of Refrigeration. 1994;90:12–22.
11. Hård M. Machines are frozen spirit: The scientification of refrigeration and brewing in the 19th century – A Weberian interpretation. Boulder: Campus Verlagand Westview Press; 1994.
12. Kirk AC. Improvements in refrigerating apparatus. British patent GB1218. April 25, 1862.
13. Pearson, A. Carbon dioxide: new uses for an old refrigerant. International Journal of Refrigeration. 2005;28:1140–48.
14. Perkins J. Improvements in the apparatus and means of produce ice and in cooling liquids. British patent GB6662. 14 August, 1835.
15. Pictet R. Improvements in producing cold, and in the manufacture of sulphurous acid, and in apparatus for these purposes. British patent GB 2727. August 3, 1875.
16. Pouillet CSM. Rapport sur un appareil de M. Carré ayant pour objet la production du froid artificial. Bulletin de la Société d'Encouragement pour l'Industrie Nationale. 1863;10:32–47.
17. Tellier C. Certain improvements in the manufacture of methylic ether, and its applications to the production of artificial ice. British patent GB387. February 15, 1864.
18. Toselli JB. Fabrication artificielle de la glace: glacière à vapeur d'eau. Le Génie Industriel. 1866;32:11–13.
19. Twining AC. Manufacturing ice. British patent GB13167. July 3, 1850.
20. Twining AC. Manufacturing ice. United States patent US10221. November 8, 1853.

CHAPTER 3.3

HIGH-PRESSURE APPARATUS

ANTHONY S. TRAVIS

The Synthesis of Ammonia

The application of high pressures to bring about chemical reactions, particularly those involving gases, was first investigated in the laboratory at the turn of the twentieth century. Theoretical developments in gas reactions had followed the work of Henri Le Chatelier (1850–1936) in France, who at the end of the 1880s formulated the principle that when an external force exists, a reversible reaction takes place in the direction that reduces the force. A special focus was on forcing the unreactive gas nitrogen to combine with hydrogen to form ammonia that could be converted into stable compounds for application as fertilizers. The ammonia synthesis is the basis of the most important early development in high-pressure chemical synthesis:[1]

$$\underset{4\ \text{volumes}}{N_2 + 3H_2} \rightleftharpoons \underset{2\ \text{volumes}}{2\,NH_3} \quad \Delta H = -92\,\text{kJ}\,\text{mol}^{-1}$$

The equation above shows that the reaction is favored by the application of pressure; the equilibrium moves in the direction that reduces the effect of the pressure, that is, to the ammonia side of the equation. However, ammonia is formed with difficulty, not only because of the inertness of nitrogen but also because the reaction is exothermic. Lowering the temperature encourages the formation of ammonia, but the rate at which the reaction takes place becomes extremely slow. It is better to use a high temperature to speed up the process, even though the yield is low. The way to reduce the barrier to reaction is to introduce an active catalyst. Le Chatelier, aware that a compromise between the rate of reaction and yield of ammonia could be obtained by adjusting the pressure and temperature and introducing a catalyst, investigated the synthesis of ammonia under such conditions in 1900. In 1901 he filed a patent for the method. Further development was brought to an abrupt halt when his laboratory apparatus blew up due to the presence of oxygen in the gas mixture.

Gases were early candidates for study under conditions of high pressure at University College London (UCL). A noteworthy UCL student who would play a critical role in bringing about the high-pressure gas phase reaction between nitrogen and hydrogen was Robert Le Rossignol (1884–1976).[5] In 1908, he joined physical chemist Fritz Haber (1868–1934) at Karlsruhe Technische Hochschule, as a research assistant. Haber, following prompting by physical chemist Walther Nernst (1864–1941), who worked with a ceramic apparatus, had set to work on inventing a system in which the synthesis of ammonia could be achieved at 200 atmospheres pressure (402 kPa) in a steel converter (reaction vessel). Building the apparatus was Le Rossignol's assignment. By mid-1909, Haber and Le Rossignol had achieved formation of ammonia in the presence of catalysts, of which osmium gave optimum results. Because the yield at around 500°C was very low (around 5 percent) and the intention was to devise a method that could be applied commercially, it was necessary to recirculate the unreacted gases in a continuous process, using a closed apparatus (Figure 3.3.1).

Haber's work was sponsored by the chemical firm BASF of Ludwigshafen, Germany, where chemist Carl Bosch (1874–1940) scaled up the laboratory process. This presented several novel technical challenges, including the need for a catalyst that had to be much cheaper and at least as efficient as Haber's osmium. Bosch's colleague Alwin Mittasch (1869–1953) came across a useful catalyst based on iron containing small amounts of potassium oxide and aluminum oxide.

No less problematic were the metallurgical and engineering challenges in design of the high-pressure converter. At 200 atmospheres,

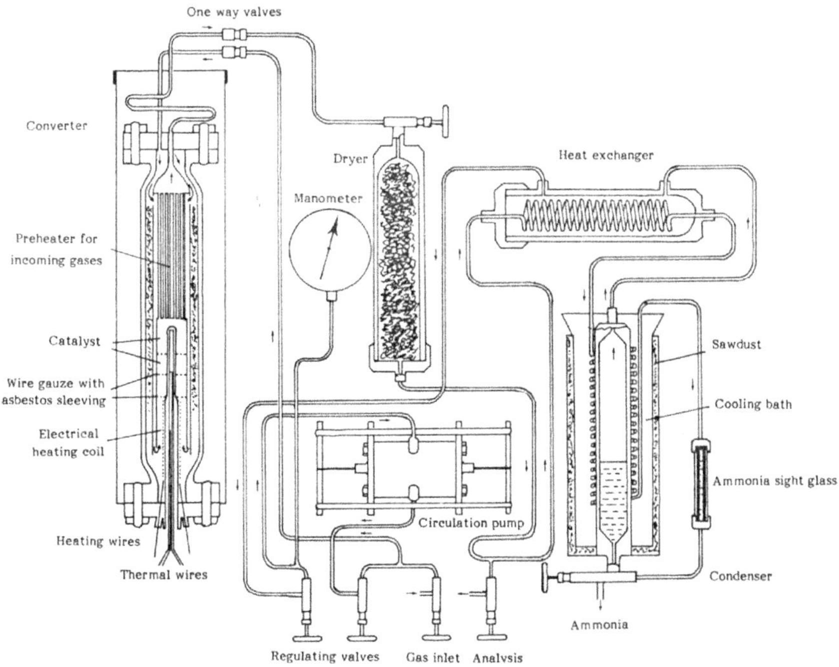

Fig. 3.3.1. Diagram of laboratory apparatus for continuous synthesis of ammonia from nitrogen and hydrogen, with complete loop, after Fritz Haber and Robert Le Rossignol (1909).

hydrogen reacted with the carbon of the steel, causing it to become brittle, resulting in failure and explosions. Bosch overcame this problem by designing a double-walled high-pressure converter. The inner wall, of low carbon steel, allowed hydrogen to diffuse through it, undergoing a loss of pressure. The outer wall, of ordinary steel, was subjected to the hydrogen at a low pressure and did not become brittle. Sufficient progress was made by 1911 to justify erecting a full scale plant, that was installed at Oppau, north of the Ludwigshafen works.[7;8] Extremely pure nitrogen and hydrogen gases required the scale-up and modification of existing coal-based processes. The gas mixture, adjusted to give seventy-five percent hydrogen and twenty-five percent nitrogen, was compressed in reciprocating compressors. When inaugurated in September 1913, the process was an immediate

Fig. 3.3.2. Erecting a high-pressure ammonia converter at Oppau, 1920 (BASF unternehmensarchiv).

success (Figure 3.3.2). However, rather than fertilizer (ammonium sulphate), during World War I much of the ammonia was converted into concentrated nitric acid, required in the production of modern explosives such as trinitrotoluene (TNT) and ammonium nitrate.[9]

The success of what was known as the Haber–Bosch process spurred inventors and entrepreneurs to devise similar synthetic ammonia processes. Most successful were Luigi Casale (1882–1927) and Giacomo Fauser (1892–1971), in Italy, and Georges Claude (1870–1960), in France. Casale and Claude worked with extreme pressures, 800 and 1,000 atmospheres, respectively, or what Claude called "hyperpressures."

The two very-high-pressure processes required less active, and less sophisticated, catalysts. Casale's catalyst was scrap iron (Figure 3.3.3). In addition, they not only gave higher conversions, but required far less catalyst than the Haber–Bosch process. One major problem, however, was the great heat generated in the converter in the area where the mixed gases were exposed to the catalyst, which, unless brought under control, caused the apparatus to explode. The effect was magnified, compared with the Haber–Bosch process, by the use of smaller converters and very high pressure. Claude estimated that his high-pressure process generated up to fifty times the heat of a Haber–Bosch converter. Casale overcame the problem by a simple and elegant appreciation of the chemistry involved. As the temperature increases, the concentration of ammonia decreases, as

Fig. 3.3.3. Assembling a Casale high-pressure converter, 1923. The lid is open, showing the massive bolts and studs required to maintain the pressure of approximately 800 atmospheres. *Credit*: Casale SA.

does, correspondingly, the quantity of heat developed in the apparatus. Thus, a convenient means of temperature regulation could be achieved by allowing some ammonia to be carried over with the recycled reactants. The very first large Casale ammonia facility to open, in 1923, was in Japan, at Nobeoka on the east coast of Kyūshū island.

From 1917, as part of his secret government wartime research, Claude investigated high-pressure ammonia synthesis using pressures of around 1,000 atmospheres at 500°C. The Claude process faced several difficulties before special steel converters were produced in Sheffield, England, during the early 1920s. Another important process was developed, in 1925, by chemical engineering contractor Friedrich Uhde (1880–1966) of Dortmund. The converter in this "Mont Cenis" process operated at just 100 atmospheres and required an extremely active catalyst. From 1923, using its ammonia converter, BASF produced methanol (methyl alcohol) from the high-pressure reaction between hydrogen and carbon monoxide.

The combination of high pressure and high temperature produced enormous stresses on the pressure tubes. This encouraged the development during the 1920s of steels based on chromium, nickel, and molybdenum alloys for ammonia converters. Now, as then, in all converters the catalyst is contained centrally in a thin vessel and the inner shell of the pressure tube is kept cool by circulating gas. After World War II, for larger plants, reciprocating compressors were replaced by centrifugal compressors. Notwithstanding the different industrial conditions for synthesis of ammonia, all processes relied, and continue to rely, on the same basic chemistry devised by Fritz Haber in 1909.[5]

Pure Hydrogen

The first high-pressure processes, for manufacture of ammonia and methanol, were reliant on the production of pure hydrogen, which from around 1920 was an industry in its own right. BASF and American manufactures relied on coal-based water gas; most Japanese and several European firms generated hydrogen by electrolysis, which was favored at sites where low-cost hydro-electricity was available.

Fig. 3.3.4. Compressor room, Casale process, Union Chimique Belge, at Ostend, Belgium, late 1920s. This facility mainly relied on hydrogen obtained from coke oven gas. *Credit*: Casale SA.

From around 1930, coke ovens became the main source of hydrogen until well into the 1940s (Figure 3.3.4). Today, natural gas is the principal source of hydrogen. However, as part of the move towards sustainability there is a renewed interest in the generation of hydrogen by electrolytic methods.

High-Pressure Hydrogenation of Coal to Oil

During World War I, Friedrich Bergius (1884–1949) worked on high-pressure hydrogenation of coal in order to produce synthetic gasoline. His process was adopted by BASF in the mid-1920s. The converters were similar to those employed in the ammonia synthesis. Significantly, the process involved the reaction between a solid and a gas. It was licensed to Standard Oil and Britain's Imperial Chemical

Industries (ICI). The Standard-IG Company was created in 1929, followed in 1931 by the International Hydrogenation Patents Company, through which coal-to-oil patents were pooled, and which included Royal Dutch-Shell.

Coal-to-oil became important in National Socialist strategic planning after Adolf Hitler came to power in 1933. High-octane gasoline made by the Bergius process was used by the Luftwaffe and RAF during World War II.

For his high-pressure ammonia synthesis, Fritz Haber received the 1918 Nobel Prize in Chemistry. Carl Bosch and Friedrich Bergius shared the 1931 Nobel Prize for their high-pressure work.

Polythene

The range of high-pressure investigations broadened considerably during the 1930s. At ICI's Winnington laboratories this was a result of speculative research and the move away from the more mature fields of synthetic ammonia and coal-to-oil conversion. The most significant outcome was the accidental discovery of polythene. This achievement relied in part on studies undertaken by Antonius M. J. F. Michels (1889–1969) of the University of Amsterdam's high-pressure laboratory, working at over 1,000 atmospheres, and funded by ICI from 1928. Two of his assistants were ICI physical chemists Reginald Oswald Gibson, and Michael Willcox Perrin (later Sir; 1905–1988), who, on returning to England, joined organic chemist Eric William Fawcett to undertake high pressure studies that might be of value to the dyestuffs division. Their apparatus was designed by ICI engineer W. R. Dermot Manning.

On March 27, 1933, the reaction between ethylene and benzaldehyde was observed to afford a white, waxy solid that lined the inner wall of the reactor. The pressure was just under 2,000 atmospheres; the temperature 170°C. In further attempts the waxy solid was sometimes obtained; at other times a black substance was deposited. In June, Gibson and Fawcett reported that the reaction giving the waxy solid, "appears of interest as a possible source of long chain compounds." The yield, however, was small and the decomposition reaction did not suggest that the method could be easily controlled. From December 1935, the reaction was repeated in an improved apparatus. Sufficient waxy product became available to confirm that it was a long-chain polymer of ethylene. It was readily molded and cast, and showed excellent electrical insulating properties, as well as being inert. In 1936, its formation was found to have been the result of the presence of a small amount of oxygen, acting as catalytic initiator.

Manufacture of the polymer polythene (poly(ethane)) began in 1938. Its use as insulated covering for underwater cables was demonstrated by a mile-long connection between the Isle of Wight and mainland England. The novel material was soon in use in high-frequency radio transmitters and receivers required in radar, since it absorbed far less radio energy than existing dielectrics.[1] Later, other olefines were used to manufacture plastics and resins. Developments in novel catalysts and technical improvements, including the use of lower pressures, gave a wide range of plastics and higher yields.

Synthetic Rubber: Acetylene under Pressure

At BASF, Walter Reppe (1892–1969) began work on synthetic Buna rubber around 1930. His research included a viable route to the monomer butadiene as a strategic priority. Reppe developed a technical route to butadiene starting with acetylene, through application of high pressures previously considered far too dangerous, since acetylene under pressure is inherently unstable and explodes with extreme violence. Acetylene's reactivity is a result of the triple bond between its two carbon atoms. In September 1937, Reppe applied pressure to a mixture of acetylene and formaldehyde in the presence of a percussive copper acetylide catalyst, to produce 2-butyne-1,4-diol (butynediol), which retained the triple bond of acetylene. The butynediol was converted into 1,4-butylene glycol, which in turn gave the required butadiene.[3] Some of Reppe's reactions, conducted under pressure in the presence of a nickel-carbon catalyst, were introduced commercially after 1945, including in the United States.[10]

Other high pressure processes include catalytic cracking and reforming (hydroforming, 1940; platforming 1949) for improvement of gasoline; manufacture of artificial ruby and diamond; and food processing by pascalization (also bridgemanization), which involves high-pressure preservation, a viable alternative to pasteurization and sterilization, without loss of nutritional value, and has been developed since the early twentieth century.[2; 4]

High-pressure industrial chemistry was a leading breakthrough technology of the early twentieth century. It completely rewrote approaches to plant design and contributed to the rapid emergence of chemical engineering. Since the 1960s, following ongoing research and new technical developments, more efficient catalysts have been introduced that enable lowering in the very high pressures previously required in many industrial chemical reactions. An important outcome is considerable reduction in energy costs, for example in the production of the bulk chemicals such as polythene and ammonia. Thus in 1983, ICI introduced an ammonia process operating at around 80 atmospheres. Finally, high-pressure nitrogen capture remains critical for feeding growing populations, particularly in India and, even more so, in China, which is the world leader today in production of nitrogen fertilizers.[6]

References

1. Harness A. The story of polythene. Brentford, Middlesex: Plastics Processing Industry Training Board; ND.
2. Hite BH, Giddings NJ, Weakly CE Jr. Certain micro-organisms encountered in the preservation of fruits and vegetables. West Virginia University. Agriculture Experiment Station, Morgantown, WV. Technical Bulletin 146. October 1914.
3. Morris PJT. The industrial history of acetylene: the rise and fall of a chemical feedstock. Chemistry and Industry. December 19, 1983;18:710–15.
4. Recio JM, Mendénez JM, de la Roza AO, editors. An introduction to high-pressure science and technology. Baca Raton: CRC Press; 2016.
5. Sheppard D. Robert Le Rossignol, 1884–1976: engineer of the 'Haber' process. Notes and Records of the Royal Society London. 2017 DOI: 10.1098/rsnr.2016.0019 [Accessed March 18, 2017].
6. Smil V. Enriching the earth: Fritz Haber, Carl Bosch, and the transformation of world food production. Cambridge MA: MIT Press; 2001.

7. Tamaru K. The history of the development of the ammonia synthesis. In: Jennings JR, editor. Ammonia synthesis: fundamentals and practice. New York: Plenum; 1991. 1–18.
8. Topham SA. The history of the catalytic synthesis of ammonia. In: Anderson JR, Boudart M, editors. Catalysis: science and technology, vol. VII. Berlin: Springer; 1985. 1–50.
9. Travis AS. Nitrogen capture: the growth of an international industry (1900–1940). Cham: Springer; 2018.
10. Travis AS. Unintended technology transfer: acetylene chemistry in the United States. Bulletin for the History of Chemistry. 2007;32(1):27–34.

CHAPTER 3.4

BIOORGANIC SYNTHESIS

VICTORIA LEE

When we think about the processes industries use to develop chemicals for food and drugs, or when these processes are portrayed in popular culture, they tend to be the sterile procedures of synthetic organic chemistry. But in fact, most of the compounds we ingest in our meals and our medicines are generated through much squishier processes: they take advantage of the laboratories that nature has built into the Earth's living organisms. From the viewpoint of the discipline of chemistry, which has long prided itself on making products "synthetically," gathering products from living things "naturally" has often appeared as the opposing approach. Scientists have begun rhetorically linking the two only recently.[2; 33] The latter, however, has a long history in the form of fermentation research. In Japan — one of the leading countries in fermentation research after World War II — the term "fermentation method" (*hakkōhō*) was long synonymous with "biosynthetic method" (*seigōseihō*), and was used directly in contrast to "synthetic method" (*gōseihō*), which employs more traditional laboratory techniques. The term "fermentation" (*hakkō*) in Japanese could also refer broadly to the chemical reactions occurring in living cells.

By focusing on fermentation research as it was performed in the 1970s and 1980s, this essay emphasizes the roots of bioorganic synthesis in applied and industrial microbiology. It offers a historical snapshot of what materials research looked like at the interface of biology and chemistry, on the cusp of the boom in recombinant DNA and genomic sequencing techniques. Today, biosynthesis work has reconfigured itself around newer fields, including biomaterials science (with methods not only from biotechnology but also nanotechnology and microelectronics) and fields at the interface of chemistry and biochemistry (such as bioorganic chemistry and chemical biology). It also continues in older fields, such as natural products chemistry and the related fields of plant and microbial biology.[14; 30; 31] Now, "soft chemistry" (*chimie douce*) whereby "reactions ... take place at room temperature in rather 'messy' aqueous environments," inspires interest as an environmentally friendly method for industrial production in contrast to traditional chemistry.[3 p602] The hope is that biosynthetically developed, "green alternative" products will help to "[replace] harsh industrial processes with milder biotechnology that does not require toxic metals or large amounts of organic solvent... [and leads to] lower amounts of side products and waste."[20 p10] The local institutional configuration varies with any given context, but in all cases the use of cellular processes to create novel materials — especially those of medicinal and pharmacological interest — finds one of its most direct precedents in antibiotic research.

In fermentation, a living organism (usually a microorganism) manufactures a desired material, which is produced as a metabolite, a result of the cell's biochemical pathways. The diversity and variability of the natural world is the main tool that researchers employ for making target materials. Researchers have used an array of methods to harness natural diversity and variability as a tool and extend its capability to work for their purposes. In the traditional food and drink fermentation industries, scientists aim to enable various microbes to produce a mixture of metabolites that act together to produce flavors or other desirable properties. Synthetic fermentation research, dating to the early twentieth century, differs in seeking to enable a particular microbial strain to accumulate one specific target substance in high quantities.[22]

The key process in laboratory bioorganic synthesis research is screening. Natural diversity presents enormous possibilities in its library of microbial strains, most of which are still unknown.

Screening is the process by which, from this diversity, researchers identify and select a microbial strain with an unusual ability to produce a substance of interest in quantities that could allow for mass production. In other words, screening is a carefully designed search system for such a microbe. It requires both biological and chemical techniques because along with the isolation, culturing, preservation, identification, and selection of microbial strains, the process involves using available tools of biological and chemical analysis to assay the substance of interest after extracting it from culture broths and purifying it. Microbial sensitivity and variability become important instruments as scientists work to optimize the strain's ability to produce the target substance by adjusting the composition of the culture medium, and by selecting and improving the strain.

This essay begins by outlining the scope of fermentation as a method of synthesis. It then concentrates on screening as the primary method of constructing and refining microbial diversity and variability for making materials, through the example of the antiparasitic drug avermectin, discovered and developed collaboratively in the 1970s by the Kitasato Institute in Japan and Merck & Co. in the United States, and for which William Campbell and Ōmura Satoshi received the Nobel Prize in Physiology or Medicine in 2015. The workflow of this process presents us with a view of older screening methods in the 1970s, which then offers us the background to make sense of the explosion in "combinatorial" approaches to materials discovery since the 1990s.

The Horizon of Biosynthetic Functions

Through the twentieth century, scientific research extended understanding of the living cell's potential to produce compounds for human uses. During World War I, glycerol fermentation for explosives and food yeast production, as well as acetone fermentation to create a biofuel substitute for gasoline, became pillars of the wartime economy in Germany and Britain respectively.[8 p36] By the interwar period, lactic acid and citric acid, particularly for food flavoring, were produced by culturing molds in European and American factories, though fermentation was still peripheral as an area of investment in large chemical firms.[8 p47; 1] The scope of biosynthetic methods grew especially after World War II following the successful wartime mass manufacture of penicillin using the mold *Penicillium* in the United States (which in turn owed much to German–Czech experience in organic acid fermentations).[6; 7; 27] Governments and businesses across the world invested in microbiological expertise for pharmaceutical research and development, as well as in the construction of large-scale fermentation tanks and fermentation engineering skills in order to produce penicillin locally.[7; 15; 18; 19; 35]

The subsequent "screening boom"[26 p439] from the 1950s to the 1970s initially focused on uncovering new lead compounds for antibiotics, but the horizon of possibility rapidly expanded to substances including anticancer drugs, agents used to inhibit disease-causing enzymes, agro-antibiotics, and industrial enzymes including those used as biocatalysts in chemical synthesis.[4] The search also expanded to primary metabolites *other* than alcohols and organic acids (including metabolites involved in making "essential macromolecules such as DNA, RNA, proteins, lipids, and polysaccharides"[12 p5] which do not typically accumulate in great quantities in the cell) like amino acids, nucleotides, and vitamins. Just a few of the goods that rely on biosynthetic production, for example, are the anti-cholesterol drugs known as statins, the flavor product monosodium L-glutamate (MSG), the plant hormone gibberellin which is used in producing seedless grapes, and various enzymes used as catalysts in sugar production or steroid hormone production.

Why did scientists look to fermentation — effectively "borrowing" the cell's metabolism — as a more competitive tool for manufacturing certain materials, instead of looking to synthetic organic chemistry, or to other natural product methods such as extraction from plant or animal preparations or decomposition of proteins (for example, by acid hydrolysis)? For drugs such as antibiotics, the diversity of the natural world is far more powerful than any computer for revealing new core structures; the vast majority of such molecules have been natural products, whereas chemical synthesis has mainly been used only to modify the branches. Whether seeking new antibiotics or enzymes, scientists came to dream of finding extremophile bacteria or fungi that survive in uncommon environments — as Canadian researcher Naowarat Cheeptham put it, "deep down in ocean sediment, or a desert, or hot spring, or going into different types of caves" — and which might therefore offer "a different and new metabolic pathway that we haven't seen before."[4; 10]

For primary metabolites, industrial fermentation can be more economical than alternative methods. In the early twentieth century, citric acid could be fermented more cheaply than it could be chemically synthesized or extracted from lemon juice. In the 1950s and 1960s, the Dutch company DSM developed a synthetic process for the amino acid lysine, only to find that it was uncompetitive with Japanese fermentative manufacturers such as Kyōwa Hakkō due to the expensive step of separating the D-isomer from the biologically active L-isomer, whereas fermentation yielded only the L-isomer.[32 p116]

Glossary

Metabolite — A *primary metabolite* is a compound derived from processes of primary metabolism. These are "chemical reaction mechanisms and pathways employed by living cells for synthesizing or modifying the basic organic compounds essential for life, such as carbohydrates, fats, proteins, or nucleic acids." They are "essentially identical in all known organisms, owing to their common evolutionary ancestry." A *secondary metabolite* is a compound derived from processes of secondary metabolism. These "employ metabolic pathways for the production of certain complex organic compounds which are unrelated to and therefore unnecessary for an organism's growth, development, or reproduction." Secondary metabolites are "usually unique to a specific organism or group of organisms."[25 pxvii–xviii]

Lead compound — A term used in drug development. It refers to a compound that shows desirable bioactive properties and is therefore selected to undergo further improvement by minor chemical modifications.

Core structure and branches — Terms used to describe the molecular structure of antibiotic compounds. Compounds belonging to one family of antibiotics share a common core structure, but their branches vary. For example, penicillins share a core structure consisting of a β-lactam ring fused to a thiazolidine ring. Modifications in the branches — in the case of penicillins, the acyl side chain — result in penicillin types that differ in their spectrum of activity (the kinds of bacteria which are affected) and pharmacokinetics (the ways in which the drug moves through the human body). Penicillins were discovered and mass produced by biosynthetic methods, but synthetic methods were then used to modify the acyl side chain to produce more penicillin types; this was known as penicillin semisynthesis.

Microbiologists such as Ōmura Satoshi argue that the diversity of the natural world is of interest to chemists because "microorganisms can perform very interesting reactions ... never known previously in organic chemistry," and "on the metabolic map, we can see many reactions which are difficult to perform chemically."[29] They emphasize both the variety of bioactivity of a single natural compound, and the variety of substances produced by a single organism.

Screening

Streptomyces avermectinius, an actinomycete and producer of avermectin (used to treat river blindness in humans and the top-selling veterinary drug for over two decades), was isolated in 1974 from the soil of a golf course in Kawana, Shizuoka Prefecture, Japan. No strain of the same species, nor any other strain that produces avermectin, has been reported to have been isolated since.[28] *S. avermectinius* was found when the team of Ōmura Satoshi of the Kitasato Institute gathered soil samples, took them to the lab, isolated strains, cultured them, and characterized many thousands of their fermentation broths. Ōmura had trained in the organic chemistry of natural products with special expertise in structural determination by NMR. After research in the brandy lab in the Department of Fermentation

Technology at Yamanashi University, and upon joining the Kitasato Institute, Ōmura gained familiarity with microbiology and became known for developing original methods of screening (he is credited with discovering fifty-three new species, thirteen new genera, and about five hundred new organic compounds).[11]

Of the *Streptomyces* isolates from the golf course, the Kitasato Institute team selected about fifty of the most unusual strains to be sent to Merck in the United States as part of a joint effort to develop products from fermentation. To Merck scientists, this partnership was a "logical sequel" to a similar microbial screening collaboration set up in 1954 between Merck and Compañía Española de Penicilina y Antibióticos (CEPA) in Spain.[34] At Merck, the strains received from Japan were extensively screened for bioactivity. William Campbell's team purified the active compound avermectin from one of the broths which showed effectiveness against parasites.[9]

Today, avermectin is mass produced as antibiotics are: it is fermented aerobically at ordinary temperature and avermectin accumulates in the *S. avermectinius* cells as they are cultivated, after which the cells are gathered and avermectin is extracted using an organic solvent such as acetone or ethanol. Since avermectin consists of a mixture of several structurally similar compounds that vary in bioactivity, the most effective constituents must then be purified in order to be used in treatment.[28] Ivermectin, the compound widely used as a drug due to its efficacy and safety profile, is a semisynthetic derivative of avermectin made by altering avermectin B_1 using selective hydrogenation.[9]

According to Ōmura, an effective screening system for new drugs can take one or two years to design and is "not just routine work." He describes it as an interdisciplinary process in which "each one of our group has a basic research assignment such as taxonomy, structure determination, mode of action, biosynthesis, or chemical synthesis, as well as participation in screening work."[29 p268] The screening system is carefully tailored to the type of microbe and target substance, and is devised to capture only a narrow slice of the existing natural diversity (Figure 3.4.1). The search might begin by looking for unknown microbes, or it might screen from a select genus and species,

TABLE 1. Screening methods and the new compounds found by Ōmura's group at Kitasato (Dec. 1985)

Screening system	Compounds (reference)
Antibacterial	A-73A, aurantinins (44), factumycin (14), kinamycins (12), KM-8 (47), LA-1 (55), leucomycins (94), takaokamycin (69)
Bacterial cell wall synthesis inhibitor	AM-5289, asukamycin (43), azureomycin (110), nanaomycins (116), setomimycin (29), vineomycins (19)[a]
Antifolate	AM-8402 (73), diazaquinomycin (39)
Antimycoplasmal	2′-Amino-2′-deoxyadenosine (26), cervinomycins (58), frenolicin B (93), OM-173 (23)
Antianaerobic	Clostomicin, luminamycin (64), lustromycin, thiotetromycin (76)
Antiviral	Virantmycin (40), virustomycin (56)
Antifungal	AF-8 (alboleutin) (61), cerulenin (119), irumamycin (92), O-2867 (104), prumycin (49)
Antiparasitic	Avermectins (3), hitachimycin (stubomycin) (46), setamycin (79)
Anticancer	OS-3256B (105), sporamycin (120)
Herbicidal	Herbimycins (70), karabemycin (75), oxetin (72), phosalacine (53)
Penicillinase inhibitor	KA-107 (45)
Adenosine deaminase inhibitor	Adechlorin (57)
Elastase inhibitor	Elasnin (41)
Antiplatelet	OM-3209
Alkaloid (chemical screening)	AM-2504 (prolimycin) (60), 1,3-diphenethylurea (22), herquline (54), NA-337 A (97), neoxaline (33), pyrindicin (25), quinoline-2-methanol (63), staurosporine (11), TM-64 (narcoactine) (34)
Hybrid biosynthesis	Chimeramycins (80), mederrhodins (18)
Chemical modification	Rokitamycin (3″-*O*-propionylleucomycin A_5) (102), 3,3″,4″-tri-*O*-propionylspiramycin I (81)
Others	Irumanolides (101), protylonolide (90)

[a] Asukamycin, nanaomycins, setomimycin, and vineomycins do not inhibit bacterial cell wall synthesis, but were isolated as by-products from this screening system.

Fig. 3.4.1. Table of some of the screening systems used and the resulting new compounds found by Ōmura's group at the Kitasato Institute.[29 p260]

such as certain molds for organic acid production or particular actinomycetes or molds for antibiotics.[22] Taxonomic studies, including studies of fermentation physiology, are important because they offer clues to what kinds of strains might be promising. Within the partnership with Merck, the Kitasato team aimed to select "unusual" *Streptomyces* strains based on their unique morphology (appearance under a microscope) and cultural characteristics, thereby "weeding out" producers of well-known antibiotics, whereupon they would send the dried fermentation broths to Merck for further assays.[9 p854]

The next step in screening consists of fermenting the isolated strains in hundreds of thousands of broths in order to evaluate the target substance and to select out a strain that produces it.[22] Merck sent shipments of microbial isolates from the Kitasato Institute to (among other assay programs) its *in vivo* mouse assay program for anthelmintic (antiparasitic) activity. It involved putting the dried fermentation broths into animal feed that had been infected with nematode parasites. Efficacy was evaluated by checking for the absence of worm eggs in the feces of animals that had consumed the infected feed, as well as checking for worms in the intestine upon necropsy of the same animals. This assay turned up anthelmintic activity in one of the *Streptomyces* broths from Japan, and the strain came to be characterized as a new microbial species, *S. avermitilis*, later renamed *S. avermectinius*. Follow-up testing showed that the previously unknown active compound, named avermectin, was also active against arthropod parasite vectors.[9]

Once activity is confirmed, the active substance must be extracted and purified from the fermentation broth — advanced tools for which include high-pressure liquid chromatography — and its structure analyzed by techniques such as NMR spectrometry. By the late 1970s or early 1980s, such techniques had "made it possible to isolate and analyze easily a very small quantity of an active substance."[29 p276] Researchers at Merck showed avermectin to have four different structural components, each with a major and minor subcomponent, and elucidated their structure in detail using chromatographic and spectroscopic tools, including mass spectroscopy and NMR. This enabled subsequent research to clarify the precise structure–activity relationship.[9]

The next stage in fermentation development is to increase the production ability of the microbe. At this point, scientists test numerous culture media per strain. In *S. avermectinius*, avermectin production occurs only in a specific medium with a particular ratio of quantities of the nitrogen source and carbon source, which consequently required extensive research to identify. Moreover, the fermentation occurs slowly, taking about a week even at the laboratory scale to reach the highest yield (and longer at the industrial scale).[28] As Campbell put it, the unmodified avermectins "are often referred to as 'naturally occurring' as distinct from the semisynthetic ones such as ivermectin. However ... we know only that such compounds occur naturally in the unnatural environment of laboratory culture."[9 p856] Fermentation physiology directs "the conditions which allow nonproducing strains to become producing strains," and is one aspect of microbial variability to be studied and harnessed.[29 p274]

The manipulation of strain variation in colony selection, cultivation, and improvement further raises production ability. Perhaps tens of colonies of the producing strain might be cultivated as a basis for selection.[22] Further random variation in the strains may also be created by inducing mutation through exposure to nitrosoguanidine or UV light — a technique that has been in use in microbial genetics as well as fermentation research since the 1940s and 1950s.[5] Although random selection of induced mutants might seem time consuming, screening systems can be made highly efficient by thoughtful design, for example by choosing the kind of culture medium or conducting bacterial assays that make colonies of the desired strains stand out visually from the rest.

Strain improvement works hand-in-hand with basic study of the biochemical and genetic mechanisms of biosynthesis of the desired material. For example, with some knowledge of the biosynthetic pathway and its regulation, researchers can screen for mutants in which the regulatory control is released to allow the desired compound to accumulate efficiently in the cell. Methods to deliberately alter the cell's biochemistry are called "metabolic engineering."[22]

Researchers at the Kitasato Institute and Merck continued researching *S. avermectinius*, using mutant strains in which part of the biosynthetic pathway of avermectin was blocked. By analyzing the intermediates that accumulated in various mutants, they could clarify the details of the biosynthetic pathway and by crossing mutants they could relate the biochemistry to mutated points on the chromosome, eventually elucidating seventeen genes controlling the biosynthetic pathway. Recombinant DNA technology was subsequently applied to heighten the production ability of the strain: by cloning the genes that govern biosynthesis of the desired compound, the activation could be amplified. With this biochemical and genetic knowledge, using gene manipulation techniques such as crossing and recombinant technology, the production ability of *S. avermectinius* for particular effective constituents of avermectin was engineered. Biochemical and genetic research has more recently been spurred on by genomic analysis tools.[28]

Combinatorial Materials Research

In the 1990s, materials scientists began to move away from natural product screening and toward screening very large libraries of synthetically generated novel organic compounds. Two new areas of technological systems development in particular enabled this trend, which was known as a "combinatorial" approach to the discovery of new materials: automated synthesis, and high-throughput screening. High-speed synthesis techniques were initially developed for electronic materials, and then were extended to organic molecules as well, including the automated production of a variety of amino acid sequences for pharmaceutical research. High-throughput screening systems, on the other hand, had their origins in chemical analysis machines developed for clinical laboratory testing in medicine. Subsequently, the pharmaceutical industry developed high-throughput systems to subject libraries created by combinatorial chemistry to various screening techniques, including optical screening (for example, using binding assays and fluorescently labeled receptors), mass spectrometry, gas and liquid chromatography, parallel calorimetry, and optical and NMR spectroscopy. By the 1990s, a number of high-throughput screening systems that could analyze hundreds of samples per day were available to laboratory researchers at affordable prices through the commercial instrumentation market. Since the "rational" design of materials based on theoretical prediction of properties appeared to be a distant dream, screening large libraries of chemicals seemed to promise the way forward.[21]

The expansion of laboratory instrumentation and methods in biomaterials research owed much to a general transformation of the research economy in the United States since the 1980s. In response to new challenges to American global economic dominance, particularly the economic growth of Japan as well as Germany and Western Europe, legislative changes including the Bayh–Dole Act of 1980 and the Economic Recovery Act of 1981 encouraged the patenting of research from academic laboratories as well as technology transfer to commercial applications, while universities courted industrial patronage and partnerships on an unprecedented scale. The development of biotechnology — the spread of recombinant DNA tools and then, in the new millennium, genomic sequencing techniques — was the most dramatic example of the resulting changes in science.[36 ch8; 17; 16] But the trend was not limited to biotechnology. A wide range of "small" sciences such as nanotechnology and microelectronics similarly saw their bench-top instrumentation revolution in this period, on the heels of (and in contrast to) the "big science" transformations exemplified by particle physics following World War II.[23 p7; 24] These changes across diverse fields manifested in shifts in biomaterials science and bioorganic synthesis, from research on nanoparticle and microelectromechanical drug delivery systems, to the systems described above for discovering new bioactive materials.

By the new millennium, it was clear that the conjunction of synthetic combinatorial chemistry and automated screening systems was not as fruitful for materials discovery as had been envisaged.[25 pxix] At the same time, techniques of *bioorganic* synthesis had *also* expanded and were used for library generation in combinatorial approaches. For example, where the genes that control the relevant enzymes for producing certain secondary metabolites were known, recombinant methods could then be used to alter the genes and therefore the enzymes, resulting in cells producing a series of novel metabolite variants that were structurally related to the original metabolite. In 2018, Frances Arnold, George Smith, and Gregory Winter received the Nobel Prize in Chemistry for their development of methods and tools for biosynthetic library generation and screening. Based on work they undertook in the late 1980s and early 1990s, Arnold mastered an efficient methodology of "directed evolution" to create DNA-sequence libraries and screen for enzyme variants with optimized desirable properties, and Smith and Winter contributed to the technology of "phage display" of peptides and antibodies respectively, which greatly simplified the screening process for these materials. According to press releases, Arnold initially attempted to design new enzymes rationally, that is by using theory to predict enzyme properties, but encountered difficulties "even with contemporary knowledge and computer power." In response, she "decided to abandon this 'somewhat arrogant approach' and instead found inspiration in nature's own method for optimising chemistry."[13 p2]

Genomic sequencing data for plants and microbes showed that many more genes for biosynthesis existed than researchers could have imagined from the metabolites so far identified, suggesting a vast world of unknown potential natural products.[30] Recombinant DNA methods made it possible to study productive microbes that could not be pure cultured in isolation from their ecosystem, by extracting from them the gene cluster controlling the biosynthetic pathway of interest and introducing it into a different host microbe. The sense of possibility broadened as interest in promising microbes shifted from bacteria to molds, with the reasoning that the materials that affect the bodily chemistry of higher organisms (such as humans) are more likely to be produced in high quantities by molds as regulatory substances in their own — similarly eukaryotic — physiology.[22] Hopes of discovery turned back to the natural world, driving the development of new tools for exploiting its diversity and variability, not just at the molecular but also the microbial level.

References

1. Benninga H. A history of lactic acid making: a chapter in the history of biotechnology. Dordrecht: Springer; 1990.
2. Bensaude-Vincent B. Discipline-building in synthetic biology. Studies in History and Philosophy of Biological and Biomedical Sciences. 2013;44(2): 122–29.
3. Bensaude-Vincent B. Chemists without borders. Isis. 2018;109(3):597–607.
4. Beppu T. Development of applied microbiology to modern biotechnology in Japan. Advances in Biochemical Engineering/Biotechnology. 2000;69:41–70.
5. Brock TD. The emergence of bacterial genetics. Cold Spring Harbor, NY: Cold Spring Harbor Laboratory Press; 1990.
6. Bud R. Innovators, deep fermentation and antibiotics: promoting applied science before and after the Second World War. Dynamis. 2011;31(2): 323–42.
7. Bud R. Penicillin: triumph and tragedy. Oxford: Oxford University Press; 2007.
8. Bud R. The uses of life: a history of biotechnology. Cambridge: Cambridge University Press; 1993.
9. Campbell WC. History of avermectin and ivermectin, with notes on the history of other macrocyclic lactone antiparasitic agents. Current Pharmaceutical Biotechnology. 2012;13(6):853–65.
10. Cham J. PHD comics: the search for new antibiotics [Internet]. Los Angeles, CA: Piled Higher and Deeper Publishing; 2013. Available from: http//phdcomics.com/comics/archive.php?comicid=1635.
11. Crump A. The Nobel Prize in Physiology or Medicine 2015: Satoshi Ōmura – biographical [Internet]. Stockholm: Nobel Foundation; 2015. Available from: https//www.nobelprize.org/nobel_prizes/medicine/laureates/2015/omura-bio.html.
12. Demain AL, Fang A. Natural functions of secondary metabolites. Advances in Biochemical Engineering/Biotechnology. 2000;69:1–39.
13. Fernholm A. The Nobel Prize in Chemistry 2018 popular science background: a (r)evolution in chemistry [Internet]. Stockholm: The Royal Swedish Academy of Sciences; 2018. Available from: https://www.nobelprize.org/uploads/2018/10/popular-chemistryprize2018.pdf.
14. Fukushima M. Between the laboratory and the policy process: research, scientific community, and administration in Japan's chemical biology. East Asian Science, Technology and Society: An International Journal. 2013;7(1):7–33.
15. Gaudillière JP, Gausemeier B. Molding national research systems: the introduction of penicillin to France and Germany. Osiris. 2005;20(1):180–202.
16. Hughes SS. Genentech: the beginnings of biotech. Chicago, IL: University of Chicago Press; 2011.

17. Johnson A. The end of pure science: science policy from Bayh-Dole to the NNI. In: Baird D, Nordmann A, Schummer J, editors. Discovering the nanoscale. Amsterdam: IOS Press; 2004. 217–30.
18. Lee V. Microbial transformations: the Japanese domestication of penicillin production, 1946-1951. Historical Studies in the Natural Sciences. 2018;48(4):441–74.
19. Lee V. Scaling up from the bench: fermentation tank. In: Bauer S, Rentetzi M, Schlünder M, editors. Boxes in action: a field guide. Manchester: Mattering Press; forthcoming.
20. Linse SS. Scientific background on the Nobel Prize in Chemistry 2018: directed evolution of enzymes and binding proteins [Internet]. Stockholm: The Royal Swedish Academy of Sciences; 2018. Available from: https://www.nobelprize.org/uploads/2018/10/advanced-chemistryprize-2018.pdf.
21. McFarland EW and Weinberg WH. Combinatorial approaches to materials discovery. Trends in Biotechnology. 1999;17(3):107–15.
22. Miwa K. Hakkō biseibutsu (Microorganisms for fermentation). In: Zaidan hōjin baioindasutorii kyōkai hakkō to taisha kenkyūkai, editor. Hakkō handobukku (Fermentation handbook). Tokyo: Kyōritsu Shuppan; 2001. 456–57.
23. Mody CCM. Instrumental community: probe microscopy and the path to nanotechnology. Cambridge MA: The MIT Press; 2011.
24. Mody CCM. The long arm of Moore's Law: microelectronics and American science. Cambridge, MA: The MIT Press; 2017.
25. Morrow GW. Bioorganic synthesis: an introduction. New York: Oxford University Press; 2016.
26. Murao S. Hakkō no rekishi (History of fermentation). In: Zaidan hōjin baioindasutorii kyōkai hakkō to taisha kenkyūkai, editor. Hakkō handobukku (Fermentation handbook). Tokyo: Kyōritsu Shuppan; 2001. 438–40.
27. Neushul P. Science, government, and the mass production of penicillin. Journal of the History of Medicine and Allied Sciences. 1993;48(4):371–95.
28. Ōmura S. Ebaamekuchin hakkō (Avermectin fermentation). In: Zaidan hōjin baioindasutorii kyōkai hakkō to taisha kenkyūkai, editor. Hakkō handobukku (Fermentation handbook). Tokyo: Kyōritsu Shuppan; 2001. 90–91.
29. Ōmura S. Philosophy of new drug discovery. Microbiological Reviews. 1986;50(3):259–79.
30. Osbourn A, Goss RJ, Carter GT, editors. Natural products: discourse, diversity, and design. Ames, IA; 2014.
31. Ratner BD et al., editors. Biomaterials science: an introduction to materials in medicine. Cambridge, MA: Academic Press; 2013.
32. Rooij Av. The company that changed itself: R and D and the transformations of DSM. Amsterdam: Amsterdam University Press; 2007.
33. Roosth S. Synthetic: how life got made. Chicago, IL: University of Chicago Press; 2017.

34. Santesmases MJ. Gender in research and industry: women in antibiotic factories in 1950s Spain. In: Ortiz-Gómez T, Santesmases MJ, editors. Gendered drugs and medicine: historical and socio-cultural perspectives. Farnham: Ashgate; 2014. 61–84.
35. Santesmases MJ, Gradmann C, editors. Circulation of antibiotics: historical reconstruction. Dynamis. 2011;31(2).
36. Wolfe AJ. Competing with the Soviets: science, technology, and the state in Cold War America. Baltimore, MD: The Johns Hopkins University Press; 2013.

CHAPTER 3.5

CHROMATOGRAPHY

APOSTOLOS GERONTAS

Chromatography was one of the most influential analytical techniques of the last century. Today it is hard to imagine a chemical laboratory that does not routinely apply it in one form or another. In contrast to the physical methods that infiltrated chemistry in the twentieth century, chromatography is native to chemistry. Yet its development has been discontinuous, with the result that it was relatively late to be canonized in the analytical laboratory. All through its development, a question mark hovered over the theoretical justification for chromatographic processes, until, eventually, the question became relatively insignificant due to the mounting successes of these processes in research.

Chromatography was created as a separation tool by Mikhail Semenovich Tswett (1872–1919) to tackle the issue of chlorophyll isolation. It was recognized and standardized as a chemical tool in the mid-1930s through the research of Richard Kuhn and Edgar Lederer on carotenoids. In other words, the then-new field of physiological chemistry, through traditional organic chemistry, generated modern chemistry; chromatography spread from organic chemistry into new fields such as radiochemistry, physical chemistry, and biochemistry.

The technique was revamped in the high-tech instrumentation era after the 1950s, evolving and branching out into a broad group of instruments and variations, the majority of which are adaptations of gas chromatography (GC) and high-performance liquid chromatography (HPLC). For these machines, an unending variation of technological accessories and specialized techniques have been developed through the decades, allowing them to enter new fields and, occasionally, to spur the creation of new specializations and communities in chemistry and materials research more generally.

The First Chromatography

Born in Asti, Italy, raised and educated in Switzerland, Mikhail Tswett moved to his paternal homeland, Russia, in 1896, only to discover that his scientific credentials from the University of Geneva were not recognized there.[1 p51; 2 p3; 22 p85] While earning a new set of degrees, he held a minor, unpaid post at the laboratory of the Russian Academy in St. Petersburg. There, he turned his attention to the extraction of plant pigments from leaves to a solution. As was well known, leaf pigments could be easily extracted with acetone and ethanol, whereas petroleum ether — which normally dissolves chlorophyll perfectly when it is already extracted — led to no extraction of a green pigment from leaves, only a yellowish one (today's "carotenoids").

Tswett reasoned that this phenomenon should be attributed to molecular forces that kept the pigments bonded to the leaf substrate and that these forces depended on the characteristics of individual molecules. That is, the strength of the bond had a higher value for one molecule and a significantly lower one for another. Tswett considered "adsorption" to be the basis of the molecular forces and decided to attempt to imitate the process using a substrate that would behave similarly to the tissue of the leaves, namely filter paper, since it also consisted primarily of cellulose.

After initial success with filter paper, Tswett proposed that it could be replaced by other more specific substrates — in powder form — leading to a totally new "physical method based on adsorption for the separation of substances."[2 p4] In 1901, Tswett moved to the University of Warsaw (then Russian). Over the next two years, he investigated a large number of potential substrates, leading him to the familiar tube structure of today's liquid-column chromatography. Adsorbents, ground to a powder, were placed in layers into a small filter funnel. A solution of chlorophyll and ligroin (an organic solvent) was introduced to the column under slight pressure, causing filtration to take place so that adsorption could be observed by the color changes — hence "chromatography." The name carries a double meaning; on one hand, it refers to the colors produced by the technique during its first applications in plant pigment research (although Tswett explicitly stated that chromatography would also work with non-pigmented substances). However, the name is also a pun: "Tswett" can mean "flower" or "color" in Russian, so "chromatography" can be also interpreted as "Tswett-ography."[7 p20–21] The system was simple and remarkably similar to modern apparatus. There was no need for elaborate chemical reactions, no need for heating and crystallization, and the system allowed the separation of all the leaf pigments in a sample at one go, without deterioration of their colors.

Yet despite his apparent success in separating pigments by physical means, Tswett faced serious problems of theoretical legitimacy, discontinuity of practice, and opposition from the bigger names in the plant pigments field. As a result, chromatography did not go far. During Tswett's time, it was reportedly used only by three persons. After his death in 1919, the method went dormant, to be revived (ironically) by Richard Kuhn, a former student of Richard Willstätter, one of Tswett's most renowned critics. Kuhn and his own doctoral student, Edgar Lederer, were researching carotenoids at the prestigious Kaiser Wilhelm Institut für Medizinische Forschung in Heidelberg. When Lederer joined the institute in September 1930, one of his first assignments was the separation of carotenoids from egg yolk for medicinal use.[24 p45–46; 4 p299; 5] The problem was difficult with traditional analytic methods because carotenoids are extremely sensitive to chemical manipulations.

The solution came from a copy of Tswett's book translated into German years before for Willstätter's personal use. Passed on to

Kuhn and then to Lederer, chromatography offered an immediately successful outcome: by December 1930, the carotenoids were successfully separated with no chemical manipulation whatsoever. The results appeared in three seminal papers, which could be considered the beginning of chromatography's new life.[17–19] After this re-introduction, the technique took off. It was re-introduced to deal with problems arising from organic chemistry research on biological materials, which were receiving increasing attention in the 1930s for many reasons, including their economic value. Even before Kuhn's and Lederer's research, organic chemists' attention had shifted toward nutrients and nutritional requirements of both laboratory animals and humans.

Chromatography's significance and entrenchment during the 1930s was a result and an accelerant of the research trends of the era. Chromatography soon showed its potential in non-biological, and indeed inorganic, fields of research too. It was used for research on radioactive substances during World War II and its initial liquid-and-tube form saw variations meant to adapt the technique for use in other fields. By 1941, chromatography was on the way to replacing a great portion of the previously dominant separation techniques.

The Chromatographic Instruments: Gas Chromatography

The German organic chemist Gerhard Hesse had proposed the idea that a gas instead of a liquid could be used as the mobile phase in chromatography in 1940. However, a paper presented by Archer John Peter Martin (1910–2002) and Richard Laurence Millington Synge (1914–1994) in 1941 is most commonly referenced as the starting point of the history of gas chromatography and of partition chromatography (as opposed to adsorption chromatography).[12; 13; 21] In partition chromatography, the solute is dissolved in a liquid phase, which is coated on a surface, as opposed to adsorption chromatography, in which the solute is adsorbed directly on the surface. Today, there are other types of chromatography as well, such as ion exchange chromatography. Since Martin and Synge also suggested some advantages of the use of gas as a carrier, they have been credited for foreseeing gas chromatography as well.

At about the same time, in Austria, Erika Cremer (1900–1996) was working out a method for determining the amounts of two gases in a mixture. Classical adsorption techniques had proven useless since the adsorption heats were too close in value, and, after reading Hesse's work, she decided to try a chromatographic process using an inert gas as the mobile phase. Her work was accepted for publication in 1945, but was not published because of the war. After the war, the work was taken further in collaboration with two of her graduate students and in 1952 her institute, the Physico-Chemical Institute of the University of Innsbruck, exhibited what must be considered the first real gas chromatograph at the world's largest chemical exposition held in Frankfurt am Main.[1 p268]

Developed in the postwar instrumental-industrial epoch, gas chromatographs represented something new in that they were among the first automated analytical instruments that could be used in non-specialist laboratories. Gas chromatography (GC) had the capacity to analyze a broad range of substances and was applicable on different scales of quantity and precision, from the miniscule to the mass industrial. In addition, by changing the columns, a GC apparatus could easily be adapted for preparative work (that is, as a tool for preparing a material for further manipulation or characterization). Perhaps even more important was that only minor adaptations were needed for the GC apparatus to perform all these functions.

Gas chromatography transcended disciplinary and sub-disciplinary boundaries. In one single apparatus, physical-chemical phenomena (partition of chemicals between two phases) were combined with chemical technology (the GC column), ultra-sensitive electronic sensors (detectors), gas pressure controllers, and advanced multiple temperature control. Eventually, electronics were introduced, followed by microprocessors and computers in the 1960s for instrument control, data collection, analysis and interpretation, and output generation.

The success of the gas chromatograph depended on more than the science of chemistry. Even if the theoretical foundations of all of the components of the gas chromatograph were not fully understood, they were sufficient to support the method's stone-upon-stone development. Yet new postwar techniques also stimulated chromatography's evolution from an empirical art into a theory-based method. As Leslie Ettre and Csaba Horváth, important actors in the history of chromatography, put it in retrospect, both "the rapid growth of this technique and the ease of obtaining quantitative results with the gas chromatographs provided strong stimuli for theoretical work."[3 p443A]

However, the popularity of gas chromatography also meant that individual practitioners knew less and less about the technique over time. The broad usefulness of the machine, its potential as a product in a more or less secure market, coupled with the possibilities of variations of technology that would act protectively against patent restrictions, attracted companies with a relative advantage in one or more of the technologies involved in the complete instrument. Thus, production was delegated to the industry, which made the machine available to a broad range of chemists. Industrialization also paved the way for standardization (of sizes and functions), which could not be reached through the alternative "do-it-yourself" strategy. Procuring instruments in this way is common in chemistry laboratories today, but until the 1950s it was an unexplored path — in part because it is accompanied by some inherent psychological distance between the researcher and their instrumentation.

Moreover "outsourcing" of construction to industry turned GC apparatus into commercial objects like every other, creating a vibrant market and giving companies motivation to compete for control of this market not only through improved technology and products, but also through aggressive commercial strategies. Patenting, exchanges of modular technology packages, service structures, advertisements, lobbying, and special relations with clients and policymakers all became characteristics of chromatography, and eventually driving factors both in the development and the dissemination of the relevant technology. Of course, all these developments occurred in parallel with an increase in the number of instruments sold, the number of suppliers of instruments, the number of technical variations, and the number of areas of research.[8; 9]

From Gas to High Performance Liquid

The most persistent shortcoming of GC was that not all analyzable substances can be readily vaporized, and even those that can cannot always be vaporized without losses, production of unwanted byproducts, or (more importantly) significant damage to their molecular structure. This holds true for organic macromolecules and smaller polar molecules, but it becomes crucial for biological substances, where maintaining the functional structure of a molecule is as important as isolating it. This is particularly true for proteins, which have a potential pharmaceutical value only in their functional form and in which a disruption of the folding structure leads to denaturation.

As mentioned earlier, chromatography as a technique that preserved molecular structures came in handy in (analytical) biochemistry in the 1930s. Interest in biochemistry, its substances, and its systems was sparked once again in the 1950s. The discovery of the structure of the DNA molecule in 1953, the introduction of antibiotics in mass production, the Green Revolution, as well as scandals and accidents involving the pharmaceutical and the agrochemical industries, offered incentives for the development of new techniques and instruments. Hunting down and investigating pharmaceutically useful proteins became a part of this, and with it a focus on protein structures in biomedical research.

The new focus on biomedical substances in general and protein structures in particular, and the newly comprehended potential economic significance of such research, elicited a speedy response from the chromatography community. The then-developing ion-exchange chromatography — which could separate ions and some polar molecules — offered the basis for the first liquid-chromatography instrument, the amino-acid analyzer of Stanford Moore, William H. Stein, and Darrel H. Spackman — a direct result of research funded, organized, and executed at the Rockefeller Institute for Medical Research. This machine was first described in 1958 and was in commercial use one year later.[23] Yet the amino-analyzer was still almost a decade removed from the possibility of an actual high-speed liquid chromatograph (LC) because it was too specialized to be characterized as a generic LC and because the appropriate technology to expand it into such an instrument was missing. Yet, this instrument did encourage researchers in academia and industry who were thinking of creating a mechanized LC. When it successfully entered mass production (and mass dispersion) it proved to the interested industrial players and the institutions cooperating with them that a lucrative market indeed existed.

The American chromatographist J. Calvin Giddings (1930–1996) made the first move in this direction. Working on the assumption that the limitation in LC is diffusion in the liquid mobile phase, Giddings came up with a series of suggestions for automated-LC-to-be in 1963.[10; 11] He concluded that a column LC with performance similar to the GC would need much smaller particles (2–20 μm in diameter) as column packing, combined with high pressure across the column body. The theory demanding great pressure drops along the columns was in place by the early 1960s, and the description of the packing materials, at least roughly, by the 1940s. However, there were no high-pressure pumps that would not overwhelm with noise and no packing material consistent with the theoretically predicted particle size.

To resolve the former problem, efforts were directed toward the design of so-called pulseless pumps or pulsation-dampening systems. In 1966, R. Jentoft and T. H. Gouw designed a pump that operated at pressures up to 1,000 pounds per square inch (psi) and flow rates up to 100 ml/min, with no detectable change in output pressure or output flow rate during its different cycles.[16] The most important and decisive steps towards the final automated LC however, happened when chromatography became embroiled in space research. In particular, the US Apollo program boosted chromatographic techniques in three distinct ways. First, a gas chromatograph for monitoring possible trace contaminants within the Apollo spacecraft was of the most urgent priority, and was indeed constructed and functioning by 1966.[15] A second effort focused on the possibility of in situ analysis of lunar soil — a project that led to the development of an ultra-portable lunar GC for this purpose. A third focus, more directly relevant to this narrative, was GC analysis of the moonrocks that were expected to be brought back to Earth for signs "of life, pre-life or post-life" (as Leslie Ettre noted in personal communication on February 14, 2010).

Prof. Seymour ("Sandy") R. Lipsky (1925–1986) became important for the latter two efforts. A medical doctor turned chromatographist, Lipsky made his name by achieving the first successful analyses of fatty acids with GC in the 1950s. An academic, businessman, and inventor, Lipsky, working in his Yale laboratory, received a substantial NASA grant in 1960 to develop a technique for moonrock analysis. Through informal encounters, he came to know and employ at his laboratory a promising Hungarian researcher, Csaba Horváth. The grant from NASA was generous, and Lipsky was trusted to the point of virtual independence from oversight by the agency, so it was possible to divert money toward an idea of Horváth's that was only vaguely connected to the targets of the grant.

Horváth had a background in biomedical research and had been disappointed by the general "primitive" state of liquid chromatography applied there. He had applied to the Harvard Medical School with a proposal for a new automated LC, but his application was rejected there (and everywhere else at Harvard). For the theoretical basis of this apparatus, he would adapt lessons from the kinetics of GC as they had been modified by Giddings in 1963. Practically, for the function of a liquid–solid chromatograph built with GC as a model, very high pressures were the first requirement. This first requirement had been easy to satisfy by the first model of his machine, built only a few months after the start of his employment at Yale.[6 p241–242]

The pellicular material that he had developed for his PhD thesis at the University of Frankfurt (under the supervision of another Hungarian sage of GC, Istvan Halász) proved useful and one more basic requirement for the existence for such an apparatus (namely, the solid material that would fill the column) was luckily readily satisfied. By the summer of 1965, Horváth had a full instrument which could have gone into production if not for the lack of critical individual parts. The machine was presented at the 6th Symposium for Gas Chromatography (held September 1966 in Rome, Italy).

Horváth named his machine HPLC, with the "P" initially standing for "pressure." Soon, it was changed to "performance." In subsequent years, Horváth explained this name change by stressing that the initial "high pressure" name gave the false impression that the only thing different from the old LC was the pressure. Instead, he said the new apparatus was characterized by "high performance" as a result of a series of factors in addition to high pressure (small diameter particles, column technology, accurate low volume injectors, sensitive low volume detectors, and good pumping systems). It is probable that the name change had more to do with the promotional needs of the new machine — and, in any case, this name change did not prevent later chromatographists from joking that in HPLC, the "P" in reality stands for "price." After all, the HPLC was for decades among the most expensive pieces of equipment for an analytical laboratory.

Horváth's colleagues referred to him until his death in 2004 as "Mr. HPLC," and his doctoral students (and their doctoral students) were members of a lineage of liquid chromatography experts active on both sides of the Atlantic. His contacts with industry only strengthened with time, to the point that his birthday celebrations were often organized by instrumentation companies. His laboratory at Yale also long remained one of the hubs of HPLC applications for biomedical studies.

Concluding Remarks

Chromatography has been one of the reforming powers for modern chemistry. It has played a major role in the shift toward separations based on physical processes and it has aided the construction and maintenance of the bridges that today connect several disciplines. Chromatography created common ways of viewing analytical practice, common vocabularies shared across disciplines, and common expectations among specialists about what a successful method of analysis should do.

These shared expectations have shaped modern analytical practice. Analysts have learned to expect direct results that are easily transferable to the next step of their analytical processes. Chromatography instrumentation after the 1950s reinforced this expectation. Furthermore, under the market pressures that shaped the instrumental revolution, chromatography and its applications were essentially, and probably irrevocably, liberated from the control of academic disciplinary boundaries. Chromatographic technology became modular. Users could absorb just the technological knowledge needed for their specific purposes, while ignoring any unnecessary super- or substructures. This allowed for faster transmission and dissemination of the technology; any user could achieve results with minimal training. In retrospect, this distancing of chromatography from its academic credentials had a positive effect by allowing a faster transfer of analytical knowledge while also sparing energy. Yet it also created theoretical gaps in the results generated by chromatographists. It further led to a redefinition of who counted as a chromatographist, with separation entrusted to lower-status technicians and sharply divided from the increasingly abstract task of analysis.

References

1. Ettre L. Chapters in the evolution of chromatography. Hinshaw JV, editor. London: Imperial College Press; 2008.
2. Ettre L. M.S. Tswett and the invention of chromatography. LC·GC Europe. 2003;16(9):2–7.
3. Ettre LS, Horváth C. Foundations of modern liquid chromatography. Analytical Chemistry. 1975;47:422A-46A.
4. Ettre LS. Professor Edgar Lederer 75 years old. Chromatographia. 1983;17(6):299.
5. Ettre LS. Richard Kuhn and the Chemical Institute: double bonds and biological mechanisms. Available from http://www.nobelprize.org/nobel_prizes/themes/medicine/states/richard-kuhn.html [Accessed October 1, 2012].

6. Gehrke CW, Wixom RL, Bayer E. Chromatography — a century of discovery 1900–2000. Amsterdam: Elsevier; 2001.
7. Gerontas A. Reforming separation: chromatography from liquid to gas to high performance liquid. PhD thesis; Norwegian University of Science and Technology; 2013.
8. Gerontas A. Creating new technologists of research in the 1960s: the case of reproduction of automated chromatography specialists and practitioners. Science and Education. 2014;23:1681–700.
9. Gerontas A. Chromatographs as epistemic things: communities around the extraction of material knowledge. Nantes: Cahiers François Viète; 2017.
10. Giddings JC. Analytical Chemistry 1963;35:439–49.
11. Giddings SC. A tribute to J. Calvin Giddings. Advances in Chromatography. 1998;38:iii–vii.
12. Hesse G, Tsachachotin B. Naturwissenschaften. 1942;30:387–92.
13. Hesse G. Adsorptionsmethoden im Chemischen Laboratorium. Berlin: De Gruyter; 1943.
14. Horváth C, Preiss BA, Lipsky SR. Analytical Chemistry 1967;39:1422–28.
15. Huebner VR, Eaton HG, Chaudet JH. A gas chromatograph for the Apollo spacecraft; Preliminary Report. J. of G.C. 1966;4(4):121–25.
16. Jentoft R, Gouw H. Pulseless high pressure pump for liquid chromatography. Analytical Chemistry. 1966;38(7):949–50.
17. Kuhn R, Lederer E. Zerlegung des Carotins in seine Komponenten. (Über das Vitamin des Wachstums, I. Mitteil) Ber. dtsch. Chem 1931;64:1349.
18. Kuhn R, Lederer E. Fraktionierung und isomerisierung des carotins. Naturwissenschaften. 1931;19:306.
19. Kuhn R, Winterstein A, Lederer E. Zur Kenntnis der Xantho- phylle. Hoppe-Seyler's Zeitschrift fur physiologische Chemie. 1931;197:141.
20. Lederer E, Lederer M. Chromatography: a review of principles and applications. Amsterdam: Elsevier; 1955.
21. Martin AJP, Synge RLM. A new form of chromatogram employing two liquid phases. Biochemical Journal. 1941;35:1358–68.
22. Senchenkova EM. Michael Tswett, the creator of chromatography. Moscow: Russian Academy of Sciences; 2003.
23. Spackman DH, Stein WH, Moore S. Automatic recording apparatus for use in the chromatography of amino acids. Analytical Chemistry. 1958;30: 1190–206.
24. Wixom WL, Gehrke CW, editors. Chromatography: a science of discovery. Hoboken, NJ: Wiley; 2010.

CHAPTER 3.6

LITHOGRAPHY

JOSEPH D. MARTIN AND CYRUS C. M. MODY

Lithography became an essential tool for materials research during the post–World War II computing revolution. Increasing computing power required shrinking circuits and packing transistors more tightly together. Lithography made it possible to write small, precise circuits on a semiconducting surface, setting the stage for modern computing and fueling Moore's Law — the observation that transistor density on chips has tended to double every eighteen months.[2; 12] But lithography was by no means a postwar development. It dates to the late-eighteenth century and is notable as a technique borrowed for materials research from the storied and ostensibly distant craft practices of ink-based printing. What ties these disparate applications together — aside from their name — is their close relationship to the commercial incentives of the times in which they developed.

"Lithography," literally "stone writing" when broken into its Greek roots, has become something of a misnomer. The name originated in the practice of rendering shapes on a flat limestone surface with a substance that would repel acid, so that soaking the stone in acid would leave the image in relief. A text or image prepared in this way could be reproduced effectively ad infinitum. The practice became a popular printing technique and artistic medium in the nineteenth century (Figure 3.6.1), shortly after the German playwright Alois Senefelder developed it in the 1790s. It then quickly migrated to non-lithic surfaces, such as metal. Lithography has since come

Fig. 3.6.1. Artists such as Eugène Delacroix were early adopters of lithography. He created this lithograph of Mephistopheles as a plate for a French translation of Goethe's *Faust* in 1828. *Credit*: Wikimedia Commons.

to encompass a collection of related techniques that involve creating patterns by changing the properties of some part of a chemically treated surface, whether on stone or not.

Senefelder invented lithography while experimenting with limestone as a replacement for expensive copper plates as an etching surface. According to his own account of the discovery, he scribbled a laundry list — for want of a piece of paper — on a stone surface he had just prepared for etching. This made him curious what would happen if he then treated the stone surface with acid. He discovered that his ink, composed of soap, wax, and lampblack, resisted the corrosive action of the acid enough to leave the inked portions of the stone in relief, providing a raised surface that was then easily re-inked for printing.

Senefelder had stumbled upon the concept of a *resist*, a substance that resists chemical action on a surface to which it is applied — just as his oily ink resisted his water-based acid. All subsequent lithographic techniques relied on identifying an appropriate resist material, the truly novel feature of Senefelder's process. Etching, a similar technique, was older, but it required creating grooves in a surface by carefully coordinating the mechanical application of acid with the lines and shapes the etcher wanted. Lithography achieved similar ends, while separating the chemical process from the design process, allowing more latitude for error and greater precision. That novel feature was important for Senefelder because of its commercial potential. He wrote of his discovery that "what was the most important for me, this method of printing was entirely new, and I might hope to obtain a franchise and even financial aid."[14 p9]

Commercial potential was similarly the catalyst for the importation of lithographic techniques into materials manufacture. The lithographic process makes clever use of the properties of materials, but it was not until the mid-twentieth century that its utility for industrial development made it an essential technique for making materials themselves. Those materials were, most prominently, semiconductor systems, which were central to electronics and computing technologies. Semiconductors offered tantalizing potential for commercial exploitation, which incentivized the development of new tools to produce them more effectively. We will consider three varieties of lithographic technique that emerged as a result: photolithography, electron beam lithography, and X-ray (and extreme ultraviolet) lithography.

Photolithography

Some substances change their chemical properties when exposed to light, a fact well known by the early twentieth century. It was, after all, the basis for photography. But early photography required cumbersome glass plates that were fragile and difficult to transport, which incentivized companies to experiment with new photographic media. This line of research opened up the possibility of photolithography, which uses a beam of light to create a pattern on a surface to which a photosensitive resist has been evenly applied. The light changes the properties of some parts of the resist, making it either more or less soluble. This allows it to be developed, leaving the pattern behind.

Historians regard the interwar period as the beginning of a golden age of American industrial research, in which corporate laboratories effectively linked fundamental research with patentable developments, often with the goal of improving materials.[4] This strategy paid off handsomely for Eastman Kodak. Decades of chemical research bore fruit in the 1950s and 1960s in the form of a plethora of patents on photoresists. Just as Senefelder's ink resist made it possible to etch letters or images on a stone surface, a photoresist made it possible to etch circuits on a wafer of a semiconductor such as silicon, known as a substrate, by exposing portions of the resist to ultraviolet (UV) light (Figure 3.6.2). Photoresists that were effective, easy to use, simple to manufacture, and, crucially, under patent, made photolithography commercially viable and encouraged its spread.

The late 1950s into the 1960s saw photolithography become a favored means by which to manufacture semiconductor devices. After a photoresist was applied to a doped semiconducting wafer, lithography could be used to create topographic patterns of hills and valleys on the wafer's surface. The acid and resist would then be washed off and new materials, such as silicon dioxide (an insulating material) or copper (a conductor), would be added to the surface in a new pattern of hills and valleys using evaporation-condensation or other processes. Then a new layer of photoresist would be applied and the process would begin again. After several (today, several dozen) such steps, a complex three-dimensional circuit made of insulating, conducting, and semiconducting regions would take shape. Transistor effects would take place in some of these regions, and the circuit as a whole would include many such transistors (today, often more than a billion). Because the transistor is a powerful amplifier or an efficient on-off switch, circuits containing many transistors are useful in digital communications, signals processing, and computing.

The transistor had been invented in 1947, about a decade before photolithography was used to create "integrated circuits" containing many transistors. Its potential to replace the vacuum tubes in giant, unreliable, slow, power-intensive digital computers such as ENIAC

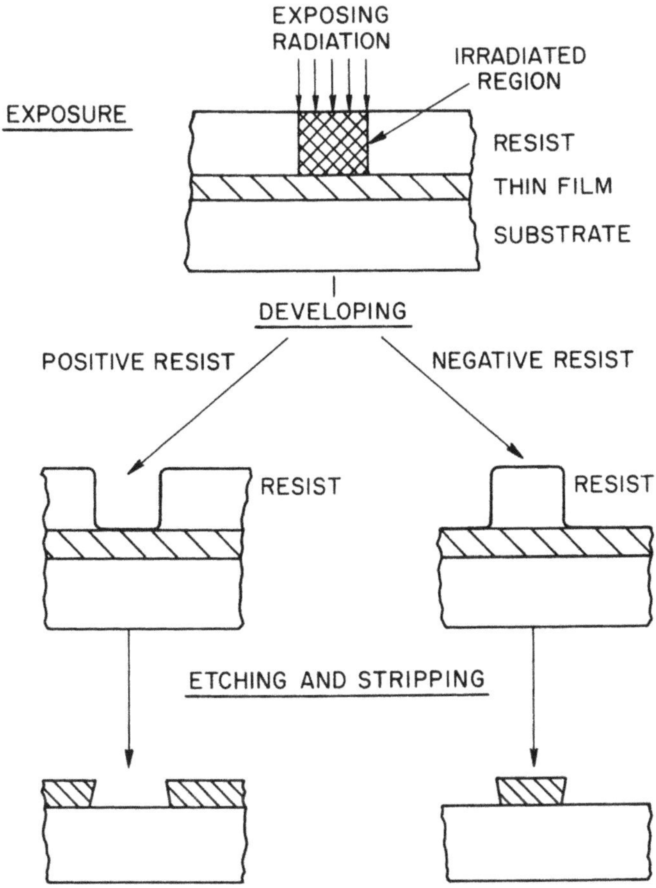

Fig. 3.6.2. The photolithographic process and the difference between positive and negative photoresists. The original caption reads: "The schematic diagram showing the formation of a polymeric relief image using lithography. The resist pattern is used to subsequently modify the underlining substrate."[17] *Credit*: American Chemical Society.

was immediately obvious. That potential was difficult to realize in practice, however, because early transistors were difficult to manufacture at industrial scales.[10] The ability to easily write transistors onto a surface using photolithography changed that. Because the transistors could now be capped by an inert "passivation layer," and because the connections between circuit components could now be written directly into the wafer (instead of soldering wires together), integrated circuits made with photolithography were much more reliable and lasted much longer than circuits made with vacuum tubes or discrete components. Reliability was paramount for the military customers who were the most important early customers for integrated circuits.

Almost from the first, though, manufacturers realized that integrated circuits manufactured using photolithography would also have tremendous commercial advantages for civilian markets. In particular, circuits written with photolithography could be made much smaller. Miniaturized circuits are faster and can be packaged into smaller devices such as today's mobile telephones. Smaller circuits are also cheaper, because it costs roughly the same amount of money to process a given area of silicon — so the more components that can go into that given area, the cheaper each component is.[9] Photolithography was therefore an essential prerequisite for the rise of Silicon Valley and the rapid increase in computing power through the second half of the twentieth century.[13]

But photolithography ran up against some fundamental limits. Typically, the procedure involved an opaque mask into which the desired circuit pattern had been cut. The mask would be placed above a semiconducting surface (usually silicon) prepared with a photoresist and exposed to a burst of UV light, which would impart the pattern to the resist for developing. But the wavelength of UV light placed a lower-bound of a few microns across on the size of the elements photolithography could produce. Higher resolutions — that is, smaller components and denser circuitry — would require other lithographic techniques.[12 p124–125; 17]

Electron Beam Lithography

By the 1960s, materials scientists had been putting electrons to work for some time in electron microscopes (see the contributions by Thomas Vogt and Pedro Ruiz-Castell in the next section). It is therefore unsurprising that electron beam lithography (EBL), which emerged in the 1960s, did so as a spin-off of electron microscopy.

EBL began, in fact, by turning a nuisance that plagued electron microscopy into an asset. Although electron beams offered unparalleled resolution for imaging samples, the electrons used to bombard those samples invariably contaminated them, changing their chemical properties and surface features. But whereas this is a problem if you are creating detailed images, it is precisely the point of a lithographic procedure.[18 p3–5] The story of how the bugbear of electron microscopy morphed into the basis for EBL takes us to Cambridge, England, and to the close university-industrial partnerships that grew after World War II.

Cambridge University in the 1960s was home to Charles Oatley (Figure 3.6.3), a pioneer of the scanning electron microscope. In addition to being a virtuoso of electron beams, he was a prodigious graduate advisor. His doctoral students, known as "Oatley's boys," were among the first to see commercial potential in the electron microscope, which they pursued through the Cambridge Instrument Company.[13 p6; 5] When Oatley's group began to pursue electron beams as a lithographic tool, therefore, a ready commercialization pathway was already in place.

Fig. 3.6.3. Sir Charles Oatley, front center, at his festschrift in 1994, backed by a number of his colleagues and former research students, left to right: Alec N. Broers; Bill Nixon; Fabian Pease; Tom Everhart; Dennis McMullan; Kenneth C. A. Smith; Oliver Wells; Chris Grigson; Gary Stewart; Philip Chang; Haroon Ahmed. *Credit*: Cambridge University.

EBL was attractive because it promised to bring circuit elements smaller than one micron — around the limit of UV photolithography — within reach, and therefore to continue the miniaturization of microchips. Through the 1960s, groups in Europe, Asia, and the United States proposed procedures that could repurpose the electron beams used in electron microscopes for lithography, but they were limited by the lack of a suitable resist — photoresists did not respond well to electron bombardment. The first breakthrough came in 1967, when a group at IBM developed poly-methyl methacrylate (PMMA), the first successful electron-beam resist.[17 p8] In EBL, the properties of the resist rather than the size of the electron set the resolution, since electrons, although very small, tend to scatter through any material they bombard. PMMA contained electron scattering sufficiently to allow circuit elements on the order of tens of nanometers.[1 p134]

But resist materials were not the only bottleneck. Drawing circuits accurately at the sub-micron scale required exacting control standards, which could only be reliably achieved by computerizing the lithograph's control systems. The collaboration between Oakley's group at Cambridge University and the Cambridge Instrument Company facilitated the development of the first computer-controlled commercial electron-beam lithograph in 1969.[17 p9; 19] This and later commercial EBL systems were expensive, especially when compared with photolithography systems, but the computing and telecommunications industries were willing to bear the expense in order to ensure that circuitry miniaturization proceeded apace. Two of Oatley's students, Fabian Pease at AT&T Bell Laboratories and Alec Broers at IBM, were particularly important in those two companies' in-house development of EBL systems used for chip manufacturing. AT&T's technology was then licensed to firms such as Perkin-Elmer, Etec, and Varian, which then made it available to the semiconductor industry.

Many proponents of EBL dreamed that it would replace optical lithography in mass production of the chips used in consumer products such as personal computers, telephones, and televisions.

The problem, though, is that EBL is a serial process, in which the beam writes only one part of the circuit pattern at a time, whereas optical lithography is a parallel process in which the pattern is written all at once. EBL is therefore much slower than optical lithography, and increasingly so as chips have become more complex. EBL is therefore only used to "direct-write" very specialized chips, for instance for military or space applications. Its more common niche today is in making the masks which are then used in optical lithography. In other words, the different varieties of lithography are partly competitors, but are also mutually reinforcing parts of a single lithography complex.

Extreme UV and X-ray Lithography

Another way to overcome the resolution limitations of UV photolithography was to use shorter-wavelength electromagnetic radiation. This approach would lead to the advent of extreme ultraviolet (EUV) and X-ray lithography (XRL) in the 1970s (ultraviolet and X-ray frequencies lie next to each other on the electromagnetic spectrum, so extreme ultraviolet lithography converges on "soft X-ray" lithography). The path-breaking XRL research was conducted at Lincoln Laboratory, a US government-funded defense lab associated with the Massachusetts Institute of Technology. In the heart of the Cold War, microelectronic components had become critical for the US military, which used them in aircraft, submarines, and missile guidance systems. The defense establishment therefore generously funded research into materials science, in the process doing a considerable amount to define the field as we know it.[11 p13–16] Indeed, defense patronage benefited materials research not just through the direct stimulation of military-relevant fields such as microelectronics, but also through indirect cross-fertilization from one military-sponsored field to another. Development of EUV, for instance, has borrowed heavily from laser technology, which the military sponsored for communications, anti-ballistic missile defense, and other applications. Naturally, once EUV and XRL moved beyond the proof-of-concept stage, they were also quickly adopted at sites like IBM and Bell Laboratories, which were always on the lookout for new ways to improve their microelectronics production processes.[16]

XRL, like EBL, required new resists. It also posed the additional challenge of developing masks that could withstand highly penetrating X-rays. But with these technical challenges solved, the technique offered several advantages. EBL was limited to negative resists, whereas XRL could be used with both positive and negative resists. Rather than relying on the precise computer control needed for EBL, XRL could be executed with a single, broad-area exposure.[15] It therefore developed alongside EBL and ion-beam lithography, a technique that, as the name implies, used ions in the place of electrons. Each technique had its advantages and drawbacks, and preferences for one over the others were often driven by local considerations, leading to what were known as the "beam wars."[17]

XRL would tie lithography, previously very much a lab-bench practice, to big science. It developed in parallel with synchrotron radiation sources. Synchrotron radiation occurs when charged particles are accelerated in curved trajectories, which causes them to give off radiation in the form of photons. The effect was problematic for high-energy particle accelerators, since the emitted radiation sapped beam energy. But it proved ideal for materials research applications (much like the damage caused by electron microscope beams ended up being ideal for EBL). Synchrotron sources could provide highly coherent X-ray beams — in which the photons in the beam all fall into a precise frequency range — making them ideal for lithography, where better coherence translated into finer control. In the 1970s, several countries dedicated new facilities producing synchrotron radiation, including the United States, which invested in the National Synchrotron Light Source at Brookhaven National Laboratory (Figure 3.6.4).[6]

X-ray lithography represents the transition of a mature technique into a new historical context. Robert P. Crease and Catherine Westfall have called the type of large-scale research that began to dominate in the 1980s and 1990s the New Big Science.[7] Like the big science projects of nuclear and high energy physics, synchrotron

Fig. 3.6.4. The National Synchrotron Light Source, one of many accelerator facilities optimized to produce highly coherent X-ray beams for materials research. *Credit*: Brookhaven National Laboratory.

sources were large, expensive installations, but they differed in a number of ways, including being designed for outside users rather than the laboratory staff. Industry therefore played a much more prominent role in the New Big Science, and XRL was one of industry's most prominent routes into it. In fact, in the 1980s, many observers, particularly in Germany and Japan, were hopeful that commercial synchrotrons would soon be built next to chip factories to facilitate mass-production XRL.[8]

Yet commercial synchrotrons have not become commonplace, and the optimistic predictions for EUV and XRL, just as for EBL, have not come true. Instead, optical lithography remains dominant, thanks to its continued improvement. The ability to use optical lithography to make commercial circuit components with features smaller than the diffraction limit of employed wavelengths is truly astounding, and quite surprising to many industrial and academic scientists and engineers who have staked their careers on the eventual demise of optical lithography. No single factor explains the surprising obduracy of optical lithography. Rather, optical lithography has survived because firms have been ingenious in coming up with incremental innovations to every single aspect of this technological complex. Photoresists, for instance, are no longer the simple lacquers of the early days. Instead, firms today use "chemically-amplified resists," initially developed at IBM in the 1990s, to sharpen the edges of the patterns cut into the wafer.[3] Materials for polishing wafers, interconnecting circuit components, packaging materials, and so on have all vastly improved. EBL, EUV, and XRL are therefore unlikely to replace photolithography, or indeed to become widespread for more than specialized applications, until their technological complexes are capable of similar rapid, incremental change along multiple dimensions simultaneously.

Conclusions

Lithography for semiconductor production developed at the confluence of a remarkable diversity of materials science techniques. It required the ability to generate and control photon, electron, and ion beams. It demanded the sensitive command of crystal growing necessary to produce substrates with the right properties. It depended upon vast chemical knowledge of resists, solvents, and washes. Moreover, mature systems were coordinated using computing technologies driven by use of the very microelectronics earlier systems had been used to manufacture. These ingredients were mixed in the crucible of commerce. Much like Alois Senefelder, who was motivated to pursue print lithography because he recognized its potential to attract investment and ultimately make money, the pioneers of semiconductor lithography proceeded with one eye squarely on the computing industry and its relentless push to miniaturize electronic components.

References

1. Broers AN, Hoole ACF, Ryan JM. Electron beam lithography — resolution limits. Microelectronic Engineering. 1996;32:131–42.
2. Brock DC, editor. Understanding Moore's law: four decades of innovation. Philadelphia: Chemical Heritage Foundation; 2006.
3. Brock DC. Patterning the world: the rise of chemically-amplified photoresists. Philadelphia: Chemical Heritage Foundation; 2009.
4. Cerveaux A. Taming the microworld: DuPont and the interwar rise of fundamental industrial research. Technology and Culture. 2013;54(2):262–88.
5. Cattermole MJG, Wolfe AF. Horace Darwin's shop: a history of the Cambridge Scientific Instrument Company, 1878–1968. Bristol: Adam Hilger; 1987.
6. Crease RP. The National Synchrotron Light Source, part I: bright idea. Physics in Perspective. 2007;10(4): 438–67.
7. Crease RP, Westfall C. The new big science. Physics Today 2016;65(5):30–36.
8. Heuberger A. X-ray lithography. Microelectronic Engineering 1986;5(1–4): 3–38.
9. Lathrop JW. The Diamond Ordinance Fuze Laboratory's photolithographic approach to microcircuits. IEEE Annals of the History of Computing. 2013;35(1):48–55.
10. Leslie SW. Blue collar science: bringing the transistor to life in the Lehigh Valley. Historical Studies in the Physical and Biological Sciences. 2001;32(1):71–113.
11. Martin JD. What's in a name change?: solid state physics, condensed matter physics and materials science. Physics in Perspective. 2015;17(1):3–32.
12. Mody CCM. The long arm of Moore's Law: microelectronics and American science. Cambridge, MA: MIT Press, 2017.
13. Schattenburg ML. History of the "three beams" conference, the birth of the information age and the era of lithography wars. 2007. Available from: http// eipbn.org/wp-content/uploads/2015/01/EIPBN_history.pdf. [Accessed June 28, 2017].
14. Senefelder A. The invention of lithography. Muller JW, trans. New York: Fuchs & Lang; 1911.
15. Smith HI, Spears DL, Bernard SE. X-ray lithography: a complementary technique to electron beam lithography. Journal of Vacuum Science and Technology. 1973;10(6):913–17.
16. Spiller E. Early history of X-ray lithography at IBM. IBM Journal of Research and Development. 1993;37(4):291–97.

17. Thompson LF. An introduction to lithography. In Thompson LF, Willson CG, Bowden MJ, editors. Introduction to microlithography. Washington, DC: American Chemical Society; 1983.
18. Utke I, Moshkalev S, Russell P. Nanofabrication using focused ion and electron beams: principles and applications. Oxford: Oxford University Press; 2012.
19. Wallman BA. From microscopy to lithography. In Breton BC, Dennis M, Smith KCA, editors. Advances in imaging and electron physics, vol. 133, Sir Charles Oatley and the scanning electron microscope, 350–86. Amsterdam: Elsevier; 2004.

CHAPTER 3.7

ION IMPLANTATION

CHRISTOPHE LÉCUYER

Ion implantation, the technique whereby one shoots charged particles (ions) at high velocity into solids in order to alter their physical, chemical or electrical properties, is widely used in the manufacture of semiconductors and the finishing of metals.[9] Alongside photolithography, it is essential for the fabrication of silicon-based integrated circuits. In the case of metals, it makes toughening them possible and it protects their surfaces. This technique originated in nuclear physics research.[6] Starting in the early and mid-1960s, it was adapted by nuclear physicists and semiconductor engineers to the manufacture of solar cells, resistors, transistors, and integrated circuits. A decade and a half later, technologists applied ion implantation to the production of cutting instruments, jet engine bearings, and medical devices. It is now employed in the production of virtually all microchips. It is also routinely used in the fabrication of high-precision devices and in consumer products such as razor blades.

Like many other tools and techniques in materials research and fabrication, ion implantation emerged out of a region-based community of practice. In the case of ion implantation, this community of practice was located in Massachusetts, especially in the region around Boston. The Boston area was an important center for nuclear physics research and instrumentation and an early hub for the semiconductor industry. The idea of shooting ions into semiconductor crystals to change their electrical conductivity did not originate in

Massachusetts, however. It was conceived by William Shockley, the co-inventor of the transistor and winner of the Nobel Prize in Physics, at the Bell Telephone Laboratories in New Jersey. In 1954, Shockley filed a patent application, in which he proposed to bombard semiconductor crystals with ions in order to place dopants with great precision inside wafers (dopants are atoms of arsenic, boron, or phosphorus that change the electrical characteristics of silicon wafers in the regions where they are inserted). In his patent application, Shockley also suggested heating crystals after implantation in order to repair damage induced by radiation. He regarded this technique as an alternative to solid-state diffusion for making semiconductor devices. Solid state diffusion was the process whereby dopant atoms migrated into crystals when placed in a high-temperature oven. Diffusion quickly established itself as the preferred technique for the manufacture of diodes and transistors and Bell Labs did not pursue Shockley's innovative idea.

But his concept was later realized by physicists and engineers working for High Voltage Engineering Corporation (HVEC) and Ion Physics Corporation (IPC), two firms located along Route 128 near Boston.[5] HVEC was a maker of accelerators for physics research. It had been established by Robert Van de Graaff and John Trump, two MIT faculty members, to design and fabricate electrostatic instruments for nuclear physics research. IPC was the firm's subsidiary specializing in the execution of R&D contracts for the federal government, especially the military services, such as an ion engine it made for the Army Ballistic Missile agency.

Responding to the industrial environment in which it was situated, IPC soon gravitated toward nuclear physics techniques for the production of semiconductors. Starting in the early 1960s, William King and other physicists at IPC implanted ions into silicon to make diodes, solar cells, and bipolar transistors. They also developed annealing techniques, where they placed the implanted wafers in high-temperature furnaces in order to restore their crystalline structure. This work was funded by the US Air Force. As IPC and its parent company ran into significant financial difficulties in the late 1960s, they spawned several spin-offs such as KeV, Extrion, and

Spire. These startups further enriched and deepened the expertise in ion implantation techniques and instruments in the Boston area. Another locus of innovation was Hughes Aircraft's central research laboratory in Southern California. There, in 1965, at the request of the firm's management that wanted to find commercial uses for their newly developed ion engines for spacecraft, Robert Bower, a physicist, devised a technique for making microchips with ion beams. As the examples of IPC and Hughes amply show, ion implantation was very much a product of the Cold War research and funding environment.

The pioneering work conducted at IPC and Hughes inspired a research group at Sprague Electric in western Massachusetts to investigate the use of ion implantation for making silicon transistors and integrated circuits. Sprague, a capacitor manufacturer, was a new entrant into microelectronics. It aimed to find alternatives to the standard processes of diffusion and photolithography that were well mastered by its competitors. In 1965, Kenneth Manchester, a physical chemist, and John Macdougall, a nuclear physicist, built an ion implanter and attempted to write microchip circuit patterns directly onto silicon wafers with ion beams (that is, without diffusion and photolithography). As it proved impractical to do so, they explored the ways in which ion implantation could supplement existing production processes.

Among other projects, Manchester and Macdougall explored how to make a new kind of transistor, the MOS (metal-oxide semiconductor) transistor, with ion beams.[3] Over the course of their research, they made two fundamental findings. By varying the quantity of ions implanted into the channel region of the transistor, they could change its threshold voltage (the voltage at which it switched on and off). This enabled the production of MOS transistors that required less power, had uniform characteristics, and matched the threshold voltage of bipolar transistors, the most widely used devices at the time. Macdougall also found that ion implantation permitted the fabrication of two different types of MOS transistors (depletion-mode and enhancement-mode transistors) on the same chip, which considerably accelerated the speed of the resulting circuit.

In 1969, Robert Palmer, an engineer working for Mostek, heard about these findings from Manchester and Macdougall. Mostek was a start-up focusing on MOS microchips that was financed by Sprague. It also produced its chips in Sprague's factory in Worcester, Massachusetts. Palmer understood the great technical and commercial potential of the new technology. In his view, it would enable the production of MOS integrated circuits with unheard-of characteristics. These chips would use little power, they would be fast, and they would be fully compatible with bipolar microchips. Soon, Palmer integrated ion implantation techniques into Mostek's manufacturing process to make high speed and low threshold-voltage MOS microchips. This bet on ion implantation established Mostek as a significant semiconductor manufacturer.[1] It became a leading supplier of memory and custom logic chips in the early and mid-1970s.

Mostek's technical achievements and its commercial success led other semiconductor firms to embrace ion implantation. They increasingly employed it to fabricate MOS chips and bipolar circuits. Four factors facilitated this rapid adoption. First, it became possible to buy ion implanters for semiconductor work on the open market. IPC sold ion implanters. Its spin-offs, KeV and Extrion, specialized in the design and manufacture of these machines. Other manufacturers also emerged, such as CGA and Accelerators, Inc. (Accelerators, Inc. commercialized Sprague and Mostek's implanter). The most successful of these firms, Extrion, sold more than five-hundred ion implanters in the 1970s. Second, consultants who had expertise in ion implantation helped semiconductor firms integrate the new technology into their production processes. For instance, after leaving Hughes in 1970, Bower built a thriving consulting business assisting corporations such as Intel, AMD, and Motorola in selecting ion implanters and establishing production lines with them. Third, several corporations started to offer ion implantation services to the semiconductor industry. Several of these firms were established by former employees of IPC and were located in the Boston area. A case in point is Spire, which offered implant services to microelectronics firms. It used Extrion's implanters to process microchips from Fairchild Semiconductor, General Instrument, and other customers.

Fourth, by the late 1960s, semiconductor engineers could rely on a solid body of scientific knowledge to perform their implants. An important site of ion implantation research was James Gibbons's laboratory at Stanford University.[4] In the mid- and late 1960s, Gibbons and his students calculated the penetration range statistics of implanted ions. Their tables were widely used in the microelectronics industry to precisely control the placement of ions into silicon substrates.

In the following decades, ion implantation gradually replaced solid-state diffusion for the doping of silicon crystals. An indication of the growing importance of this technique is the rapid increase in the number of implantation steps used in the manufacture of integrated circuits. In the mid-1970s, in most factories, ion implantation was used once or twice in the manufacture of microchips, the other doping steps being performed with solid-state diffusion. Twenty years later, microchips went through thirty to thirty-five implanting steps. Ion implantation offered key advantages over diffusion. It provided tight dose control (the dose being the number of ions implanted per square centimeter). It also gave excellent control of the vertical depths at which dopants were introduced into substrates. Thereby, it facilitated the formation of very thin layers with the desired ion concentrations within silicon crystals. Along with advances in photolithography, the growing reliance on ion implantation enabled the fabrication of smaller and smaller transistors and the extraordinary increase in the complexity of microchips. The densest chips of the mid-1970s, had 60,000 transistors; two decades later the densest chips had more than five million transistors. In other words, ion implantation became a crucial enabler of Moore's Law, the trend whereby the number of transistors per integrated circuit doubled every two years. Moore's Law has been the defining characteristic of semiconductor technology and industry since the 1970s.[7]

Ion implantation also became an important process for the treatment of metals. The introduction of techniques first developed for microelectronics into metal finishing was spearheaded by another Boston-based corporation, the Gillette Company. Gillette, the manufacturer of razors, was interested in producing blades that had harder edges and were more corrosion resistant. In the early 1970s, it commissioned a group of British engineers to perform research on the use of ion implantation for the treatment of cutting edges.[2] They found that the implantation of elements such as nitrogen, oxygen, and argon into razor blades made them harder, protected them against corrosion, and improved the adhesion of coatings on the blades. Going one step further, Gillette implanted the molds used to make the plastic components of razor blades. This manufacturing step strengthened the molds and extended their useful life. Following Gillette's lead, other firms based in Massachusetts moved into the new field. For example, in the early 1980s, Spire, which had previously sold implant services to semiconductor makers, expanded into metal treatment. Among its clients was Johnson & Johnson, for which Spire implanted artificial hip and knee joints with ions in order to make them more resistant to wear and corrosion. Another customer was the Navy, which wanted to make the bearings used in planes and helicopters impervious to saltwater.[8] Since then, ion implantation has been extensively employed by aerospace corporations, tool and die makers, medical device manufacturers, and the producers of cutting blades.

Ion implantation grew into a powerful and flexible materials fabrication technique. But it was also a hazardous one. The implanted elements came in a gaseous form. In the case of semiconductor implants, these gases were highly toxic. Some gases, such as arsine and phosphine, were so lethal that they had been used in trench warfare during World War I. Ion implanters also created significant radiation hazards. Finally, they were dangerous to service and repair. The main cause of deaths in the semiconductor industry in the 1970s and 1980s was the electrocution of technicians repairing ion implanters and sputterers. Accidents involving operator exposure to toxic gases were common. It took at least two decades for the manufacturers of ion implanters and the safety departments of semiconductor firms to tame ion implantation and reduce the dangers it posed to the people working with it.

From the early 1960s to the mid-1980s, ion implantation, a technique of nuclear physics research, became a critical manufacturing process used in a wide range of industries, including microelectronics, aerospace, medical devices, and consumer products. Much of this transformation occurred in Massachusetts, especially the Boston area, and it relied on close collaborations among nuclear physicists, solid state physicists, materials scientists, chemists, and electrical engineers. It is likely that without Cold War funding, ion implantation would have remained a nuclear physics technique. Pioneering research on the adaptation of ion implantation to semiconductor manufacturing occurred in military-oriented firms such as IPC and Hughes. But it was at Sprague and Mostek, two commercially oriented chipmakers, that ion implantation was established as a critical manufacturing process, as it allowed the production of advanced microchips that could not be fabricated in other ways. From that point on, ion implantation became widely used in microelectronics and it attracted the attention of other industries interested in improving the quality, durability, and reliability of their products. Research techniques have surprising lives indeed.

References

1. Brock D, Lécuyer C. Digital foundations: the making of silicon-gate manufacturing technology. Technology and Culture. 2012;53:561–97.
2. Dickson KE. The development of ion implantation technology in the UK semiconductor industry. Phys. Techno. 1985;16:171–76.
3. Fair R. History of some early developments in ion-implantation technology leading to silicon transistor manufacturing. Proceedings of the IEEE. 1998;86(1):111–37.
4. Lécuyer C. What do universities really owe industry?: the case of solid state electronics at Stanford. Minerva. 2005;43:51–71.
5. Lécuyer C, Brock C. From nuclear physics to semiconductor manufacturing: the making of ion implantation. History and Technology. 2009;25(3):193–217.
6. McKenna C. A personal history of ion implantation equipment for semiconductor applications. IEEE Conference on Ion Implantation Technology, 2000. 2000:1–7.
7. Mody CCM. The long arm of Moore's Law: microelectronics and American science. Cambridge MA: MIT Press; 2016.
8. Pollack A. Making metals wear longer. New York Times. May 8, 1986.
9. Yarling CB. History of industrial and commercial ion implantation, 1906–1978. Journal of Vacuum Science and Technology A. 2000;18:1746–50.

CHAPTER 3.8

POWER SOURCES

MATTHEW N. EISLER

Contemporary power-source research is interdisciplinary. It originates from several fields of physical science, including those traditionally associated with battery technology, notably chemistry and electrochemistry. Since the 1960s, solid-state materials science has become increasingly important to power-source technoscience, with important implications for how users employ tools and instruments in this field. In solid-state microelectronics research, the line separating devices and instruments from the materials that compose them is blurry.[2; 4] Measuring a quality of a device (such as the switching speed of a transistor) also measures the characteristics of the material that device is made from (such as carrier mobility in silicon that has been doped in a particular way). And the intertwining of devices and materials has important implications for knowledge production and technological innovation.

Material–device blurring is also present in solid-state power-source research. However, a single, standardized material does not dominate batteries as silicon dominates semiconductors. Nor has there been a trend in exponential scaling of components and performance in the manner of Moore's Law. In the twentieth-century United States, power source technoscience gradually declined in social importance after World War I as fossil-fueled internal combustion power dominated. After World War II, only federal institutions were willing to procure powerful new batteries, mainly for specialized military applications. In this period, the vanguard of power source technoscience shifted from the traditional field of electrochemistry to the new field of materials science.

The reason that conventional delineations of devices, instruments, and materials blur in materials research is that, at the nanoscale, the distinction between experimental conditions (including devices and instruments) and unknown or vaguely known material objects of investigation can collapse.[6; 7] As practitioners study materials in ever-smaller increments, instruments can provoke physical phenomena and themselves become objects of research, an effect known as "instrument knowledge."[5] In the power-source sector, the gulf between test and real-world environments has been especially large thanks in good measure to the historical contingencies of scaling in this field. Predicting how reactions and side reactions will unfold over time becomes increasingly difficult as electrodes and cells are clustered into batteries and integrated into applications, and that problem has in turn been greatly complicated by the longstanding estrangement of power source engineering from electronics engineering. The history of lithium-ion batteries illustrates the exigencies of instrument knowledge and the entanglements of material, device, and instrument in power-source research.

Material as Tool

In accounts of materials research, X-ray diffraction is often framed as a classic instrument, and the technology was no less vital in the discovery of compounds central to the lithium-ion battery.[1] For John B. Goodenough, a solid state physicist and inventor or co-inventor of several of the most important of these materials, X-ray diffraction was the "principal tool" for determining crystal structure in the early 1950s, at the outset of his career. Goodenough was motivated primarily by materials design and basic questions of solid state science. Nevertheless, he developed problems (or "engineering targets") issuing from devices, which became the basis of experiments he designed for chemists to execute.[3]

Some of the most important tools in the co-construction of materials science and modern electrochemistry derived from battery components and batteries. One such device, the sodium–sulfur battery, yielded important engineering targets. A product of the Ford Motor Company's Scientific Laboratory, invented in the late 1960s by Joseph T. Kummer and Neill Weber, the battery was impractical owing to its high operating temperature and volatile materials, which caused durability and safety problems.

Nevertheless, the device's components aroused considerable interest, particularly its ceramic electrolyte beta-alumina, which stimulated study of the reversible insertion of ions inside solids. Essentially, this was blue-sky materials research. Kummer and his colleague Y.-F. Yu Yao used a powder X-ray diffractometer to characterize the structure of beta-alumina and also employed a crystal scintillation detector to count the diffusion coefficient of sodium ions using radioactive isotopes of the diffusing ions. Goodenough credited his role in investigating the sodium-sulfur battery with helping bring about the convergence of solid state physics with electrochemistry, resulting in the new field of solid state ionics.

As electrodes, electrochemical cells, and batteries became both tool and object of solid state ionics, so did the toolkit of the electrochemist become relevant to solid state researchers. Foremost among this equipage was the potentiostat. A relatively simple bench-top instrument, the potentiostat was invented by the electrochemist Archie Hickling at University College, Leicester, in 1941 to measure and control voltage in an electrochemical cell. Together with the galvanostat, a variant used to control current, the potentiostat became a standard tool in fundamental studies of electrochemistry and electrophysiology, potentially allied fields that, like electronics and power sources, were socially alienated from one other.

In 1970s and 1980s, Goodenough's chief engineering targets were supplied by the lithium-titanium disulfide battery. Invented by the chemist M. Stanley Whittingham in the laboratories of Exxon Research and Engineering, this device represented another milestone in solid-state power-source technoscience. It operated on the principle of the insertion and extraction of lithium ions (a process known as intercalation) from layered sulfides. To understand the rate of lithium diffusion, Whittingham employed data from a colleague using nuclear magnetic resonance technology to measure the magnetic interaction between lithium ions and these host structures and the resulting change in the electronic properties of these structures. Like the sodium-sulfur battery, the titanium-disulfide battery proved impractical as a power source. Whittingham had focused on the cathode, pairing it with an interim test anode made of metallic lithium, a dangerous combination when the cell was subject to repeated recharging.

Nevertheless, titanium disulfide chemistry posed interesting research questions for Goodenough. Believing metal oxides offered superior performance as a host structure for lithium ions, he and a team of researchers proceeded to develop the lithium-cobalt oxide cathode. The project illustrates the convergence of the tools of solid state materials research, chemistry, and electrochemistry. Goodenough's team used X-ray photography and ethylenediaminetetraacetic acid (EDTA) to characterize a sample of the material. Then they tested a lithium-cobalt oxide cathode in a potentiostat set up in a vacuum glove box; to complete the circuit, the researchers initially paired the cathode to a "counter" electrode made of lithium metal. It yielded an astounding four volts. They used coulometry, a set of techniques of analytical chemistry to understand the transformation of matter in electrolysis reactions, determining that lithium ions could be reversibly removed from the cathode. And in order to confirm that lithium and cobalt had been ordered into the stable layers necessary to host lithium ions, the team employed X-ray powder diffractometry.

Goodenough's team did not invent a practical battery, since a suitable anode remained to be developed. From 1985, Sony's Energytec division worked on this problem, developing a safer graphite anode that enabled the company to commercialize the lithium-cobalt-oxide battery in 1991, initially for use in a cordless telephone. The results would revolutionize consumer electronics, but would also create new problems for practitioners that would require new tools, techniques, and instruments.

Batteries and Surface Science

Along with the nickel–metal hydride battery, the commercialization of the lithium-ion battery signaled a rebirth of electrochemistry. In the realm of consumer electronics, designers assumed that lithium cells only had to last as long as the devices they powered, which was only a few years. In electric vehicles, on the other hand, batteries had to last much longer and had to be manufactured to much higher quality standards, imposing new demands on diagnostic and production equipment. Fortunately, the commercialization of lithium-ion power coincided with rapid new developments in materials science instrument technology. From the late 1980s, many new instruments and techniques capable of engaging phenomenon *in situ*, such as probe microscopes and high-resolution transmission electron microscopes, became available.

Surface analysis instruments like X-ray photoelectron spectroscopy and infrared spectroscopy were particularly relevant to commercial lithium-ion batteries. They were crucial in characterizing a structure essential to battery lifetime known as the solid electrolyte interphase (SEI), a layer of decomposed solvents, salts, lithium ions, and impurities from the electrolyte that formed on the graphite anode mainly in the first charge cycle. Ideally, these materials completely coated the anode and provided electronic insulation, enabling ionic conductivity and intercalation while preventing decomposition of electrolyte and loss of lithium ions and power density.

In the years following the discovery of the SEI layer in 1979, researchers became aware that this structure was vital to lithium battery performance and expended much energy in developing charging procedures and additives that promoted its proper formation. In this and other efforts to understand and increase the lifespan of lithium ion batteries for electric vehicles, the federal government played an important role, notably through the Department of Energy's Advanced Technology Development program, launched in 1998. By the late 2000s, the question of how SEI layers formed in the five or so major lithium ion chemistries then on or nearing the marketplace had become of crucial significance to battery manufacturing.

If the revolution in materials science helped beget the revolution in power sources technoscience, the knowledge problems of the commercial lithium-ion battery had in some respects outpaced the capabilities of the tools of materials science. Especially nettlesome was the question of how to prove battery lifetime. The problem with empirical testing regimes was that they approximated neither the way electric vehicles were driven in real-world circumstances, nor battery behavior over time. Electrolyte and charged electrode materials constantly undergo parasitic reactions whether batteries are being cycled or not.

One instrument solution was the ultra-high precision charger, a device that uses high-rate cycling to "beat the clock" on problematic side reactions. Pioneered by Dalhousie and Kyoto universities, the technology was transferred to instrument manufacturers in the 2010s, a classic example of how academic research and development sometimes addressed innovation in niche fields that established industry had hitherto ignored.

On the other hand, there were limits to the ability of tools and instruments to solve problems in this context. Most *in situ* instruments were not well suited to characterizing how fully integrated batteries aged over time because they were invasive, aging and damaging the device in the act of investigating it.

Economic pressures, moreover, created technical problems that instrumentation alone could not resolve. Competition and narrowing margins led Japanese manufacturers of lithium-ion batteries to move production to South Korea and China in the early 2000s in operations that were initially characterized by extensive bugs and high cell-scrap rates.

Electric-vehicle battery manufacturing faced different sorts of economic pressures relating to the question of battery ownership, one that has been obscured by the US federal government's subsidization of battery and electric vehicle production. The question of how to reconcile the cost, reliability, and safety of lithium power for electric traction will be a key driver of solid-state ionics and electrochemistry, ensuring that this sector is an important locus of innovation of the instruments and tools of power source technoscience.

References

1. Bensaude-Vincent B. The construction of a discipline: materials science in the United States. Historical Studies in the Physical and Biological Sciences. 2001;31(2):223–48.
2. Brock DC, Lécuyer C. Digital foundations: the making of silicon gate manufacturing technology. Technology and Culture. 2012;53(3):561–97.
3. Goodenough JB. Rechargeable batteries: challenges old and new. Journal of Solid State Electrochemistry. 2012;16(6):2019–29.
4. Lécuyer C, Brock, DC. The materiality of microelectronics. History and Technology. 2006;22(3):301–25.
5. Marcovich A, Shinn T. How scientific research instruments change: a century of Nobel Prize physics instrumentation. Social Science Information 2017;56(3):348–74.
6. Mody CCM, Lynch M. Test objects and other epistemic things: a history of a nanoscale object. British Journal for the History of Science. 2009;43(3): 423–58.
7. Nordmann A. Molecular disjunctions. In Baird D, Nordmann A, Schummer J, editors. Discovering the nanoscale. Amsterdam: IOS Press; 2004, 51–62.

CHAPTER 3.9

MOLECULAR BEAM EPITAXY

BRIAN TYRRELL

John R. Arthur Jr., then a research scientist at Oregon State University interested in surface physics, remembered listening to a radio program about nanotechnology in the early 2000s. The program focused on the fantastical possibilities of science at the small scale. Applications such as invisibility and space elevators dominated the discussion. Visions of the "flamboyant future" powered by nanotechnology left Arthur feeling "impatient." In all the buzz about nanotechnology, it seemed nobody talked about the workaday breakthroughs in fields like surface science. Broadly defined, nanotechnology refers to the precise control of matter at the near-atomic scale. Researchers use nanotechnology to create devices or novel structures. Under the auspices of the National Nanotechnology Initiative, the US government has invested around $21 billion in nanotechnology research since 2001. Arthur can be forgiven his impatience. As a pioneer of molecular beam epitaxy (MBE), Arthur had been building nanostructures "one atom at a time" for more than thirty years. He was a nanotechnologist *avant la lettre*.[1 p190]

Molecular beam epitaxy refers to a process for growing thin films of a variety of materials ranging from "oxides to semiconductors to metals" in a manner which the *New York Times* likened to "spray painting ... with atoms."[1 p189; 4] Telecommunications companies applied the process to build commercially successful compound semiconductors, and semiconductor fabrication still dominates MBE.

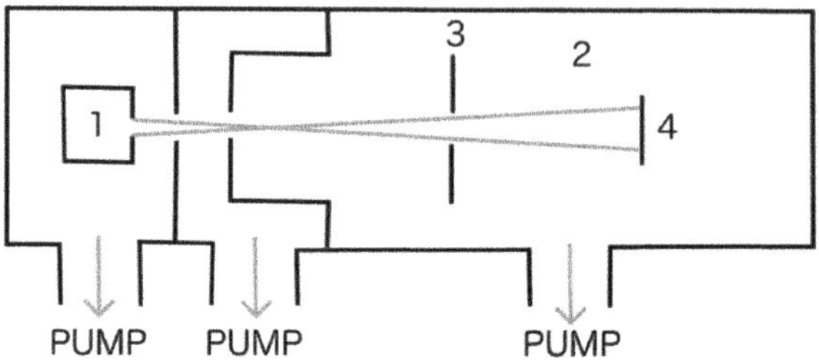

Fig. 3.9.1. This image depicts the basic apparatus used to form a molecular beam. An oven[1] vaporizes a material and the gas molecules effuse through a small aperture into an isolating chamber. The "beam" travels through the isolation chamber into the observation chamber.[2] The observation chamber contains another aperture,[3] which defines the beam toward a detector.[9 p38; 4]

The invention of the transistor and the dawn of the computer age placed a premium on precise construction of semiconductor crystals, and MBE proved to be an effective tool for doing so.

To continue with the *Times*'s spray paint analogy, MBE (Figure 3.9.1) brings a molecular "beam" (the paint) in contact with a "substrate" (the wall). In the first step of the process, pure samples of beam materials are vaporized in separate ovens. The atoms then travel in a focused pack — hence, a "beam" — through an ultra-high vacuum chamber toward the slightly heated crystal substrate. The ultra-high vacuum environment minimizes contamination of the growing surface of the substrate. Beam atoms and molecules travel nearly collision free until they reach either the growing surface or the chamber's chilled walls, at which point the atoms condense.[1]

This process continues, atomic layer by atomic layer, until the desired structures are obtained. Building crystalline structures by MBE is painstakingly slow. Material accumulates at a rate of less than half a micron or so per minute. Such a pace, combined with the high vacuum and purity requirements, results in precision control of matter. After all, the word "epitaxy" comes from the Greek words for "above" (*epi*) and "in an ordered manner" (*taxis*).[8 p259]

In the early 1950s, researchers from several scientific disciplines, including solid state physics, surface physics, and materials science, made contributions that would lead to MBE. Before MBE was used to grow semiconductor crystals, atomic and molecular beams had already been employed in the development of nuclear magnetic resonance technology and the maser. Indeed, by 1955 beam research had developed so far that Kenneth F. Smith published a monograph, *Molecular Beams*, to appraise and catalogue the field.[9 p37] The first step toward MBE occurred in 1954, when Herbert Kroemer, two years removed from receiving his doctorate at the University of Göttingen and working at the Central Telecommunications Laboratory of the German Postal Service, suggested a "non-stoichiometric mixed crystal of different semiconductors with different energy gaps."[7] Working later at American firms, Kroemer theorized about the benefits of using heterostructures, structures made of two different semiconducting materials. But a process for building heterostructures eluded researchers.[8 p259]

The first step toward beam-built crystals came about a decade after Kroemer's theoretical intervention. Between 1966 and 1968, Bruce Joyce and Bob Bradley of the Plessey Allen Clarke Research Centre in Northamptonshire, England, published a series of papers on making silicon thin films. In 1968, Joyce outlined their progress in a paper he delivered at the Second International Conference on Crystal Growth (ICCG2) in Birmingham, England. Yet at that point, MBE had no viable commercial application. Although the speed, or lack thereof, of deposition enabled greater control over crystal growth, the pace at which Joyce and Bradley's structures assembled, about one micron per week, was simply too slow.[9 p47; 50]

Much of the research that made MBE commercially viable occurred in the private laboratories of US-based communications companies. The Bell Telephone Company established their laboratory in Murray Hill, New Jersey, in 1941. At Bell Labs, researchers applied basic scientific knowledge to the growing communications industry. In 1947, researchers at Bell Labs invented the transistor.[9 p58] International Business Machines Corporation (IBM) started their corporate research laboratory at Columbia University in 1945, and later expanded to sites in nearby Westchester County. In 1961, the firm opened the Eero Saarinen–designed Thomas J. Watson Research Center in Yorktown Heights, New York. Semiconductor research was always an important area at both Murray Hill and Yorktown. Thus, both IBM and Bell Laboratories took up the challenge of building semiconductor heterostructures, leading to a series of advances between 1968 and 1973. John Arthur, whose reaction to nanotechnology began this piece, made particular progress, as laid out in a 1968 paper that described the epitaxial growth of gallium arsenide (GaAs). With its wide band gap and high electron mobility, Arthur's film impressed upper management at Bell Labs and IBM Research who thought the structure was a good candidate for high-speed electro-optical devices.[2]

For MBE to develop, researchers needed to focus as much attention on instrumental engineering as they did on basic science. For expertise in that area, Arthur relied on Alfred Y. Cho (Figure 3.9.2), a Beijing-born scientist who accepted a position at Bell Labs in 1968. Before coming to Bell, Cho had researched ion-beam propulsion, which helps explain why all his "MBE systems all look like ion engines."[3] The equipment that Cho and Arthur developed produced

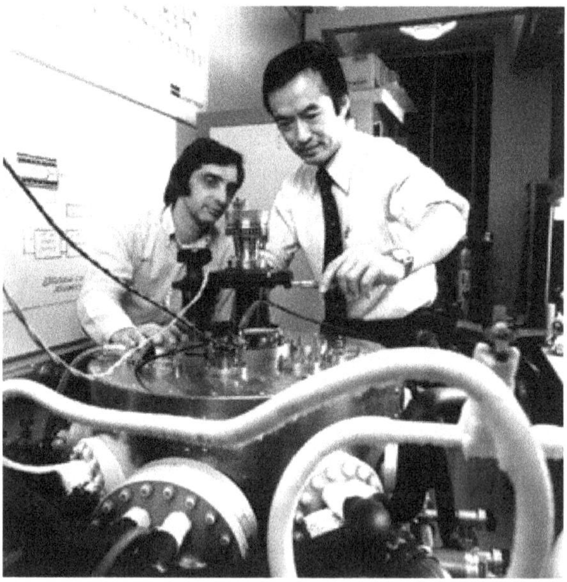

Fig. 3.9.2. Alfred Cho (right) and Charles Radice at work on an early MBE machine at Bell Labs in the 1970s. Image: Bell Labs. *Credit*: Reused with permission of Nokia Corporation.

only a few hundred angstroms of GaAs per day. Although the layers were too thin to accurately characterize, Arthur and Cho's work inspired increased interest in molecular beam crystal growth.[8 p260]

Sixty miles away at IBM's Yorktown Heights lab, Leo Esaki and Raphael Tsu built on Arthur and Cho's work by proposing, in 1970, that epitaxial methods could be used to construct a "superlattice." In a theoretical paper published while the pair worked at IBM, Esaki and Tsu speculated that by alternating ultra-thin layers of AlGaAs and GaAs they could produce a superlattice with novel features. As electrons approached the boundary of the superlattice mini-zone, they would encounter a region of negative resistance and electrons would slow. The structures that Esaki and Tsu postulated had complex features — they were far more complicated than the heterostructures proposed by Kroemer. Within their superlattices, Esaki and Tsu predicted that they could observe previously unknown physical phenomena which would provide "a valuable area of investigation in the field of semiconductors."[6] Years later, Esaki described their work as "do-it-yourself quantum mechanics."[8 p260] Initially Esaki's group tried to grow superlattices using liquid phase epitaxy (LPE), but the method proved lacking. Esaki's superlattices needed layers of thickness measured in angstroms instead of microns.[9 p64]

A Primer on Solid State Physics

To understand the mechanics of molecular beam epitaxy, a primer on solid state physics is in order.

Molecular beam epitaxy is used widely in the construction of semiconductor devices, including transistors. Semiconductors are substances whose electrical properties are intermediate between those of insulators, such as rubber, and conductors, like copper. A substance's conductivity — the ease at which an electric charge travels through it — is largely determined by its electronic band structure. The "band gap" refers to the energy differences between the top of the valence band and the bottom of the conduction band. Electrons can jump from one band to another, but this requires a specific minimum amount of energy.

Engineers use semiconductors in electric circuits because they can fine-tune semiconductors' electrical properties by introducing impurities to the semiconductors' crystal structures. Such impurities change the band gaps. By adding small amounts of new substances to a semiconductor, engineers can control the flow of electrons. This process is called "doping," and engineers create diodes by arranging doped semi-conductors in such a way that charges will flow in one direction but not the other. These properties make semiconductors, like silicon, especially useful in modern electronics.

Engineers often use superlattices, periodic structures consisting of layers of two or more materials. Superlattices were discovered in the 1920s using X-ray diffraction techniques. Superlattices made of two semiconductor materials with different band gaps, affect the way charges flow through the structure.

While writing the 1970 paper, Esaki began working with Leroy Chang to build their own MBE machine. They had learned about the technique from Cho, and they believed the new technique might be appropriate for the level of control they needed.[9 p64] Esaki's winning of the 1973 Nobel Prize in Physics probably convinced IBM to provide financial support for his research. The Nobel's persuasive force was needed, at least initially, because ultra-high vacuum environments and pure elemental sources made MBE research expensive. Cho later recalled that coworkers joked that the acronym MBE really stood for "mega buck equipment." Another factor that may have influenced Esaki's managers was the competition between IBM Research and Bell Labs. Despite the costs, both IBM and Bell funded MBE groups who raced each other to improve MBE technology and to produce new materials.

By the mid-1970s, though, MBE had spread widely enough that a community of researchers formed. Even Herbert Kroemer — who had vowed in his Nobel lecture to research epitaxial growth if it ever became practicable — began experimenting with MBE after the US Army funded the purchase of a machine for his lab at the University of California, Santa Barbara, in 1976.[8 p260] More commonly, though,

MBE attracted young researchers like physicist Arthur C. Gossard, who was in his mid-30s when MBE caught on in the 1970s. For his research on semiconductors, Gossard needed a method to build samples. Despite his manager's protests, this need drew him toward MBE. Like most before him, Gossard's group struggled with technological obstacles in getting MBE to work. Yet despite their slow start, Gossard's group eventually developed a technique known as "modulation doping," with which they could construct nanostructures that varied from layer to layer. By switching between doped and undoped layers, researchers tuned their structures' electrical properties.[5]

All of the MBE research conducted before 1975 used jury-rigged homemade equipment. Most researchers paired a vacuum system with quadruple mass spectrometers, but the individual MBE apparatus varied from laboratory group to laboratory group. At Bell Labs, for instance, John Arthur and Alfred Cho joined Morton Panish's group — according to Arthur, Panish coined the name "molecular beam epitaxy" — where they developed the MBE process. Arthur constructed a stainless steel ultra-high vacuum system for the group. Firms like Varian in the United States, Vacuum Generators in the United Kingdom, and Riber in France, sold ultra-high vacuum equipment for surface researchers, and researchers using molecular beam epitaxy paired these vacuum systems with monitoring devices. When Cho joined the Panish group in 1968, he brought experience building the "first quadrupole mass spectrometer."[9 p58–67]

By the 1980s, MBE had developed into a proper research community, which was no longer dominated by the premier corporate laboratories. The *New York Times* reported in 1982 that more than one hundred groups globally, including labs at Massachusetts Institute of Technology, Cornell, Illinois, and the Army's research laboratories at Fort Monmouth, NJ, engaged in MBE research.[4] By that point, MBE researchers had organized a biannual conference series where they could share their findings. What is more, as the community grew, researchers no longer needed to build their own MBE systems. By the early 1980s, Riber offered one of the first off-the-shelf MBE systems. The proliferation of MBE machines helped move the

Fig. 3.9.3. A researcher stands next to an MBE system in 2013. *Credit*: US Department of Energy.

process out of the laboratory and into the mainstream. MBE has even achieved a role in industrial manufacturing: for instance, MBE machines (of the type seen in Figure 3.9.3) helped produce the lasers used in compact disc readers.[8 p261]

The more visible contribution of the instrumental community that formed around MBE, though, has been new scientific knowledge. In October 1981, Horst Störmer and Daniel Tsui, physicists working at Bell Labs, began characterizing samples Gossard's team fabricated with his MBE machine. The pair discovered the fractional quantum Hall effect for which they shared the 1998 Nobel Prize. In their Nobel lectures, both physicists acknowledged the role of Gossard's instrumentation. Störmer even called "creators of materials ... the true heroes of our trade."[10]

References

1. Arthur JR. Molecular beam epitaxy. Surface Science. 2002;500:189–217.
2. Arthur Jr. JR. Interaction of Ga and As_2 molecular beams with GaAs surfaces. Journal of Applied Physics. 1968;39:4032–34.
3. Bell TE, Cho AY. [biography]. IEEE Spectrum. 1994;31:70–73.
4. Boffey PM. Atom by atom, physicists create matter that nature has never known before. New York Times. June 1, 1982.
5. Dingle R, Gossard AC, Wiegmann W. Direct observation of superlattice formation in a semiconductor heterostruture. Physical Review Letters. 1975;34:1327–29.
6. Esaki L, Tsu R. Superlattice and negative differential conductivity in semiconductors. IBM Journal of Research Development. 1970;14:61–65.
7. Kroemer H. Nobel lecture, NobelPrize.org, December 8, 2000. http://nobelprize.org/nobel_prizes/physics/laureates/2000/koremer-lecture.html [Accessed April 3, 2019].
8. McCray WP. MBE deserves a place in the history books. Nature Nanotechnology. 2007;2:259–61.
9. Orton J, Foxon T. Molecular beam epitaxy: a short history. New York: Oxford University Press; 2015.
10. Störmer H. Nobel lecture, NobelPrize.org, December 8, 2000. http://nobelprize.org/nobel_prizes/physics/laureates/1998/stormer-lecture.html [Accessed April 3, 2019].

CHAPTER 3.10

VAPOR DEPOSITION

JORIJN VAN DUIJN

Vapor deposition entails the growth of solid materials on the top of a surface. A vapor is directed over the surface, called a substrate, that needs to be coated. The substrate can be anything, from the glass of a lightbulb to a silicon wafer. The vapor bonds with the substrate and forms a thin film, the thickness and properties of which depend on the duration of the process and the quantity of processed gasses. In essence, it is a chemical reaction, which results in a solid and strong thin film of material.

It becomes more interesting and complex when we look at the technique's applications. Vapor deposition comes in numerous varieties. The application of the grown material determines the time the procedure takes, the temperature at which it is conducted, the tolerances required to ensure the desired quality, and which variant of the deposition process is used. For instance, physical vapor deposition (PVD) — also known as sputtering — allows the growth of a film independent of whether it reacts with the substrate or other processing gasses. However, compared to chemical vapor deposition (CVD), the coverage might be less optimal on an uneven or trenched surface. With CVD, almost any material can be deposited, depending on the tolerances that the desired application permits. The realm of CVD contains quite a few variations, which will be further explored in this exposition.

A material grown by means of chemical vapor deposition is the consequence of a chemical reaction between the incoming vapors (gases) and the substrate. By altering the environment in the reactor chamber, all sorts of materials can be grown. Multiple variations are possible, each with its own characteristics. These properties determine the suitability for the growth of a particular material on a particular substrate. A few variations include: atmospheric pressure (APCVD), low pressure CVD (LPCVD), atomic-layer deposition (ALD), plasma enhanced CVD (PECVD), and photo induced CVD (PICVD). By varying the pressure, the gas flow changes. This can have consequences for the growth rate of the material. By activating the vapors by means of a plasma, less thermal energy is required, which enables the growth of films at lower temperatures. Sequencing the influx of gasses enables the growth of films atom by atom.

An Illustration of Vapor Deposition: Plasma Enhanced CVD

I will illustrate the domain of vapor deposition by focusing on one particular example: the main plasma deposition furnace that was used in semiconductor manufacturing from the late 1970s through the mid-1980s in all major semiconductor markets worldwide. Though the tool was principally intended for high volume manufacturing — which poses different requirements for the machine than pure laboratory research — it was also used for research and development. Research using this machine was aimed at expanding the applicability of the process to new materials in high-volume manufacturing. The history of this plasma CVD machine nicely encapsulates the characteristics of a vapor-deposition process, its application, and configuration. Multiple components of vapor deposition come together in the plasma CVD machine, like the importance of pressure, temperature, alternative activation methods, the influx of gases, and the positioning of the substrates.

The machine in question is called the Plasma I (Figure 3.10.1). It was developed by George Engle in cooperation with two companies, Pacific Western Inc. and ASM America — the latter a subsidiary

Fig. 3.10.1. ASM America Plasma I — Horizontal hot walled PECVD reactor. The load section where the substrates are placed in the boat is on the left. The closed part on the right is the actual furnace. The far right of the machine is the controller and computer.

of ASM International.[1;2] Engle owned the intellectual property of the plasma deposition furnace and was therefore free to move from Pacific Western to ASM in 1979. As a consequence, both companies offered basically the same technology, though the systems differed in throughput, gas control, and automation. For semiconductor applications, plasma CVD was originally discovered by engineers Richard Swann and Henley Sterling in 1964, at the Standard Telecommunication Laboratory, a British subsidiary of International Telephone and Telegraph Corp.[3] Texas Instruments was the first to use a plasma CVD reactor for chip manufacturing internally since 1971.[5] A variation of this configuration was commercialized by Applied Materials in 1976, merely reverting the gas flow of the Texas Instruments machine.[6;8] The Texas Instruments variation opened the application of plasma CVD processes in chip production, but was not of sufficient quality to arouse interest in the wider industry. In the end, the variant developed at ASM America dominated the market for plasma CVD until the late 1980s.

In semiconductor manufacturing, plasma CVD is used to grow materials at low temperatures.[3–6; 8] In regular CVD furnaces, thermal energy determines the gas flow and deposition rate. This means that the higher the temperature, the higher the velocity of the vapors — depending on the materiality of the gases. However, high temperatures are not always an option. For instance, an integrated circuit — a computer chip — consists of multiple layers comprising the semiconductor device (transistors, diodes, and other components), interconnections, and dielectrics. When the final films are grown on top of the integrated circuit — in particular the passivation layer — the temperature during the growing process cannot be too high. A high temperature affects the structures below the final film and threatens to undo previous work. Manufacturing computer chips is a commercial business, which means that components that are damaged due to high temperatures in the manufacturing process undermine return on investment. Therefore, the final layer needs to be grown at the lowest temperatures possible, while simultaneously maintaining quality and throughput of the system. Using plasma-enhanced CVD, this can be achieved. The plasma, rather than thermal energy, activates the gases. When igniting a plasma, the temperature is no longer crucial. It might even be decreased from the already relatively low temperatures (for thermal deposition) of 400°C to room temperature, depending on the specifications of the user.

A plasma CVD machine incorporates various components designed to help it achieve the described process (Figure 3.10.2). The Plasma I consisted of a horizontal tube made of quartz around which a heating element was wound. This formed the reactor chamber. Vapors flowed from one end of the tube to the other. One end of the tube could be opened to load a boat containing the substrates — in case of the Plasma I, these substrates were silicon wafers. After closure, the gases were inserted through the door. The other end of the tube contained a vacuum pump, filters, and an exhaust through

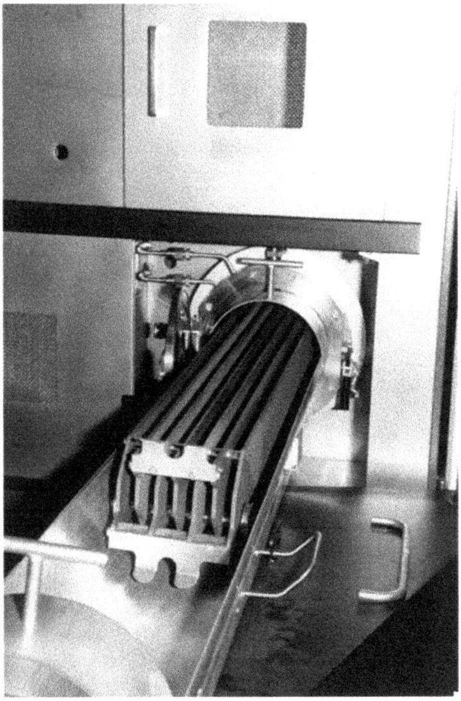

Fig. 3.10.2. ASM America Plasma I — Horizontal Hot Wall PECVD Furnace [photograph]. *Credit*: ASM America (US); 1980.

which the vapors left the reactor. The boat containing the substrates consisted of vertical plates, which were positioned lengthwise and in parallel to each other through the tube (Figure 3.10.3). The plates held the substrates vertically. Each plate had an electrode used to ignite a plasma in between them.

At the start of the deposition process, the heated chamber was purged by nitride or another gas in order to enable a vacuum, removing all oxides and other undesired elements. Next, the plasma was ignited and the reactant gases were inserted in the chamber. To prevent an immediate reaction occurring everywhere except on the desired substrate, the vapors were diluted with a so-called carrier gas. The flow of gases continued until the film had the desired thickness, an assessment that must be made by skilled operators on the basis

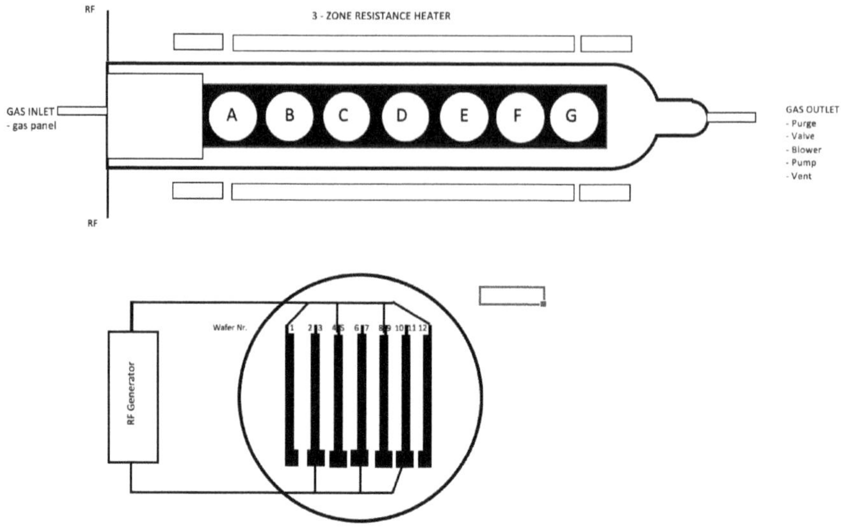

Fig. 3.10.3. Horizontal parallel boat of ASM Plasma I — Horizontal Hot Wall PECVD Furnace. Visible are the vertical plates positioned lengthwise into the furnace tube. In between the slots, the substrates are placed vertically. After loading, the boat was slide into the furnace. Boat loader ASM America Plasma I Horizontal Hot Wall PECVD Furnace [photograph]. *Credit*: ASM America (US); 1980.

of hands-on experience. Finally, the boat could be removed and the substrates covered with the desired films could be collected.

The temperature was higher near the end of the tube. This was needed to ensure that the vapors not only reacted at the front of the tube, but also at the back. The goal was to affect a uniform deposition rate across the tube. To control the process and make it reproducible, thermocouples measured the temperatures in the chamber, mass flow controllers measured the inserted gasses, and a computer kept track of the process in line with the recipe prepared by the engineers.

Research with Plasma I concerned both the expansion of applicable materials and the characterization of the materials grown by the machine — the latter was done by means of evaluation equipment. Equipment suppliers and their customers studied whether other materials could be grown by means of the hot-walled PECVD process, and for which purposes these materials might be suitable. In addition to the previously mentioned silicon nitride, the process was used for silicon oxide, but also for polysilicon, titanium silicide, and others. Although processes existed to grow these substances, substantial commercial interest hardly existed in new deposition methods. A film of, for example, polysilicon grown by low pressure CVD is not the same as a film of polysilicon grown by means of plasma CVD. The growth process contributes to the characteristics of the film, and therefore to its applicability. Some parts of chips are more vulnerable to oxides or carbon than others and need a different deposition process for higher quality. Often, higher quality means a slower deposition rate. Depending on the functionality of the film, engineers have to consider the pros and cons of a growing process like horizontal hot-walled plasma CVD (Figure 3.10.4). Therefore, by means of plasma CVD, many materials can be grown, but the materials grown this way are not practical for every potential application.

This horizontal furnace was the first configuration applied by the industry worldwide. The design lost its prominence in chip fabrication due to the increased diameter of substrates in the early 1990s. The bigger silicon wafers of about 8 inches, and later 12 inches, and the subsequently wider diameter of the tube made it more difficult to control quality across the tube, since the gas dynamics changed. This resulted in the single wafer system, concurrently developed by Applied Materials and Novellus in 1987. Instead of processing multiple substrates in one reaction chamber simultaneously, one substrate was processed horizontally at a time. Especially for laboratories, the single wafer configuration was more practical if a new material had to be grown by means of CVD. It entailed less tweaking and tuning of the process to transfer it from laboratory to an industrial environment. Clustering these single-wafer chambers together secured the desired industrial throughput to some extent.

There are more configurations, such as continuous reactors — where the substrate moves from chamber to chamber on a belt. At the advent of the bigger substrates, still some alternative horizontal batch configurations have been developed, like shorter tubes to have better control of the quality across the tube. Understanding which configuration will produce results that fall within the desired parameters requires great resourcefulness on the part of the engineers,

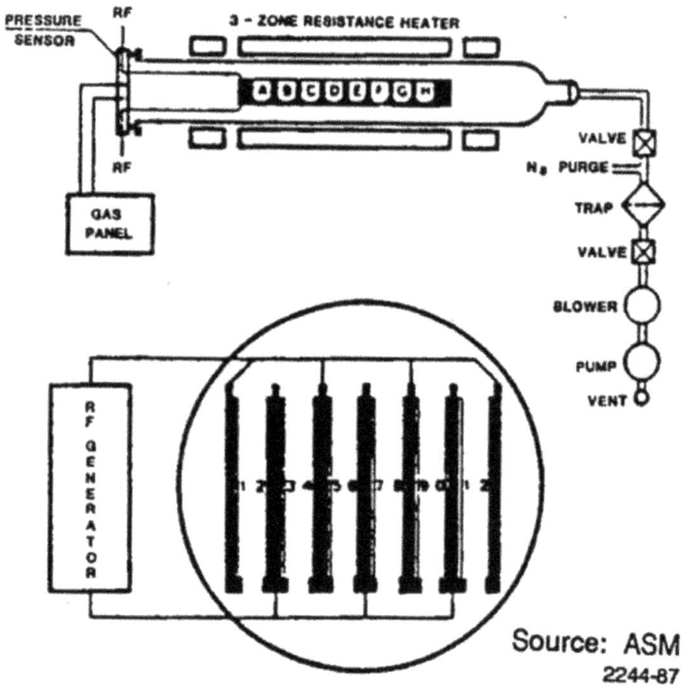

Fig. 3.10.4. Lay-out of ASM America Plasma I — Horizontal Hot Wall PECVD Furnace. Upper drawing shows the horizontal tube. Left is loadlock to load the boat containing substrates, and where the gas is inserted. Right shows the valves and pumps that pump away the gasses. In between is the tube with heating wended around it. The letter indicates the number of wafers that can be positioned in the system. The other drawing is a cross section of the tube, showing the various longitudinal plates holding the wafers, and the connections with the RF-generator required to ignite the plasma between the plates.[7 p88]

who are charged with aligning the requirements of the material with the properties of the material itself. The goal is to develop a tool suitable for as many applications or materials as possible. As discussed throughout the text, the parameters are growth rate, temperature, pressure, throughput, uniformity, film density, and others. These parameters define the suitability of vapor deposition processes. Through its adherence to these parameters, the single-wafer configuration definitively replaced the horizontal system for industrial plasma CVD processes from the early 1990s onwards.

References

1. Engle GM. Interview with Van Duijn J. May 2015.
2. Engle GM, Pacific Western Systems. Plasma enhanced chemical vapor processing of semiconductive wafers. United States Patent 4,223,048, September 16, 1980.
3. ETHW.org [internet] Swann RCG; 2015. First-hand: the birth of glow discharge chemistry (aka PECVD). Available from: http://ethw.org/First-Hand:The_Birth_of_Glow_Discharge_Chemistry_(aka_PECVD) [Accessed December 2, 2016].
4. Gorzyca TB, Gorowitz B. Plasma enhanced chemical vapor deposition of dielectrics. In: Einspruch NG, Brown DM, editors. VLSI electronics microstructure science, vol. 8: Plasma processing for VLSI. London: Academic Press; 1984. 69–87.
5. Reinberg A, inventor. Texas Instruments Incorporated, assignee. Radial Flow Reactor. United States Patent 3,757,733. September 11, 1973.
6. Rosler RS, Benzing WC, Baldo J. A Production reactor for low temperature plasma enhanced silicon nitride deposition. Solid State Technology. 1976;19(6):45–50.
7. Rosler RS, Engle GM.LPCVD-type plasma-enhanced deposition system. Solid State Technology. 1979;22(12):88–92.
8. Tolliver D. The history of plasma processing. In: Einspruch NG, Brown DM, editors. VLSI Electronics Microstructure Science, vol. 8: Plasma processing for VLSI. London: Academic Press; 1984. 2–24.

CHAPTER 3.11

NUCLEAR RESEARCH AND TEST REACTORS

MARK D. BOWLES

After World War II, with fears of weaponized nuclear power escalating daily, a science-based atomic initiative saw a proliferation of reactors built throughout the world, many of which were located on university campuses. Scientists and engineers developed these nuclear reactors as research tools to probe the secrets of materials in revolutionary new ways. Called "research and test reactors," they produced streams of neutrons for experimentation, not power for energy generation or weapons for nuclear destruction. They were little-known among the public, and so did little either as antidotes for or stimulants of nuclear fear, but instead became significant materials research tools for scientists. Since the completion of the first nuclear research reactor in 1942 at the University of Chicago, scientists and engineers have built 783 facilities throughout the world. The United States was the most active research reactor–building nation, with 264 sites, followed by the former Soviet Union with 118.[3] Today, the most operational sites by country reside in the Russian Federation, with the United States second.

It is important to differentiate between test, research, and power reactors. A test reactor is one that operates above the 10 megawatt thermal power range, whereas a research reactor is less powerful, generating between 10 kilowatts and 10 megawatts. Power reactors operate at much higher power levels for producing electricity and are conspicuous because of their size and hourglass-shaped cooling towers. Research and test reactors, like those which supported university laboratories, can be quietly tucked away.

Research and test reactors provided invaluable opportunities for researchers to use neutrons and radiation for diverse fields of experimentation. The 1950s and 1960s were the golden age of research and test reactors in the United States, when they became important experimental tools at national laboratories (Argonne, Oak Ridge, Brookhaven, and Los Alamos), on university campuses (such as North Carolina State, Penn State, Massachusetts Institute of Technology), in private industry (Westinghouse, General Electric, and other corporations), and within military research facilities.[6] Even the US space agency, NASA, built a research and test reactor at Plum Brook Station, in Sandusky, Ohio, in order to study materials for nuclear rockets (Figures 3.11.1 and 3.11.2).

Scientists and engineers promoted research and test reactors with enthusiasm in the 1950s for a variety of reasons, which combined to create a surge in momentum. First, in the United States, regulations allowed reactor owners to reap the financial rewards from any discoveries made with a reactor, while the Atomic Energy Commission (AEC) relaxed many of its previously strict standards for organizations interested in building and operating a reactor. Second, new and more abundant supplies of uranium-235 became available for use as reactor fuel. Third, there was a great desire to learn more about nuclear fission so that nations could take full advantage of the capabilities of the nuclear reactors. Fourth, President Eisenhower's "Atoms for Peace" speech in 1953 called for mobilizing nuclear experts to improve agriculture, medicine, and other civilian domains. This proved to be a stimulus for the creation of university research reactors because "nuclear engineering programs and university reactors were essential to winning the Cold War and fulfilling the promise of Atoms for Peace."[12 p165] Finally, and perhaps most importantly, there was a remarkable optimism about the potential for these reactors to improve basic science research.

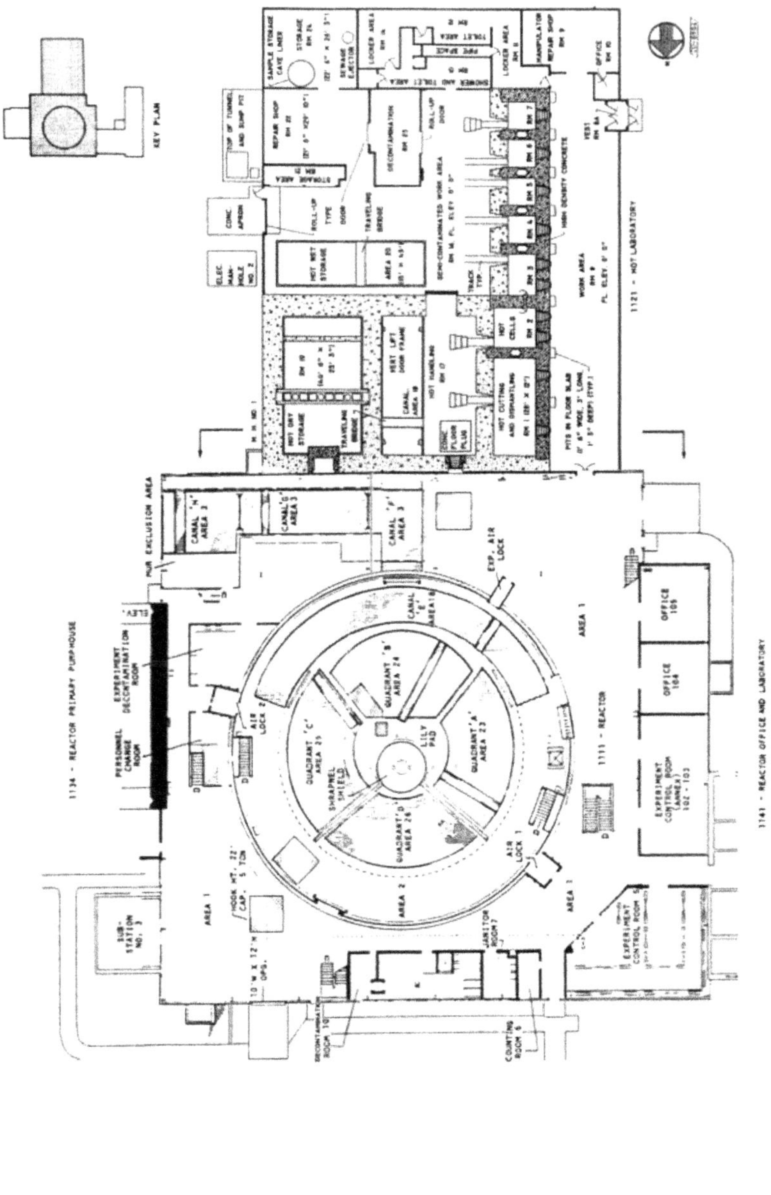

Fig. 3.11.1. First-level floor plan for the reactor building at NASA's Plum Brook Reactor Facility. *Credit:* Courtesy of NASA, Image #: 28Reactor_FloorPlan.

Fig. 3.11.2. NASA's only research and test reactors were located at Plum Brook Station in Sandusky, Ohio. Harold Geisler takes Plum Brook reactor critical for the first time on the evening of June 14, 1961. This first self-sustaining chain reaction was conducted at very low power. It was not until the following April that the reactor reached its full potential of sixty megawatts. By July 1963, the reactor had completed its first experimental cycle while critical. *Credit*: Courtesy of NASA, Image #: C-1961-56899.

Nuclear reactors had research relevance to a wide assortment of scientific specializations because they created "neutron fluxes." These were intense beams of radiation that experimenters directed at materials to better understand their atomic makeup. Researchers exposed test objects to radiation near the reactor core for a specified length of time and then moved them via underwater canals to "hot laboratories" for investigation.[2 p77] The so-called "hot labs" were shielded cells aligned in a row where the engineering operators stood side by side, peering into them through thick plate-glass, oil-filled windows. They interacted with the radioactive materials

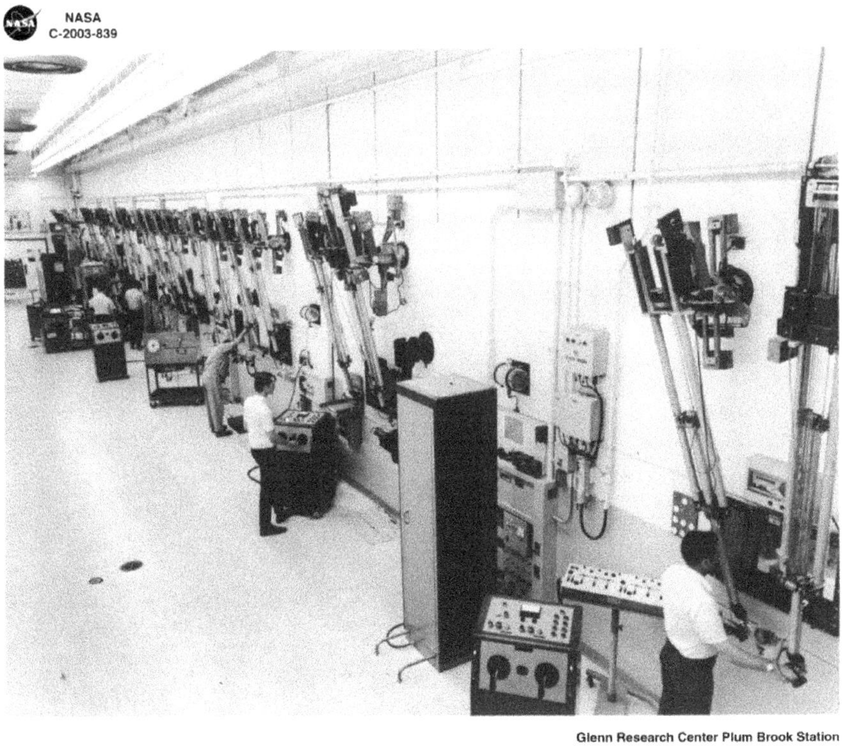

Fig. 3.11.3. The hot laboratory's safe work area at NASA's Plum Brook Reactor Facility. Operators are using manipulator arms to work with irradiated experiments in the cells. The hot lab also contained an office, manipulator repair shop, and a decontamination room that connected this "clean" operating area with the radioactive area behind the cells. *Credit*: Courtesy of NASA, Image #: C-2003-839.

through the deft use of robotic manipulator arms, which were huge claw-like devices that enabled engineers to perform experiments on materials without exposing themselves to high levels of radiation (Figures 3.11.3 and 3.11.4).

Nearly every material on earth became a candidate for "irradiation," that is, placement near a radioactive source such as a nuclear reactor. One observer wrote: "After the war [World War II], the development of reactors, with their ability to produce neutrons in unheard-of quantities, even *beams* of them, promised to revolutionize nuclear

Fig. 3.11.4. Two technicians clad in anti-contamination clothing manipulate a shim safety control rod in a water canal in the hot laboratory. The twenty-five-foot deep water provided shielding from the radiation, yet still enabled visible contact with the research experiments. The water also allowed the underwater transfer of irradiated materials from the reactor to the hot laboratory for inspection. Moving materials by canal reduced the need for lead transfer casks, though they were still needed when the radioactive materials were taken out of the water. *Credit*: Courtesy of NASA, Image #: C-1961-55808.

physics, solid-state physics, and any field in which radioisotopes could be used."[5 p11,153] In an age of increasing specialization, both in scientific and engineering professions, research and test reactors became unusually versatile instruments.

Nuclear physicists used the reactors' neutrons to help understand subatomic behavior. Solid state physicists and chemists used neutrons to discover the properties of matter and materials. Biologists and physicians used neutrons to study the effect of radiation on tumors and organic tissues.[4] Engineers used neutrons to design more efficient nuclear reactors and electrical components. For example, they could expose materials to radiation for a period of time and then assess how this affected strength and performance. This enabled safer construction of nuclear-power reactors.

Research and test reactors were used in three main ways. Using a technique called "neutron radiography," which is similar to a dental X-ray, engineers determined the structural integrity of materials. They used this tool for quality control purposes in automotive, aerospace, and medical organizations. Environmental scientists and geologists used a powerful technique called "neutron-activation analysis" to measure trace elements of pollutants in air, water, and soil. Finally, "neutron scattering" enabled researchers to explore the ways that radiation bounced off and interacted with other materials. This yielded important analytical information and enabled them to analyze "molecular structure, surfaces and interfaces, measure electronic and magnetic properties, stress and strain conditions, and gauge other characteristics."[11]

Just as with many other emerging technologies during this period (space satellites, nuclear weapons, nuclear aircraft), Cold War competition boosted both superpowers' nuclear research programs. Although the United States dominated in the sheer number of research and test reactors it built, the Soviet Union made their own important advances in the area of high-flux research reactors. The first went critical at the Russian Research Center at the Kurchatov

Institute on December 25, 1946. The Soviets founded this institute to develop nuclear weapons, yet it also became the leading post-war Russian research and development organization in the field of nuclear energy.[1 p564] In the 1960s the *New York Times* reported on a new Soviet research reactor that would test materials for long-range space voyages with its "intense flux of neutrons."[10 p71] The United States worked to keep pace and eventually surpass these early Soviet achievements with the construction of research sites like the Materials Test Reactor at Idaho Falls.

There are far fewer of these reactors today than in the 1950s and 1960s, based on data at the International Atomic Energy Agency's (IAEA) Research Reactor Database (see https://nucleus.iaea.org/RRDB/RR/ReactorSearch.aspx). The greatest number of operational sites reside in Eastern Europe with seventy-three reactors, primarily in the Russian Federation (fifty-four sites). The next largest concentration is in North America with forty-nine research reactors operational (forty-two in the United States) out of the 283 built. In Western Europe, of the 191 research reactors built, thirty-one remain active. The third most populated region is in East Asia where sixty-eight reactors have been built. All are now decommissioned or shut down except for seventeen in China and one each in Japan, Taiwan, North Korea, and South Korea.

Other geographical regions have far fewer concentrations of research reactors. Middle Eastern and South Asian countries have constructed twenty-six reactors, with fifteen still operational, including four in India, four in Iran, two in Israel, two in Pakistan, and one each in Bangladesh, Syria, and Jordan. Engineers in the African continent have constructed fourteen reactors, with six still operational. South East Asia and the Pacific countries have four if its twelve research reactors remaining (Australia, Indonesia, and Malaysia). Latin America is the only region with more operational facilities (sixteen) than decommissioned sites (seven). Two more are planned, in Argentina and Brazil (Figure 3.11.5).

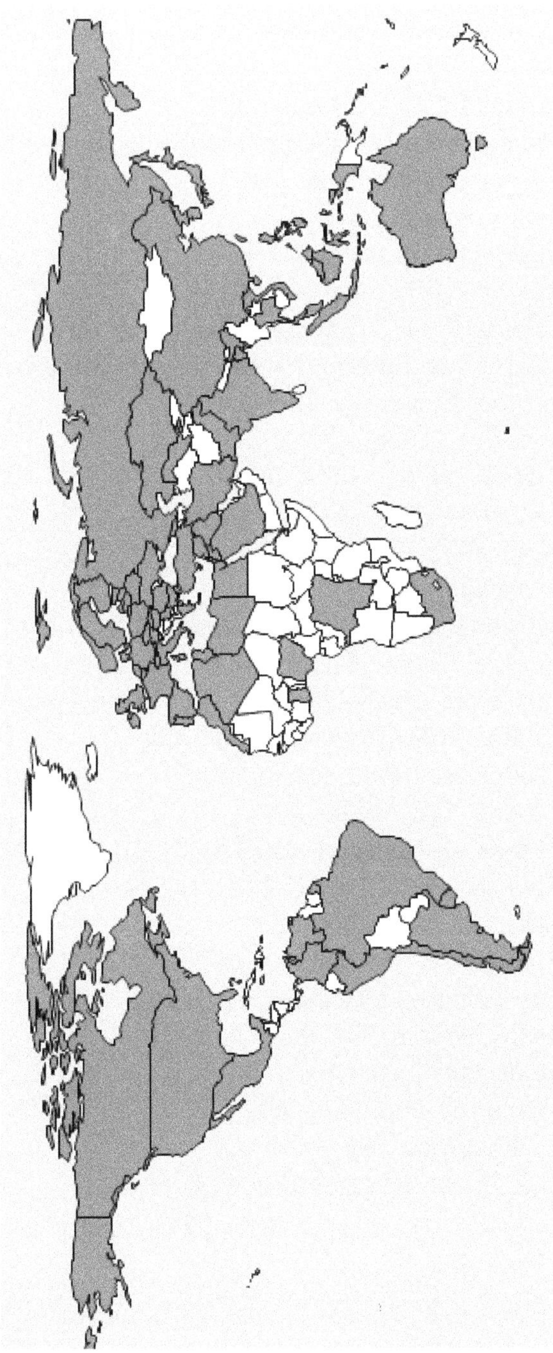

Fig. 3.11.5. Shaded countries shaded currently have an operational research or test reactor. *Credit:* ©IAEA, https://nucleus.iaea.org/RRDB/RR/ReactorSearch.aspx.

31 Operating Research and Test Reactors in the United States

Aerotest Operations Inc., San Ramon, CA
Armed Forces Radiobiological Research Institute, Bethesda, MD
Dow Chemical Company, Midland, MI
General Electric Company, Sunol, CA
Idaho State University, Pocatello, ID
Kansas State University, Manhattan, KS
Massachusetts Institute of Technology, Cambridge, MA
National Institute of Standards and Technology, Gaithersburg, MD
North Carolina State University, Raleigh, NC
Ohio State University, Columbus, OH
Oregon State University, Corvallis, OR
Penn State University, University Park, PA
Purdue University, West Lafayette, IN
Reed College, Portland, OR
Rensselaer Polytechnic Institute, Schenectady, NY
Rhode Island Atomic Energy Commission, Narrangansett, RI
Texas A&M University, College Station, TX (two reactors)
University of California-Davis, Sacramento, CA
University of California, Irvine, CA
University of Florida, Gainesville, FL
University of Maryland, College Park, MD
University of Massachusetts, Lowell, MA
University of Missouri, Columbia, MO
University of Missouri, Rolla, MO
University of New Mexico, Albuquerque, NM
University of Texas, Austin, TX
University of Utah, Salt Lake City, UT
University of Wisconsin, Madison, WI
U.S. Geological Survey, Denver, CO
Washington State University, Pullman, WA

Data from: http://www.nrc.gov/reading-rm/doc-collections/fact-sheets/research-reactors-bg.html

Despite the decline in new construction and operation, reactors are still important tools at university research facilities. Consider North America as an example. The Nuclear Regulatory Commission (NRC) currently licenses forty-two research and test reactors, with thirty-one operating, twenty-five of which are located at universities (Figure 3.11.6). Test and research reactors have also gained a newer function in terms of training personnel, as evidenced by the name of the professional organization serving this research community in the United States and Canada: National Organization of Test, Research, and Training Reactors. As Thomas Newton, Director of Reactor Operations at the MIT Nuclear Reactor Laboratory,

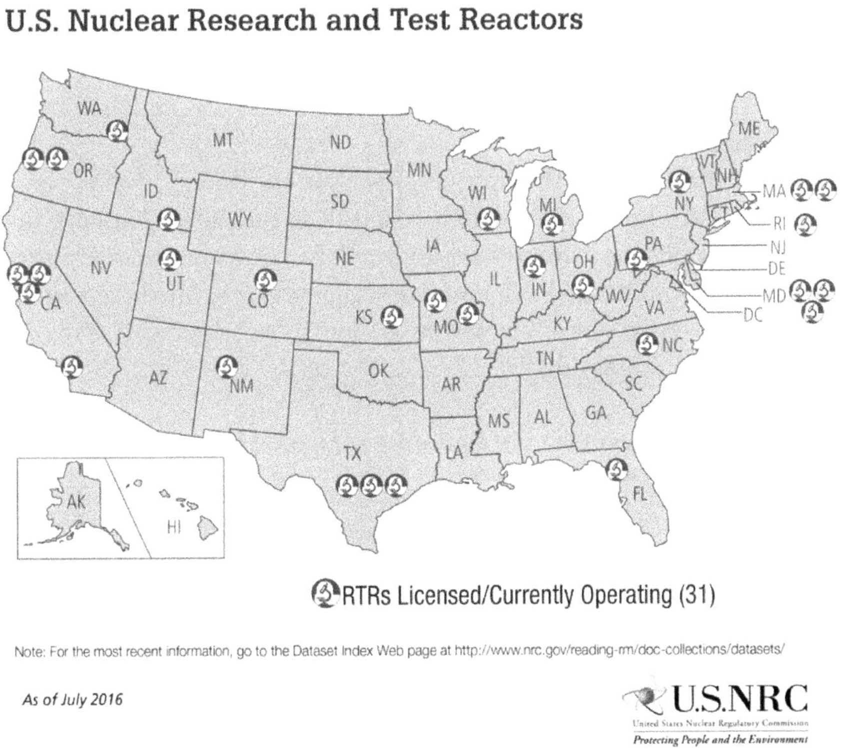

Fig. 3.11.6. Nuclear research and test reactors currently in operation in the United States. *Credit*: Courtesy of the US Nuclear Regulatory Commission, http://www.nrc.gov/reactors/operating/map-nonpower-reactors.html.

put it, these reactors were a "unique hands-on education tool for the next generation of engineers and scientists." They also play a role in "educating the public and demonstrating the safe and effective use of nuclear reactors."[8]

In 2005 the NRC noted that research and test reactors have made significant experimental contributions to almost every scientific discipline, including physics, chemistry, biology, medicine, geology, archaeology, and the environmental sciences.[11] At least in the United States, however, university test and research reactors face increasing challenges, including deteriorating equipment, increasing cost of new equipment, improved computer simulations (which can substitute for experiments run in real-world reactors), a preference to use higher powered test reactors at a national laboratory, and a "negative perception by the campus community."[9] As a result, leadership in academic reactors has now passed to the European University Research Reactors in terms of productivity and number of researchers.

According to a 2016 International Atomic Energy Agency publication, "the future of research reactors is radically changing in a more economically competitive and safety conscious marketplace." Despite their continued value in materials research, for these reactors to survive they require renewed attention to management, economic support, and marketing.[7 p15]

Table 3.11.1. Data from the International Atomic Energy Agency on every research and test reactor (774 total).

Status	Developed Countries	Developing Countries	All Countries
PLANNED	3	7	10
UNDER CONSTRUCTION	4	4	8
OPERATIONAL	154	90	244
TEMPORARY SHUTDOWN	13	5	18
PERMANENT SHUTDOWN	113	21	134
DECOMMISSIONED	327	25	352
CANCELLED	4	4	8

Data from: https://nucleus.iaea.org/RRDB/RR/ReactorSearch.aspx

References

1. Angelo J. Nuclear technology. Westport, CT: Greenwood Press; 2004.
2. Bowles MD, Arrighi R. NASA's nuclear frontier. Washington, DC: Government Printing Office; 2004.
3. Bowles MD. Science in flux: NASA's nuclear program at Plum Brook Station. Washington, DC: Government Printing Office; 2005.
4. Creager ANH. Life atomic: a history of radioisotopes in science and medicine. Chicago: University of Chicago Press; 2013.
5. Crease RP. Making physics: a biography of Brookhaven National Laboratory, 1946–1972. Chicago: University of Chicago Press; 1999.
6. Holl JM, Hewlett RG, Harris RR. Argonne National Laboratory, 1946–96. Urbana: University of Illinois Press; 1997.
7. International Atomic Energy Agency. Research reactors: purpose and future. Vienna, Austria: International Atomic Energy Agency; 2016. https://www.iaea.org/OurWork/ST/NE/NEFW/Technical-Areas/RRS/documents/RR_Purpose_and_Future_BODY.pdf [Accessed November 10, 2016].
8. Newton Jr. TH. Research and test reactor briefing. Washington, DC: United States Nuclear Regulatory Commission; 2014. Available from: http://www.nrc.gov/reading-rm/doc-collections/commission/slides/2014/20141216/newton-12-16-2014.pdf. [Accessed November 10, 2016].
9. Perez PB, Gallagher P, Bernard J, Bobek LM, Vernetson WG. University research reactors: a brief overview. September 2000. Office of Science and Technology Policy. Available from: https://web.archive.org/web/20160303183848/ http://www.trtr.org/Links/OSTP%20Briefing%202000.pdf [Accessed November 10, 2016].
10. Shabad T. Russian reactor to aid space trip. New York Times. December 10, 1967.
11. United States Nuclear Regulatory Commission. Backgrounder on research and test reactors. Washington, DC: United States Nuclear Regulatory Commission; 2014. Available from: http://www.nrc.gov/reading-rm/doc-collections/fact-sheets/research-reactors.html. [Accessed November 10, 2016].
12. Wellock TR. Triumph and trouble in the first temple of the atom. In Walker JS, editor. Nuclear energy and the legacy of Harry S. Truman. Kirksville, MO: Truman State University Press; 2016.

CHAPTER 3.12

CENTRIFUGES

ANDREW EDE

The centrifuge is a device for separating particles in a fluid by spinning the fluid at high speed. The different types of particles in the fluid separate into layers based on their particle density. In scientific terms, this behavior is based on the *volumetric mass density* of the particles, which represents the mass of an object per unit of volume. Centrifuges take advantage of centripetal acceleration that causes particles of higher density to move farther away from the center of rotation than particles with lower density. This process, also known as sedimentation or fractionation, allows centrifuges to separate materials with different characteristics such as cream from whole milk or plasma from red blood cells. The centrifuge can be used to collect materials that are difficult to separate using other methods, such as different isotopes of an element.

The earliest use of spinning to separate liquids was the creation of the first milk separator. Antonin Prandtl wrote a paper outlining the principle of centrifugal separation and suggested some design ideas around 1859.[9] Antonin's brother Alexander Prandtl, working at the Weihenstephan Central School of Agriculture in Freising, spent five years developing a working separator which he showed at the 1875 World Fair in Frankfurt am Main. The earliest models were designed for individual farm use and operated by turning a hand crank, but the principle was easily scaled up. Alexander (who was a mathematician) also did theoretical work, which he published in the journal *Weihenstephaner Milchzeitung* (*Weihenstephan Milk Journal*).[6] Although the Prandtl brothers were likely the first to create a milk separator, the first successful commercial exploitation of the process was by the Swedish engineer Karl Gustaf Patrik de Laval, who may have heard or read about the earlier invention. When de Laval showed his separator in England in 1879, he was awarded a silver medal by the Royal Agricultural Society of England.[7] De Laval began marketing his separator in 1883.

Another form of centripetal farm equipment were honey extractors (Figure 3.12.1). Beekeepers wanted to extract honey from the honeycomb without damaging the individual wax cells created by the bees, so dozens of machines were invented to spin the frames rather than remove all the cells and squeeze them in a press. One of the earliest patents was issued to G. W. Williams of San Diego, California, in 1879.

Although the milk separator quickly became a widely used device, the development of centrifuges for other commercial applications did not become commonplace until the development of the electric motor around 1900. With the availability of electric motors, the application of centrifuges to a variety of industrial applications became possible. In the sugar industry, centrifuges were used to quickly separate sugar crystals from the sugar liquor with the added advantage that the machines could be set to collect crystals of a specific size. In the petroleum industry, centrifuges were used to remove water and non-organic solids from bitumen prior to refining. Large centrifuges continue to be used for water treatment for many industrial processes such as smelting, municipal wastewater treatment, and mining. Centrifuges are useful anywhere solids need to be removed from water or other liquids. Although they are often overlooked as centrifuges, one of the largest applications of spinning technology is in the processing and cleaning of textiles. Everything from dyeing to removing water and chemicals in the process of fulling (cleaning and thickening) cloth was sped up by the introduction of power-driven spinning drums. Today, every home clothes dryer that spins is a type of simple centrifuge.

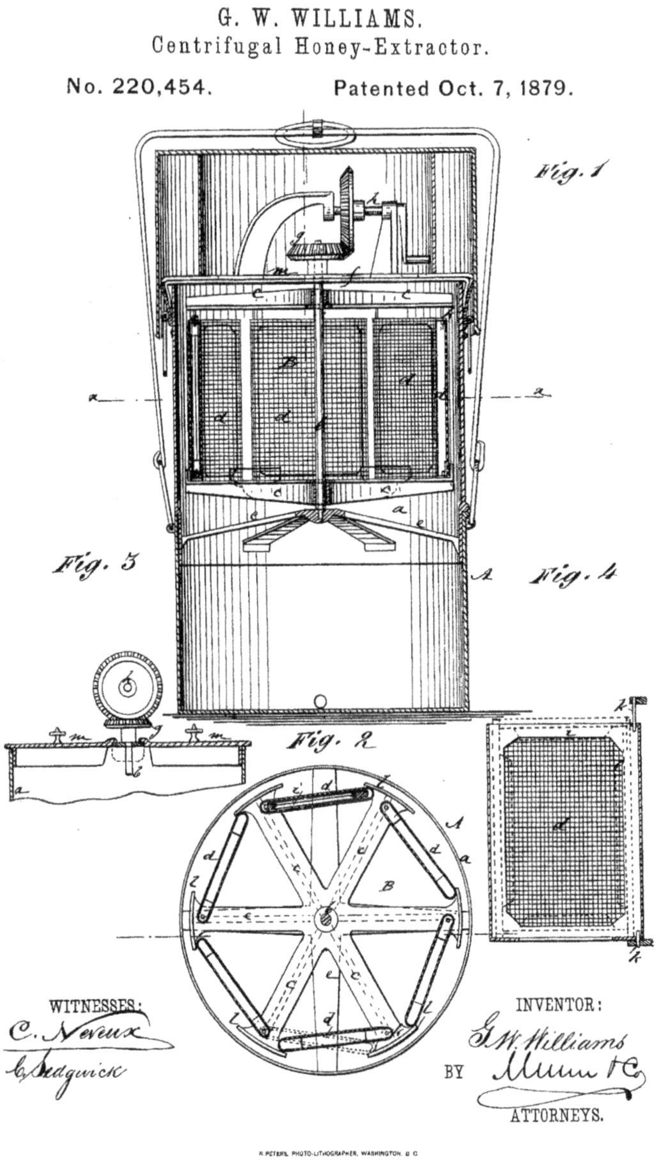

Fig. 3.12.1. Centrifugal honey-extractor, United States Patent 220,454. *Source*: Google Patents.

Most commercial or industrial centrifuges operate at relatively low speeds (a home clothes washer dry cycle spins at between 600 and 1,600 rpm), but the principle of sedimentation offered the possibility of using high-speed centrifuges for scientific research. In the early twentieth century there was a great deal of interest in the composition and structure of matter, but few laboratory tools were available to investigate materials at the size of atoms and molecules. In physics and physical chemistry, a debate raged about whether atoms even existed, while in biology and organic chemistry the composition of the cell was a mystery because it was difficult to measure or isolate the component parts. Scientists recognized that one of the ways to solve this problem was by separating particles in fluid, but there was a problem because of Brownian motion. Sedimentation in a test tube only works if the particles you want to isolate are more affected by the downward pull of gravity than the friction and kinetic behavior of the fluid. Sand will sediment out of water quickly, but most of the particles of coffee in a cup of coffee are so small they will not separate under normal conditions.

Brownian Motion

Brownian motion is the random motion of small particles in a fluid such as pollen in water or dust in air. This motion prevents small particles from ever falling to the bottom of a container under normal conditions. The strange motion of tiny particles was first noted by botanist Robert Brown in 1827, but it was not until 1905 that Albert Einstein worked out the physics of Brownian motion and demonstrated the motion was caused by the collision of atoms or molecules with the particles.

The solution to this problem was theoretically simple, but technically complex: increase the sedimentation rate by using spin to simulate gravity. In 1924, the team of The Svedberg, Herman Rinde, and A. G. Bergström working at the University of Uppsala, Sweden, built the first ultracentrifuge to scientifically study particle distribution and help to separate materials in fluids (Figure 3.12.2). The first

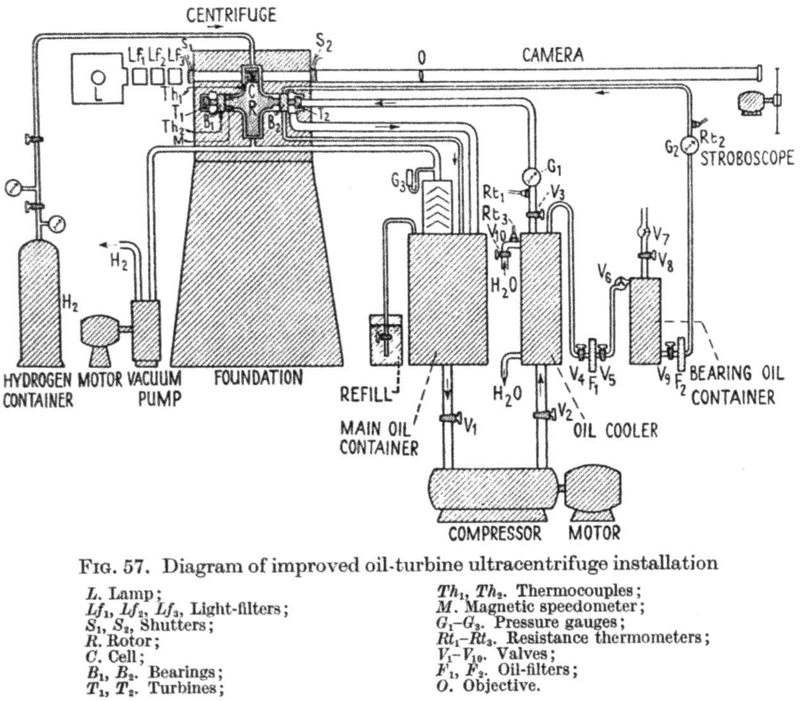

Fig. 3.12.2. Svedberg's oil turbine drive ultracentrifuge (1927). *Source*: Svedberg T. The ultracentrifuge and its field of research. *Industrial and Engineering Chemistry Analytical Edition*. 1938;10(2):114. Reproduced with permission.

version used a 1.5 hp electric motor and spun at 5,400 rpm, generating the equivalent of 150 units of gravity (g).[3 p110] The test material was placed in a glass tube inside a metal disk and there was an optical system that allowed the experimenters to measure where the particles were located in the tube while the disk was spinning. By 1925, Svedberg had increased the rotations to 8,700 rpm and produced the equivalent of 5,000 g.[8] In 1927, using a much more powerful oil turbine engine, hydrogen cooling, and spinning the sample containers in a vacuum, a speed of 84,000 rpm was achieved, but for practical purposes the ultracentrifuge worked best at around 42,000 rpm which produced 104,000 g.[3 p115] At this speed, the ultracentrifuge became a useful laboratory instrument. The machine occupied a small room with most of the engine parts in a second room under the floor of the laboratory. By the 1970s, ultracentrifuges were capable of producing 900,000 g, but new tools and research topics led most laboratories around the world to retire their experimental ultracentrifuges by the 1980s.

In physics, the ultracentrifuge helped to confirm the reality of molecules and atoms by giving scientists a much more precise examination of particle behavior, one of the keys to understanding Brownian motion and kinetic theory. Svedberg's work confirmed experimentally ideas set out by Albert Einstein in 1905.[4] It was Svedberg's work on organic chemistry, however, that did the most to change scientific ideas. Organic materials such as blood and proteins were very difficult to study because they were not composed of a single thing but were a mixture of different substances. The ultracentrifuge separated materials so that all the particles with the same characteristics would form a band at a specific point in the glass tube. By knowing the sedimentation velocity created by the ultracentrifuge, the position of the band of material could be used to calculate the molecular weight of the test material. In the 1920s scientists believed that the highest weight for a single organic molecule was 17,600 g/mol. Using the ultracentrifuge to examine hemoglobin (the metalloprotein containing iron in red blood cells that carries oxygen) showed it to have a molecular weight of 70,400 g/mol, and hemocyanin (similar to hemoglobin, but using copper rather than iron) from the Busycon snail had a molecular weight of 10,000,000 g/mol.[5 p406a] At first, Svedberg and other scientists assumed that they were seeing substances made up of smaller molecules (the molecular weight 70,400 is $4 \times 17{,}600$), but it was quickly proven that these were in fact giant molecules. This led German chemist Hermann Staudinger to formulate macromolecule theory, which described how organic molecules could be made of very long chains or polymers. This theory remains one of the fundamental ideas in organic chemistry and opened the door to technical applications such as plastics and scientific work such as discovering the structure of DNA. Svedberg won the Nobel Prize for Chemistry in 1926 for his groundbreaking work using the ultracentrifuge and Staudinger would win the Nobel Prize for Chemistry in 1953 for his work on macromolecules.

Although the original ultracentrifuge was a research instrument to physically evaluate materials (such as finding atomic weight and particle size), a number of people recognized that the centrifuge could be used as a tool for other kinds of research and laboratory work. At the University of Virginia, Jesse W. Beams and Arthur J. Weed created a much simpler (and less expensive) centrifuge that worked by compressed gas and could quickly and economical separate materials for collection and study.[5 p406a] One of Beams's students, Edward Greydon Pickels, saw the need for commercial centrifuges and founded Spinco (Specialized Instruments Corporation) in 1946. Spinco made centrifuges that could be used to separate materials for collection such as nucleic acids, viruses, and different types of proteins. Laboratory centrifuges, which typically operate at between 15,000 to 30,000 g were much more compact and could be set up on a workbench. These devices played a key role in the discovery and decoding of DNA and are still used genetics laboratories around the world.

One of the most important uses for modern centrifuges is in medical analysis. Hematocrit centrifuges spin blood samples to separate the plasma, leukocytes (white blood cells), and erythrocytes (red blood cells). This allows the different blood fractions to be collected and analyzed. On a larger scale, blood collected for medical use such as transfusions can be used as whole blood, but is often fractionated by spinning it in a centrifuge and collecting the plasma. Plasma accounts for about 55% of blood volume and it remains usable longer than whole blood, so it is often used in medical emergencies or surgery to maintain blood volume. In addition to plasma, other blood components such as factor VIII (used to treat hemophilia) and immunoglobulin (used to treat immune deficiency) can be extracted from centrifuged blood.[1]

The most technically complex types of centrifuge are gas centrifuges used to collect isotopes. This type of centrifuge was developed by the Soviet Union after World War II and is known in the west as a Zippe-type centrifuge (Figure 3.12.3) after one of its builders, Gernot Zippe, an Austrian engineer who was captured at the end of the war and later released from the Soviet Union in 1956.[2] The primary use of gas centrifuges is to prepare enriched uranium for reactors and nuclear weapons. In naturally occurring uranium, more

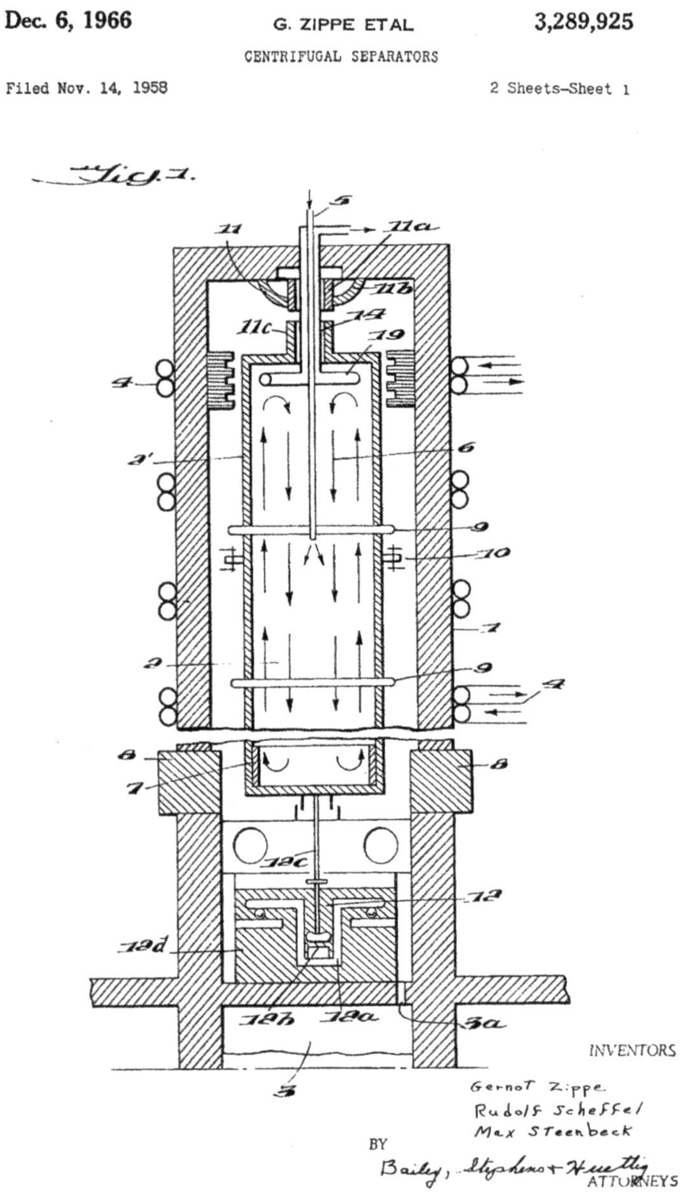

Fig. 3.12.3. Zippe gaseous centrifuge, United States Patent 3289925A. *Source*: Google Patents.

than 99% of the uranium atoms is the uranium-238 isotope. Reactors and weapons, however, require uranium that has been "enriched" in the uranium-235 isotope. Enrichment can be achieved by turning uranium into a gas (uranium hexafluoride) and pumping it into a centrifuge which is rotating at 90,000 rpm, so that the heavier U_{238} will move further away from the center of rotation than the U_{235}. The lighter gas is collected and centrifuged again, hundreds of times, until the enriched uranium contains the desired quantity of U_{235}.

Centrifuges played an important role in the development of modern science and continue to be valuable tools for medicine, industrial systems and nuclear power. They evolved from simple, hand-operated farm equipment to the cutting edge of nuclear engineering. Our clothes are cleaner and dairy products are reliable because of simple centrifuges. More sophisticated centrifuges allow modern healthcare to detect illness and produces life-saving blood products. Centrifuges were a key device for decoding genetics and today criminals are caught because centrifuges can isolate DNA collected from crime scenes. Any time we want to separate different materials, a centrifuge is frequently the device to do the job quickly and efficiently.

References

1. Brodniewicz-Proba T. Human plasma fractionation and the impact of new technologies on the use and quality of plasma-derived products. Blood Reviews. 1991;5:245–57.
2. Burkharin O. Russia's gaseous centrifuge technology and uranium enrichment complex. Princeton: Princeton University; 2004.
3. Ede A. The rise and decline of colloid science in North America, 1900–1935. Aldershot: Ashgate; 2007.
4. Einstein A. Über die von der molekularkinetischen Theorie der Wärme geforderte Bewegung von in ruhenden Flüssigkeiten suspendierten Teilchen (On the motion of small particles suspended in a stationary liquid, as required by the molecular kinetic theory of heat). Annalen der Physik. 1905;17(8): 549–60.
5. Pedersen O. The protein molecule. In Svedberg T, Pedersen KO, editors. The ultracentrifuge. Oxford: Clarendon; 1940.

6. Prandtl A. Über den theoretisch zu erwartenden Effekt der Aufrahmung durch Zentrifugalkraft. (On the effect of creaming from centrifugal force that may be expected on theoretical grounds). Weihenstephaner Milchzeitung; 1877.
7. Silver Medals. Journal of the Royal Agricultural Society. 1879;15:691.
8. Svedberg T. Low-speed ultracentrifuge with gear drive. In Svedberg T, Pedersen KO. The ultracentrifuge. Oxford: Clarendon; 1940. 80–100.
9. Vogel-Prandtl J. Ludwig Prandtl: a biographical sketch, remembrances and documents. Trieste, Italy: International Centre for Theoretical Physics; 2004.

PART IV

TOOLS FOR CHARACTERIZING MATERIALS

TOOLS FOR CHARACTERIZING MATERIALS: INTRODUCTION

JOSEPH D. MARTIN

"If you can spray them then they are real" is the philosopher of science Ian Hacking's pithy answer to the question of when we should believe in the existence of those microscopic entities we cannot see.[10 p23] Much history and philosophy of science has concerned the second half of Hacking's slogan. Historians have investigated how scientists came to believe that things like electrons, neutrons, and photons are real. Philosophers have wondered what it means to build science around the belief in entities that are unobservable, or only indirectly observable. But the first half of the quote hints at other, more rarely told stories.[7] When we spray electrons, or neutrons, or photons, how do we spray them? At what? To what end? The history and philosophy of science have said a great deal about the things we spray, but much less about how and why we spray them.

The fourth and final section of this volume concerns tools used to understand the properties of materials. Many of these tools emerged because it is possible to learn a great deal about substances by spraying things at them and then watching how those things are absorbed and reemitted, scattered and reflected, or diffracted and refracted. Like a handful of pebbles, which when tossed into a well sends back sounds that can help ascertain the well's depth and wetness, various kinds of particles, when directed at materials, return information about the unseen properties of those materials. Even while some scientists were struggling to comprehend the basic properties and behavior of electrons, neutrons, photons, ions, and other particles, another group seized on these same particles as the basis for new techniques to probe more complex matter. Almost all of the basic phenomena that interest physicists and chemists are relevant to characterizing materials. Physicists who study heat, light, sound, force, motion, electricity, magnetism, and quantum behavior can examine all of these as they behave in materials. Materials offer the chemist an equally productive means of exploring bonding, valence, affinity, reactions, and structure. Investigating why materials behave the way they do is therefore relevant both to finding new things to do with them, as we have seen in part 3 of this book, and to developing a better foundational knowledge of the world around us.

Scientists and engineers have been characterizing materials for hundreds of years. No era in this branch of study has been so lively, however, as the twentieth century, when a new understanding of the microstructure of matter, itself guided by inventive new tools, made it possible to explain easily observable phenomena like magnetism, electrical and thermal conductivity, hardness, and brittleness in terms of the arrangement and interactions of atoms and molecules. The tools described in this section are products of the fruitful give and take between better theoretical understanding of the microstructure of matter and laboratory techniques that generated new ways of seeing materials.

To understand the world that made these tools possible and gave them their purpose, we must step back a little over a century to consider how they fit with two related developments. The first is the growing interest in radiation of all types in the late-nineteenth century. New tools and detection techniques, such as discharge tubes and phosphorescing screens, made it clear that certain substances could be induced to emit some kind of energy, which could then be detected at a distance. We now understand those and other emissions as photons, or electrons, or protons, or neutrons, or helium nuclei, depending on the particles that compose them. But at the time they had names like "cathode rays," "rays of positive electricity," and "X-rays" (the last of which has stood the test of time). We also now understand many them to be exceedingly hazardous, a realization that would lead to the safety regimes described in Amy Slaton's contribution to part 1, but which arrived too late for many early experimenters.[30] These rays, which scientists sought to understand for their own sake (at their own, unwitting peril) in the late 1800s, would become the sprayed entities that made new tools to understand materials possible in the early 1900s.

Second, the 1910s and 1920s would see the advent of quantum theory. Before it was possible to fully interpret the data generated when bombarding materials with submicroscopic particles, scientists would have to come to grips with the strange and counterintuitive way that matter and energy behave at very small scales. These two developments were closely connected. Studying the rays that had piqued scientists' interest in the late-nineteenth century generated a host of problems and questions that led to the development of quantum mechanics in the early twentieth century, which in turn created the theoretical foundation that would allow a bounty of new tools based on bombarding matter with rays of some kind throughout the remainder of the twentieth century.

Both of these stories are standard fare in the history of science, where they are usually presented as a prologue to later developments in nuclear and high energy physics. As the study of rays evolved into the study of the particles that composed them, physicists embarked on a journey toward smaller and smaller length scales that took them inside the atom and then inside the nucleus. By the same token, the rise of quantum theory established the mathematical language that was necessary to talk about subatomic and sub-nuclear phenomena. But these early developments were just as important as precursors of materials research. Because of the tremendous influence, prestige, and visibility fields like nuclear and high energy physics gained later in the twentieth century, historians have naturally tended to look at earlier developments and see in them the seeds of this success. However, we should not be so quick to privilege this view over the perspective that shows these activities making equally important contributions to materials science. Early twentieth-century physicists, such as Hans Bethe and Arnold Sommerfeld for example, often made contributions both to the quantum mechanics of single particles and to the characterization of complex matter.[11; 19] The intellectual traditions we now call high energy physics and condensed matter physics were closely intertwined for much of the twentieth century, and so we must understand earlier contributions that formed the roots of one as giving rise to both.

Efforts to decipher the mysteries of cathode rays and X-rays were certainly prerequisite to later investigations of elementary particles. But they also led to valuable tools for investigating complex matter. Quantum mechanics was similarly critical for interpreting the information those tools returned. This introduction describes how late nineteenth- and early twentieth-century developments in physics and chemistry created the conditions that allowed the proliferation of tools for characterizing materials later in the twentieth century. The contributions that follow describe many of these tools, and collectively reinforce the point that the legacy of early microphysics and quantum theory is richer and more extensive than its evident contributions to nuclear and high energy physics.

The Days of Rays

The late-nineteenth century was a heady time to be a scientist in Europe. The *fin de siècle* — French for "end of the century" — was not only an era of great political and economic change; it also brought exciting developments in the scientific understanding of matter. In the 1890s, it was still scientifically respectable to deny the existence of atoms.[1] By the early 1910s, it was possible to "see," in a manner of speaking, atoms and their arrangement in materials with the help of new X-ray diffraction techniques.

How did scientists move from uncertainty about the microscopic world to confidently manipulating it in a search for new knowledge about materials? Important background conditions include nineteenth-century imperialism and the Industrial Revolution. The pursuit of empire by the major European powers encouraged exploration and exploitation of natural resources, the construction of communication networks, and forms of travel that brought the growing scientific community together.[12] News traveled fast on railroads and steamships, and through telegraph cables, and new discoveries quickly generated excitement and elicited responses. After Wilhelm Conrad Röntgen discovered X-rays, he published his results in Austria in December 1895. The news reached the English journal *Nature* in January 1896. *Science*, the journal of the American Association for the Advancement of Science, reported Röntgen's discovery in February, as did the French *L'Eclairege Electrique*.[8 p28] Industrialization encouraged scientific investigations of a host of materials, both new and old. When Marie Curie discovered the element radium in 1896, she did it by analyzing pitchblende, a uranium-rich mineral that was a byproduct of the Eastern European silver mining industry, which served the hunger in Europe for precious metals and took advantage of steam-powered mining equipment.[9]

But novel tools and techniques were perhaps the most important of all for the rise of the rays. Glassblowing and vacuum technology had become ubiquitous by the late 1900s (see contributions by Catherine Jackson in part 1 and Cyrus C. M. Mody in part 2). Through the mid-1800s, partially evacuated glass tubes, rigged with wires to allow an electrical current to be pumped through them, became parlor curiosities for the strange luminous glow they would produce. In the 1860s, William Crookes (Figure 4.0.1), working in Britain, transformed these "discharge tubes," as they were known, into more flexible tools for studying electrical phenomena by vastly improving the vacuum that could be obtained in them, creating what became known in the English-speaking world as the Crookes tube. (In the Germanic world, similar apparatus was more commonly named for Johann Hittorf.)

Crookes tubes quickly became a favored tool for investigating electricity. "The phenomena attending the electric discharge through gases are so beautiful and varied that they have attracted the attention of numerous observers," wrote the Cambridge physicist J. J. Thomson, who explained that "while the passage of this agent [electricity] through a metal or an electrolyte is invisible, that through a gas is accompanied by the most brilliantly luminous effects, which in many cases are so much influenced by changes in the

Fig. 4.0.1. Sir William Crookes, depicted with his eponymous tube in *Vanity Fair* in 1903. *Credit*: Wikimedia Commons.

conditions of the discharge as to give us many opportunities of testing any view we may take of the nature of electricity, of the electric discharge, and of the relation between electricity and matter."[27 p189]

The Crookes tube was responsible for two discoveries that would enable a wide range of future tools for materials research, electrons and X-rays, and Thomson was responsible for the former. Before there was the electron, there were cathode rays. One of the first things Crookes observed when he began evacuating his tubes to produce higher and higher vacuums was a dark patch near the cathode in what had previously been a fully illuminated tube. At better and better vacuums, the dark spot grew, until the glow was reduced to a luminous spot at the end of the tube. Because this spot could be found directly opposite the cathode, or negative electrode, the phenomenon was presumed to be produced by rays emanating from the cathode, leading the German physicist Eugen Goldstein to coin the term "cathode rays" (Figure 4.0.2).

The precise nature of cathode rays remained mysterious through the late-nineteenth century. Experiments showed that cathode rays were deflected by magnets and that two cathode rays mutually repelled each other, supporting the conclusion that they were negatively charged. Crookes's own theory was that the gas molecules remaining in the tube — even the very high vacuums Crookes could produce were far from perfect — were ionized by the electrical current and, at very low pressures, could be accelerated across the tube without losing too much momentum through collisions with others. Thomson, however, was unsatisfied with this explanation.[27 p119] He measured the charge-to-mass ratio of cathode rays, which he judged to be incompatible with Crookes's theory of ion beams, and speculated instead that cathode rays consisted of "corpuscles," which he supposed to be component parts of atoms that also bore some responsibility for their chemical characteristics. For this leap, we credit Thomson with the discovery of the electron, a term that George Johnstone Stoney had proposed for an elementary unit of electricity in 1891.[22]

Thomson proposed his "corpuscles" in 1897. Thirty years later, his son, George Paget Thomson would pioneer electron diffraction techniques.[22 p1] At that point, by Hacking's criterion, electrons were firmly real. More precisely, they were sufficiently understood to be used as the basis for new tools and techniques to characterize materials. The Crookes tube can therefore be regarded as the grandparent of the range of electron diffraction and microscopy techniques that have provided some of our most vivid images of materials at the atomic scale (Figure 4.0.3). Thomes Vogt and Pedro Ruiz-Castell

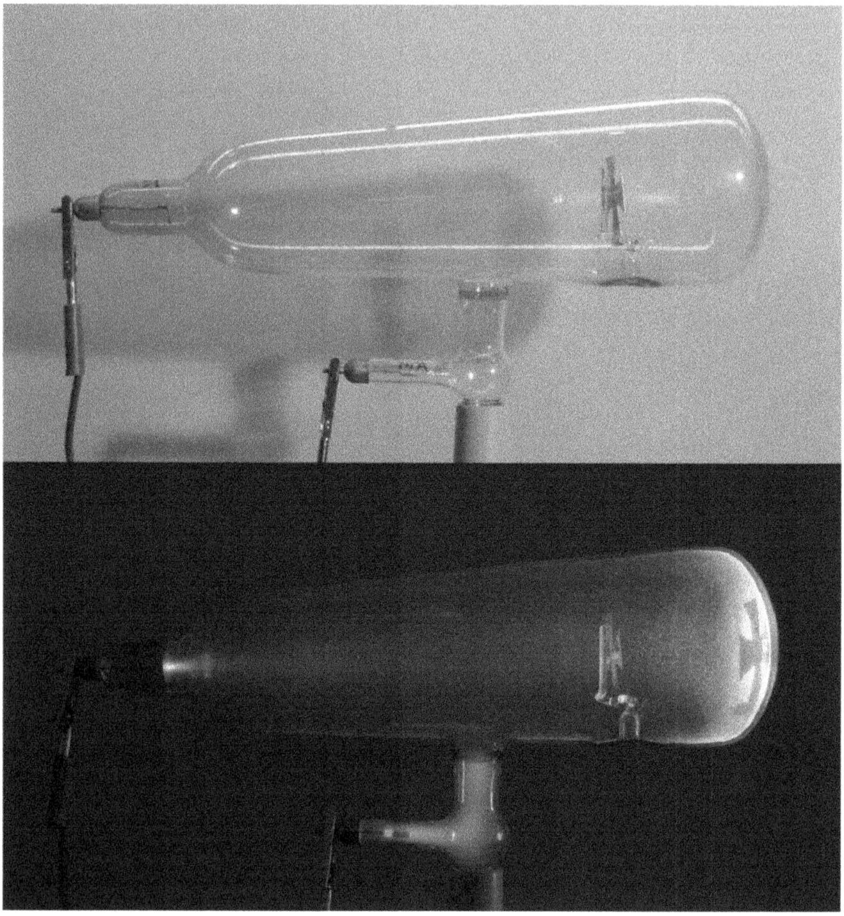

Fig. 4.0.2. A modern Crookes tube, before (top) a current is applied and after (bottom). The shadow cast by the Maltese cross provided evidence that the phosphorescence was produced by some type of emanation from the cathode. *Credit*: Wikimedia Commons.

offer further detail on these developments in their contributions to this section.

A long and steady process of research into the nature of cathode rays yielded the electron; the discovery of X-rays was somewhat more abrupt. In November 1895, Wilhelm Röntgen stumbled upon X-rays when investigating the source of an unexpected glow on a

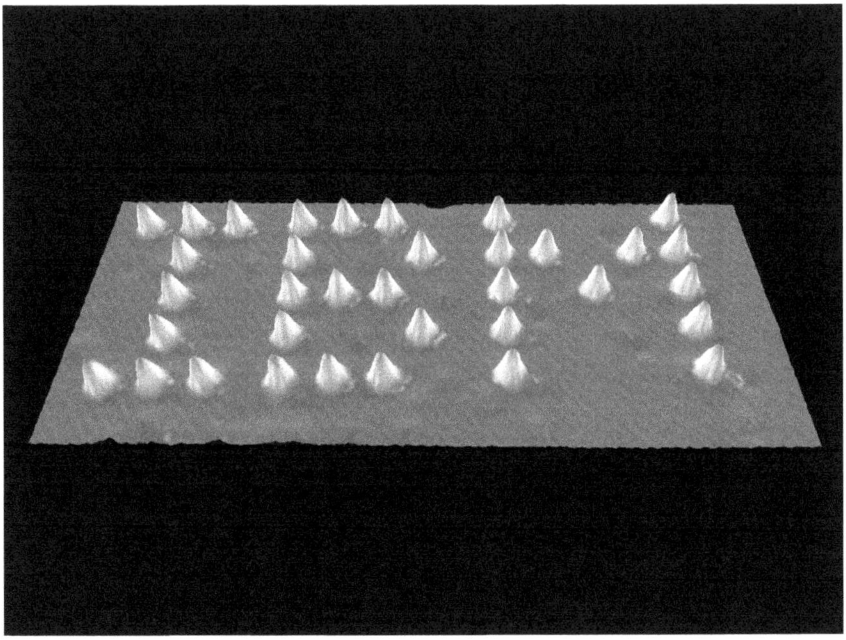

Fig. 4.0.3. On November 11, 1989, IBM researcher Don Eigler and his team used a custom-built scanning-tunneling microscope to spell out IBM with 35 xenon atoms. *Credit*: IBM News Room.

phosphorescing screen — a chemically coated sheet that emitted a telltale incandescence in the presence of light or radiation, and which could be used to determine its presence, source, and intensity. The phosphorescence Röntgen observed appeared in the vicinity of an operating Crookes tube, which was covered by an opaque box. Röntgen was not the first to notice this phenomenon, but he was the first to become curious enough to investigate it systematically.

Like any competent experimental physicist of the 1890s, Röntgen had a good idea of what sorts of properties to look for when he suspected he was dealing with a new type of ray. After ascertaining that they originated from his discharge tube, and noticing that they were highly penetrative compared with normal light, he placed a number of objects between the source of the X-rays and the screen to determine what types of materials they were capable of penetrating and to what degree. He experimented with photographic plates as

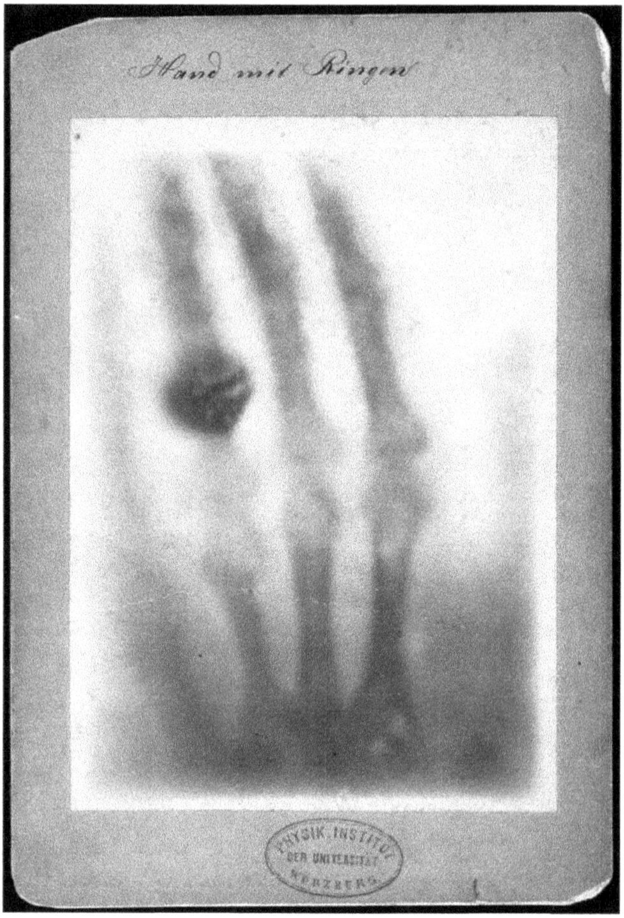

Fig. 4.0.4. Anna Bertha Ludwig's hand, as seen by X-ray. *Credit*: The Wellcome Collection.

well as phosphorescing screens, producing the now-iconic image of his wife's hand (Figure 4.0.4). He subjected them to magnetic fields and determined that they did not undergo magnetic or electrostatic deflection. He attempted to interfere them, but observed no discernable interference patterns.[23 p147–150]

After working feverishly for several weeks, Röntgen sent a short letter, "On a New Type of Rays" (*Ueber eine neue Art von Strahlen*), to the Würzburg Medical Physics Society (*Würzburger Physik.-medic. Gesellschaft*) detailing his investigations.[25] Because the set of properties he observed did not align with any other type of emanation known at the time, Röntgen named them "X-rays" (*X-Strahlen*), highlighting their mysterious nature and the provisional state of his understanding. Further investigations, in particular those by Hermann Haga, Cornelis Wind, and Charles Barkla, indicated that X-rays exhibited behavior such as diffraction and polarization, which were more commonly associated with light. By 1912, Max von Laue's theoretical prediction that X-rays would behave like light waves with a much shorter wavelength than visible light had been confirmed.[23 p148–149]

The medical applications of the new rays were the most immediately obvious. The ability to peer inside the body was a boon for doctors, who were beginning to standardize their practices and to embrace new technologies, of which X-rays were one. But many doctors were slow to adopt X-rays, in part because they had already developed other ways to view the interior of the body.[15] In contrast, X-rays represented a strikingly new possibility to peer into the submicroscopic interior of matter.

Max von Laue's astute prediction of the wavelength of X-rays relied on the additional assumption that their wavelengths would be on the order of the spacing of an atomic lattice. If that were the case, he realized, X-rays should diffract through crystals — that is, they should behave like waves and produce interference patterns. Most naturally occurring, inorganic solids are crystals, which means that the atoms that make them up are spaced out regularly, in a recurring pattern. In the early twentieth century, physicists and chemists had begun to understand solids as crystals with just such a regular structure. The insight von Laue achieved was that an experiment demonstrating X-ray diffraction would, in one fell swoop, indicate both the wavelength of X-rays and the lattice spacing of crystals, and provide a flexible tool for investigating crystal structure.[5; 6 p59–61] His lecture on the occasion of winning the 1914 Nobel Prize in Physics ended by pointing out the potential of X-ray diffraction for probing the nature of crystal lattices.[29]

Aus den Sitzungsberichten der Würzburger Physik.-medic. Gesellschaft 1895.

W. C. Röntgen: Ueber eine neue Art von Strahlen.

(Vorläufige Mittheilung.)

1. Lässt man durch eine *Hittorf*'sche Vacuumröhre, oder einen genügend evacuirten *Lenard*'schen, *Crookes*'schen oder ähnlichen Apparat die Entladungen eines grösseren *Ruhmkorff*'s gehen und bedeckt die Röhre mit einem ziemlich eng anliegenden Mantel aus dünnem, schwarzem Carton, so sieht man in dem vollständig verdunkelten Zimmer einen in die Nähe des Apparates gebrachten, mit Bariumplatincyanür angestrichenen Papierschirm bei jeder Entladung hell aufleuchten, fluoresciren, gleichgültig ob die angestrichene oder die andere Seite des Schirmes dem Entladungsapparat zugewendet ist. Die Fluorescenz ist noch in 2 m Entfernung vom Apparat bemerkbar.

Man überzeugt sich leicht, dass die Ursache der Fluorescenz vom Entladungsapparat und von keiner anderen Stelle der Leitung ausgeht.

Röntgen opened his report of his new discovery by writing:

> "If one passes the discharges of a large induction coil through a Hittorf vacuum tube, or a sufficiently evacuated Lenard or Crookes tube or other similar apparatus, and the tube is covered by a thin, black cardboard, one sees that, in the completely darkened room, a paper screen coated with barium platino-cyanide located near the apparatus, will illuminate brightly at every discharge, whether the coated or the other side of the screen is facing the discharge apparatus. The fluorescence remains visible at a distance of 2 m from the apparatus.
>
> "It is evident that the source of the fluorescence originates from the discharge apparatus and from nowhere else in the path."

The Nobel committee would continue the X-ray theme in 1915, awarding the physics prize to the father-and-son team of William Henry and William Lawrence Bragg. With his collaborators, von Laue had observed the diffraction patterns from the passage of X-rays through crystals; the Braggs had figured out how to read them. Even relatively simple crystals, such as sodium chloride (NaCl), produced intricate diffraction patterns that could be difficult to interpret (Figure 4.0.5). By figuring out how to reverse engineer these

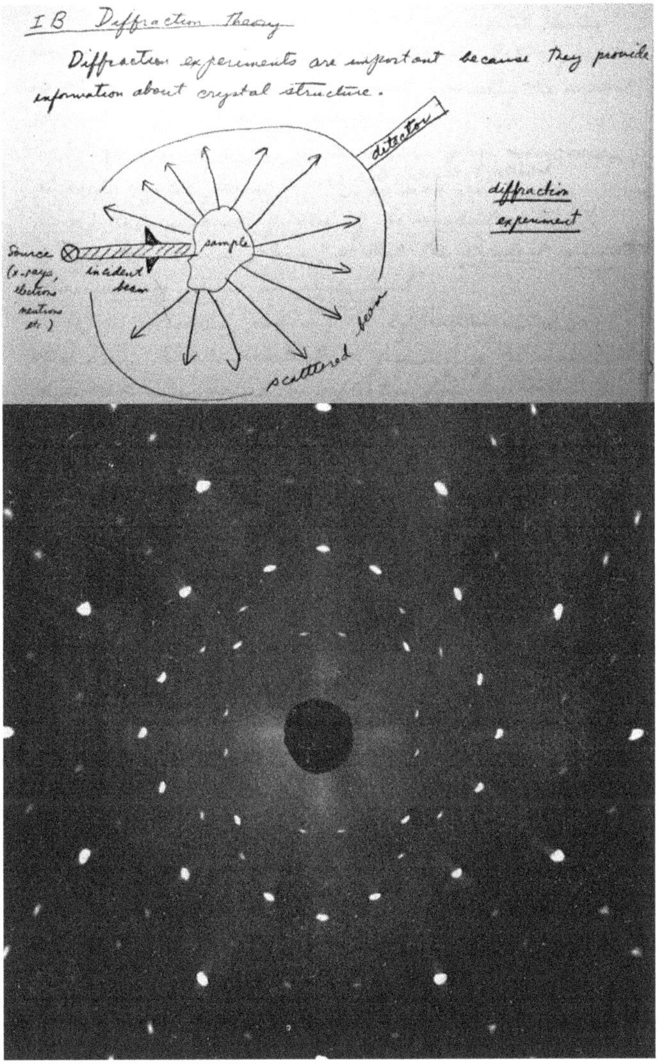

Fig. 4.0.5. Top: schematic depiction of a diffraction experiment from course notes written by Mildred Dresselhaus for her MIT course on solid state physics. *Credit:* Courtesy of Randall Richardson. Bottom: X-ray diffraction pattern of NaCl. *Credit:* Robert Eisberg and Robert Resnick, *Quantum Physics of Atoms, Molecules, Solids, Nuclei, and Particles.* New York: Wiley; 1974, 68, Copyright ©1974 by John Wiley & Sons, Inc.

patterns to infer the atomic structure of crystalline solids, the Braggs opened new frontiers in the understanding of material structure.[13]

As Robin Scheffler discusses in his contribution to this section, X-ray diffraction would eventually become a prominent tool in biological as well as in physical and chemical research. Many of the tools discussed here followed a similar trajectory. Originating as a way to arbitrate a specific question, they later became tools for materials research when scientists recognized their flexibility. Often, that flexibility became clear as new tools helped to drive new theoretical understanding. In the early twentieth century, that understanding came in the form of quantum physics, which proved critical for making sense of the data new tools were making available.

Materials and the Quantum Revolution

Optical dispersion, the tendency of material arrangements such as prisms and diffraction gratings to decompose white light into a spectrum, has been studied since the seventeenth century. Dispersion by glass prisms was, for example, one of the major subjects of Isaac Newton's *Opticks*. It was not until the nineteenth and twentieth centuries, however, that optical spectra became intimately relevant to the study of matter. In the early nineteenth century, Joseph von Fraunhofer, using diffraction gratings to produce cleaner spectra, observed a series of dark lines in the spectrum of the sun, which came to be known as Fraunhofer lines (Figure 4.0.6). Persistent efforts to understand these lines would lead to the quantum theory of matter and energy.

The realization that the dark Fraunhofer lines corresponded to bright bands of color in the spectra of light that substances gave off when they burn connected spectral lines to the study of matter. The dark lines in a spectrum of white light (an absorption spectrum) and the bright lines in a spectrum of colored light (an emission spectrum) were two sides of the same coin. Two German natural philosophers, Robert Bunsen and Gustav Kirchhoff, formed a collaboration that would cement the connection between absorption/emission spectra and the science of matter. They built their research program on new

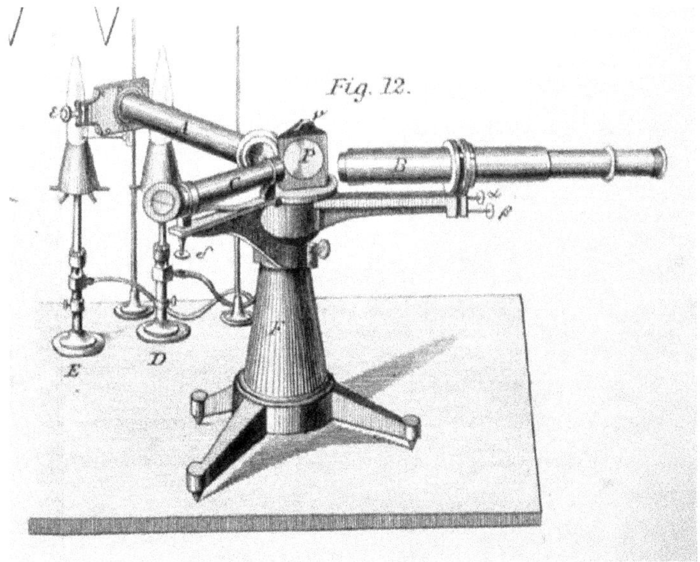

Fig. 4.0.6. Bunsen and Kirchhoff's spectroscope. The light produced by the lamps or burners E and D is decomposed by the prism P. C is a lensed tube with a scale inserted at the end. The user observes the spectrum superimposed on the scale through the scope B.[17]

tools. An improved form of heating apparatus, now known as the Bunsen burner, permitted them to burn materials in such a way that they would give off clear spectra, and a new piece of optical equipment, the spectroscope, made analyzing those spectra simpler and more precise.[20]

Their results led to the identification of a number of new materials, and their methods established a line of research that would foundationally change the scientific understanding of matter. Through the late nineteenth century and into the early twentieth, dispersion was understood to result from the interaction between light and matter, with phenomena like emission and absorption lines explained by some kind of resonance, or co-vibration, between light and matter. Investigating the nature of this interaction, and thereby explaining emission and absorption spectra, was instrumental to early quantum theory.[16]

Quantum physics, therefore, was a child of the spectroscope. Niels Bohr developed his model of the atom in 1913 to account for the Balmer series, the spectral lines that constitute the absorption/emission spectrum of hydrogen. Spectral analysis was equally important to the realization that Bohr's atomic model, which neatly described hydrogen, could not account for the structure of other atoms, and so motivated the development of a full-fledged quantum mechanics theory. As the American quantum physicist John Hasbrouk Van Vleck described the situation in 1928: "The chemist is apt to conceive of the physicist as some one who is so entranced in spectral lines that he closes his eyes to other phenomena."[28 p493]

Van Vleck's quip, although it indicates the extent to which spectral data were critical for opening new lines of research, elides the extent to which the matter responsible for spectral lines was also a subject of physical fascination. One critical example is the Compton effect. In 1922, Arthur Holly Compton observed a correlation between the wavelength of radiation produced in X-ray scattering and the angle at which it was scattered. The change in wavelength implied a corresponding loss of energy, thus softening the scattered secondary rays. This correlation posed a problem from the classical wave-theory perspective in which such effects would have been expected to disappear at the low intensities Compton was using, and he was not immediately sure how to explain his results. His crucial insight came when he recognized that the process could be treated as an inelastic collision between the scattering electron and the scattered X-ray quantum.[3 p221] His insight was possible because of apparatus descended from the Crookes tube that allowed him to generate X-rays with very precise wavelengths.

To many in the physics community, Compton scattering, and his explanation in terms of light quanta published in a 1922 *Physical Review* paper, provided the smoking gun needed to validate the reality of light quanta. Compton described the process like this: "Any particular quantum of x-rays is not scattered by all the electrons in the radiator, but spends all of its energy upon some particular electron. This electron will in turn scatter the ray in some definite direction, at an angle with the incident beam. This bending of the path of the quantum of radiation results in a change in its momentum. As a consequence, the scattering electron will recoil with a momentum equal to the change in momentum of the X-ray."[4 p485] In the face of Compton's strong experimental evidence, the bulk of the physics community finally accepted Albert Einstein's 1905 light quanta hypothesis.[26]

As the Compton effect demonstrated, understanding light to be composed of individual quanta of electromagnetic energy was central to analyzing light-matter interactions. This is one of the reasons Einstein's Nobel Prize came not for the theory of relativity, for which he is best remembered, but for his explanation of the photoelectric effect, which postulated light quanta. Quantum packets of light energy, and their interactions with the electrons confined within materials, would become the basis for a range of new spectroscopic techniques through the mid-twentieth century (see contributions to this section by Loïc Petitgirard, Catherine Westfall, Robin Scheffler, Indianara Lima Silva, and Benjamin Wilson).

Historians have often looked at quantum mechanics as a theory that was mostly worked out with appeal to elementary particles, such as the electron and the photon, and very simple systems such as the hydrogen atom, before it was applied to understand more complex physical and chemical systems. But recent scholarship has shown how the investigation of materials was critical to the articulation of quantum mechanics itself. Concepts like tunneling, resonance, exchange, and others, which feature centrally in our contemporary understanding of quantum mechanics, especially its role in determining the properties of materials, were worked out in the late 1920s and early 1930 as the theory of quantum mechanics itself was maturing.[14]

Investigations of the quantum behavior of solids and molecules were in turn critical for interpreting the results returned by new tools. After World War II, neutron bombardment became a powerful way to understand materials (see Hallonsten and Kaiserfeld in this section). Neutrons, which are electrically neutral, are less likely to interact with or be deflected by charged electrons and can penetrate materials deeply, providing glimpses of properties and behavior that photons or electrons are likely to ignore. But interpreting the results of neutron bombardment requires an understanding of quantum features of the solid state such as phonons — quantized, vibrational normal modes in a crystal lattice, which carry energy through it.[18 pIII-51–52]

Materials helped build quantum mechanics, and quantum mechanics in turn motivated a better understanding of materials. Tools for characterizing materials likewise evolved as part of a rich give-and-take between theory and experiment. Devices like the spectroscope generated the rich data set on spectral lines that made it possible for Bohr and others to imagine what the internal structure of the atom might be like. The quantum theory that emerged from their speculations matured into quantum mechanics as physicists and chemists grappled with the challenges posed by understanding the quantum properties of materials. That understanding in turn opened up a range of new possibilities for using the various species of rays discussed above to further probe the structure, behavior, and properties of matter.

Making Materials Real

We began with Hacking's suggestion that things are real if we can spray them, and considered in turn what scientists were spraying (photons, electrons, neutrons, and other particles) and at what they were spraying them (matter, predominantly regular solid crystals, which were instrumental to both developing and testing quantum theory, but later in the century including amorphous solids and biological matter as well). It remains to consider the broader purposes for which material characterization tools were deployed. As the twentieth century matured into middle age, so did the scientific understanding of materials. It would be fair to extend Hacking's observation that we take the reality of unobservable entities seriously when we spray them in service of another end to say that understanding materials provided that end. But the things scientists sprayed were far from the only thing whose reality was at issue.

A striking example comes, in fact, from the use of new tools for materials research to demonstrate an *un*reality. In 1912, just as X-ray diffraction was emerging as a technique and quantum theory was beginning to take hold, a new discovery shook the world of anthropology as well. Bone fragments discovered near Sussex, England, pointed toward a new human ancestor, dubbed Piltdown Man. The bones — a partial skull, jaw, and some teeth — threatened to upend the standard understanding of human evolution. The find was so surprising, in fact, that some anthropologists immediately suspected an elaborate and sophisticated hoax.

Despite longstanding suspicions, it was not until 1953 that the hoax was demonstrated conclusively. And an electron microscope would play a critical role in uncovering the Piltdown Man hoax. Anthropologists Kenneth Oakley, of the British Museum of Natural History, and Joseph Weiner, of Oxford University, both fascinated with the four-decade old case, joined forces to test the Piltdown artifacts using a range of methods that had emerged since 1912. These included inspection with an electron microscope, which showed the presence of organic material in the jawbone, indicating that it must be a modern specimen. Oakley and Weiner also used radiography, an X-ray imaging technique, to indicate the presence of modern dyestuffs in the bones. When combined with other physical and chemical investigative methods, these tests proved conclusively that the bones were both too young to be of evolutionary interest and most certainly the product of chicanery.[24]

Oakley and Weiner could make this determination in 1953, whereas the hoax could not be proven in 1912, because of the greater understanding of materials the characterizing tools that emerged in the interim allowed. The importance of spraying things, in other words, is not merely, or even mostly, that it lets us know that the things we are spraying are real. It also represents a level of control over materials that opens up new ways of knowing their structure, properties, and in some cases their histories. It was precisely that level of control that made tools of characterizing materials so integral to Cold War–era research and development. The development of the tools discussed in this section, indeed, entered its most urgent phase after World War II.

In 1951, the US National Research Council (NRC) formed a Materials Advisory Board (MAB) to evaluate how advances in materials research might address strategic bottlenecks, particularly in military development. In 1967, MAB issued a report, *Characterization of Materials*, which suggested: "Attempts to provide the superior materials that are critically needed in defense and industry are usually empirical and often wasteful of efforts and funds. That is so, chiefly because we do not yet have a fully developed science of materials that affords predictable and reliable results in devising and engineering new materials for specific tasks." In order to achieve this goal, the report insisted that "a wider dissemination and a wider availability of the tools of characterization are needed."[21 pI-ix, II-41] Successfully characterizing materials was essential to strategic development aims and attention to the tools that made those characterizations possible was a central part of achieving those aims.

The tools described in the following essays were responsible — largely within the context of the Cold War and building on early twentieth century developments in microphysics and quantum mechanics — for redefining what scientists mean when they talk about materials. Materials are often distinguished from other forms of matter because they can be or have been turned to human purposes. Nothing about that definition requires a robust scientific understanding of why the materials in question have properties that people find useful. The proliferation of tools for characterizing materials brought that knowledge within grasp. These tools helped fuse the many traditions of materials research into a new interdisciplinary fields of materials science.[2] In doing so, they made knowledge of the inner workings of matter essential to the concept of materials as the class of substances that humans use to achieve their aims and desires.

References

1. Bächtold M. Saving Mach's view on atoms. Journal for General Philosophy of Science. 2010;41(1):1–19.
2. Bensaude-Vincent B. The construction of a discipline: materials science in the United States. Historical Studies in the Physical and Biological Sciences. 2001;31(2):223–48.

3. Brush SG. How ideas became knowledge: the light-quantum hypothesis 1905–1935. Historical Studies in the Physical and Biological Sciences. 2007;37(2):205–46.
4. Compton AH. A quantum theory of the scattering of X-rays by light elements. Physical Review. 1923;21(5):483–502.
5. Eckert M. Max von Laue and the discovery of X-ray diffraction in 1912. Annalen der Physik. 2012;524(5):A83–A85.
6. Eckert M, Schubert H. Crystals, electrons, transistors: from scholar's study to industrial research. New York: American Institute of Physics; 1990.
7. Franklin A. The neglect of experiment. Cambridge: Cambridge University Press; 1986.
8. Glasser O. Wilhelm Conrad Röntgen and the early history of the Röntgen rays. San Francisco: Norman; 1993.
9. Goldsmith B. Obsessive genius: the inner world of Marie Curie. New York: W. W. Norton; 2005.
10. Hacking I. Representing and intervening. Cambridge: Cambridge University Press; 1983.
11. Hoddeson LH, Baym G. The development of the quantum mechanical electron theory of metals: 1900–28. Proceedings of the Royal Society of London Series A, Mathematical and Physical Sciences. 1980;371(1744)8–23.
12. Hunt B. Pursuing power and light: technology and physics from James Watt to Albert Einstein. Baltimore: Johns Hopkins University Press; 2010.
13. Hunter GK. Light is a messenger: the life and science of William Lawrence Bragg. Oxford: Oxford University Press; 2004.
14. James J, Joas C. Subsequent and subsidiary? rethinking the role of applications in establishing quantum mechanics. Historical Studies in the Natural Sciences. 2015;45(5):641–702.
15. Jamieson A. More than meets the eye: revealing the therapeutic potential of "light," 1896–1910. Social History of Medicine. 2013:26(4):715–37.
16. Jordi Taltavull M. Transmitting knowledge across divides: optical dispersion from classical to quantum physics. Historical Studies in the Natural Sciences. 2016;46(3):313–59.
17. Kirchhoff G, Bunsen R. Chemical analysis by spectrum-observations — second memoir. Philosophical Magazine, suppl. 4. 1861;22(148):329–49.
18. Materials Advisory Board. Characterization of materials. Washington, DC: National Academy of Sciences; 1967.
19. Joas C, Eckert M. Arnold Sommerfeld and condensed matter physics. Annual Review of Condensed Matter Physics. 2017;8:31–49.
20. Morris PJT. The matter factory: a history of the chemical laboratory. London: Science Museum; 2015.
21. National Academy of Sciences — National Academy of Engineering, Characterizing materials. Publication MAB-229-M. Washington, DC: National Academies Press; 1967.
22. Navarro J. A history of the electron: J. J. and G. P. Thomson. Cambridge: Cambridge University Press; 2012.

23. Nye MJ. Before big science: the pursuit of modern chemistry and physics. Cambridge, MA: Harvard University Press, 1996.
24. Oakley KP, Weiner JS. Piltdown man. American Scientist. 1955;43(4): 573–83.
25. Röntgen WC. Ueber eine neue Art von Strahlen. Sitzungsberichten der Würzburger Physik.-medic. Gesellschaft. 1895;S1–9.
26. Stuewer R. The Compton effect: turning point in physics. New York: Science History Publications; 1975.
27. Thomson JJ. Recent researches in electricity and magnetism. Oxford: Clarendon; 1893.
28. Van Vleck JH. The new quantum mechanics. Chemical Reviews. 1928;5: 467–507.
29. von Laue M. Concerning the detection of X-ray interferences. Nobel lecture, November 12, 1915. NobelPrize.org. https://www.nobelprize.org/nobel_prizes/physics/laureates/1914/laue-lecture.pdf [Accessed March 12, 2017].
30. Womack JC. Uncertainty medicine: the development of radiation therapy, 1895–1925 [PhD dissertation]. Houston, TX: University of Houston; 2016.

CHAPTER 4.1

X-RAY CRYSTALLOGRAPHY OF BIOMATERIALS

ROBIN WOLFE SCHEFFLER

Among the methods developed for the analysis of matter at the beginning of the twentieth century, X-ray crystallography (XC) promised the tantalizing ability to envision the structure of substances at the atomic level. This vision proved especially alluring for the application of XC to biomaterials. Biomaterials encompass those materials often used in medical devices, such as ceramics and plastics. These materials are distinct from "biological materials" — substances in nature such as enzymes and fibers whose structure is of interest to biologists but which have usually fallen outside the scope of materials science.[18] By the late twentieth century, however, XC provided a new means of engineering biological materials themselves as biomaterials for the delivery of pharmaceuticals and other uses, opening up the potential for molecules as biomaterials.[9] This essay will follow the role that XC played in the "molecularization" of life and the transformation of biological materials into biomaterials — the visualization of molecular biological structures provided a precursor for manipulating those structures as biomaterials.[8; 16]

The biological applications of XC were not immediately evident in 1912 when the German physicist Max von Laue observed that crystals served as a diffraction grating for X-ray radiation, nor when the father-son team of William Henry and William Lawrence Bragg worked out the formulae that could be used to calculate the dimensions of crystal unit cells — the basic arrangement of atoms that repeats in a regular solid — based on the diffraction patterns produced by XC shortly thereafter. "Bragg's Law," which gives the angles from which incident radiation is scattered by a crystal, suggested the possibility of using XC in the same manner that optical studies of diffraction had been used to establish the symmetry group of crystal unit cells. Given that the wavelength of X-ray radiation is much smaller than that of visible light, its diffraction pattern could reveal details about crystal structure that optical crystallography could not. Following Bragg's law, an X-ray beam passing through a crystal would be deflected at angles, θ, by an atom (or more accurately the electrons surrounding an atom's nucleus), satisfying the relationship $2dsin(\theta) = n\lambda$, determined by λ, the wavelength of the X-ray radiation, and d, the spacing between atoms in the crystal lattice. This one-dimensional case could be extended to all three dimensions of spacing in a crystal lattice. Therefore, with knowledge of the diffraction pattern and the wavelength of the X-ray radiation used, it was hypothetically possible to work backwards to determine the three dimensional structure of the lattice, although this became more and more complicated the more atoms that there were to diffract the X-ray beam — the diffraction pattern was the total sum of all of these interactions (Figure 4.1.1).[2]

In the United Kingdom, the analysis of biological substances gained impetus from the decision by William Lawrence and William Henry Bragg to divide the field of XC between their two laboratories.

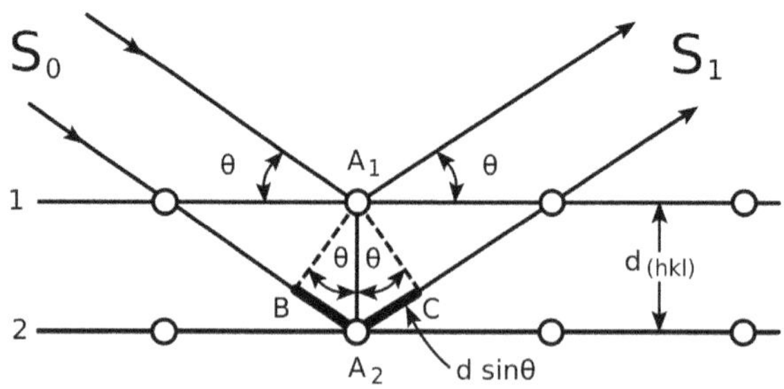

Fig. 4.1.1. Illustration of Bragg's Law.

W. H. Bragg moved to London to continue the analysis of organic materials while W. L. Bragg remained in Manchester to study inorganic materials, with an emphasis on their potential industrial applications. For researchers in Manchester, XC was of the greatest use for the analysis of inorganic crystalline structures, such as those found in mineralogy and metallurgy. The popularity of X-ray powder crystallography addressed these applications, often in connection with industrial applications in which alloys endured extremes of temperature. Affiliation with industrial firms also aided XC by allowing access to more powerful X-ray sources.[20] The XC analysis of organic substances presented possible industrial uses — such as the work of W. H. Bragg's student William Thomas Astbury on the structure of the protein keratin, which he started in conjunction with the ongoing efforts of the Department of Textile Industries at the University of Leeds to explore the chemistry of wool fiber.[13]

Another student of W. H. Bragg in London, the crystallographer and physicist J. D. Bernal, also became interested in the structure of proteins in the late 1920s, moving to study the XC of organic molecules from his mathematical studies of crystalline space groups. Bernal moved to the Cavendish Physics Laboratory at Cambridge University, where his proximity to the Sir William Dunn Institute for Biochemistry ensured him steady access to a range of purified hormones, vitamins, and other biologically active compounds.[6] Informed by his Marxist politics, Bernal hoped that the study of the structure of organic materials through XC would provide a template for the creation of new "molecular materials" to aid humanity — a remaking of the world to match a socialist remaking of society.[4] Between their two laboratories, Bernal and Astbury trained a generation of X-ray crystallographers interested in biological macromolecules. Working with Dorothy Hodgkin, Bernal produced the first XC diffraction pictures of protein crystals in 1934. However, the use of XC on proteins faced ongoing challenges: determining the appropriate humidity for protein crystals (which expanded their molecular structure), calculating the phases of refracted X-ray beams, making thousands of intensity measurements, and handling the immense amount of data generated by these measurements.[14]

Especially for proteins and other macromolecules, the large numbers of intensity readings and the corresponding complexity of the diffraction calculations made the practice of XC a problem of computation as well as instrumentation and measurement. Starting in the 1930s, X-ray crystallographers started to develop mathematical methods to address the computational question of data analysis. The practice of using Fourier transformations on data to reveal structural regularities — introduced by W. L. Bragg in 1928 and famously used in the Patterson function — offered a means of converting data on thousands of diffraction readings into a map of electron densities that was far easier to interpret.[3; 19]

Although it is most famously remembered in the biological realm for its contribution to the elucidation of the structure of DNA, XC in the 1950s and 1960s proved essential to the development of what the historian of science Robert Olby termed the "protein school" of molecular biology.[17] XC grew as a product of three mutually supporting sets of practices: improvements in the technique of producing XC diffraction images, model building, and computing power. Famously, Rosalind Franklin manipulated the humidity of DNA crystals to produce high quality XC images, a feat which played a critical role in the solution of DNA's double-helical structure in 1953 (Figure 4.1.2).[15]

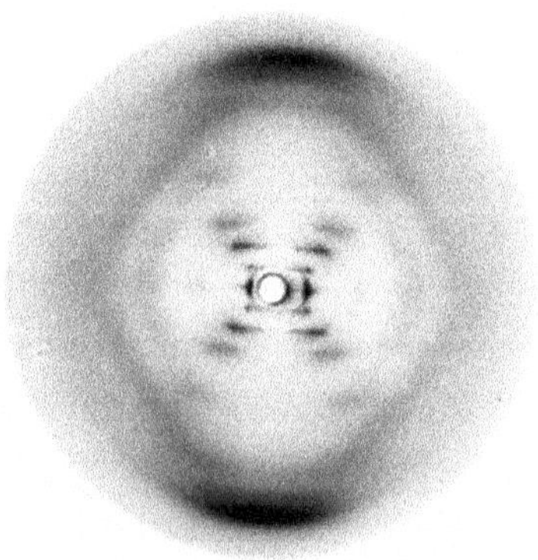

Fig. 4.1.2. "Photo 51" by Rosalind Franklin — the X-ray crystallograph image that provided suggestive evidence for the double-helical structure of DNA. *Credit:* Image used by permission of King's College London.

Even as this announcement came, Franklin had already turned to the analysis of viruses using XC methods, another productive field for X-ray crystallographers.[10] At Oxford, Dorothy Hodgkin, a collaborator of Bernal in the 1930s, provided a structure for vitamin B12 in 1955 — a structure that traditional biochemical methods would have failed to identify.[11]

Meanwhile, at Cambridge University, Max Perutz and Joseph Kendrew used a combination of new computational resources (from the Mathematical Laboratory, an early center of digital computing), refined XC methods such as heavy-atom substitution in crystals, and model-building to propose the first structural models for proteins — myoglobin and hemoglobin — in 1959 (Figure 4.1.3).[7] These structures marked major advances in biology — Perutz, Hodgkin, and Kendrew were all awarded Nobel Prizes, and controversy continues as to the question of whether Franklin's early death deprived her of the opportunity to be considered for a share of the Prize given for solving DNA's structure.[15] Interest in the structure of macromolecules

Fig. 4.1.3. Max Perutz Engaged in Model-building for the structure of hemoglobin, 1959. *Credit*: Image courtesy of the Laboratory of Molecular Biology, Cambridge University.

deepened after 1961 with growing awareness of the importance of protein folding to biochemical reactions.

Protein XC became a key component in the intellectual program of molecular biology as it sought to create a chain of connections from DNA "code" to amino acid sequence to three-dimensional protein form to biological function (Figure 4.1.3). From the perspective of biology, understanding the three-dimensional structure of molecules was important because of the connection between the chemical activity of compounds and their specific structure. An analysis of chemical composition alone could not provide this information. By the 1970s, X-ray crystallography, buttressed by expanding computing power, became integral to organic chemistry and molecular biology. Protein crystallographers developed into a semi-autonomous community within XC — one marked by a focus on using XC diffraction data for exact structures rather than the classification of crystalline materials based on space groups. Although the creation of structures remained laborious, their generation became a matter of routine laboratory work. Advances in experimental techniques and technical apparatus aided the routinization of the use of XC in structural biology. The ability to dehydrate crystals faster, improvements in computing and collection methods, and advances in X-ray radiation sources to produce beams of constant frequency have applied to the studies of many materials; their impact has been most profound on approaches to biological materials as biomaterials (Figure 4.1.4).[1; 21; 22]

In the 1980s, protein crystallography, which had earlier failed to catch industry's attention, drew intense interest from pharmaceutical companies and biotechnology firms due to the widespread use of recombinant DNA, which allowed the production of a wide range of proteins, and a growing interest in the possibility that slight structural differences in proteins might offer promising targets for new drugs.[4] This ability to examine protein structure in three dimensions using advances in XC was integral to ambitions that computer-aided design would create small-molecule drugs that could modify the biological activity of enzymes or cell receptors.[17; 23] Alternately,

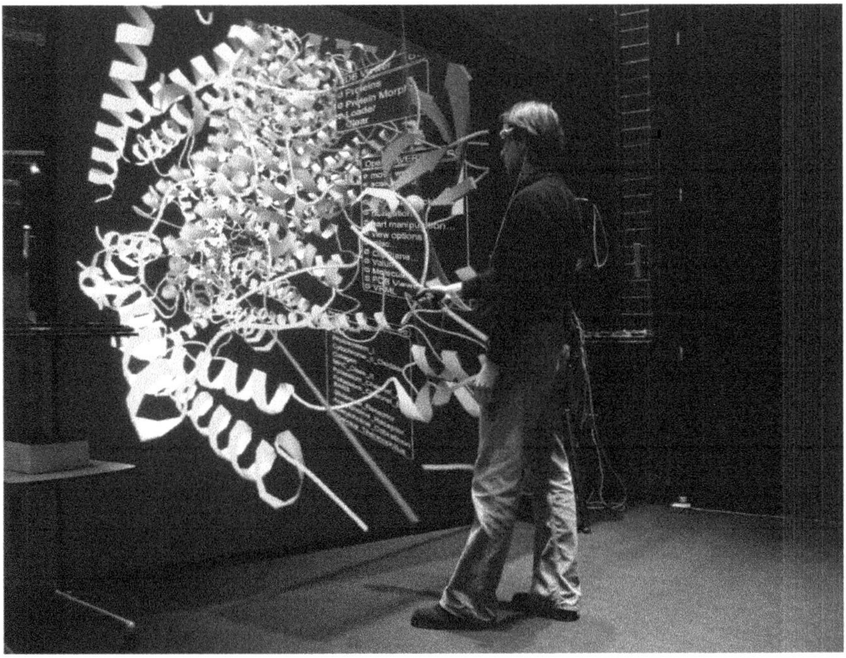

Fig. 4.1.4. Protein crystallographer working with a Protein Data Bank software interface "C-Wall" (virtual reality wall). *Credit*: National Science Foundation, 2006.

the structure of peptide chains and the proteins they help constitute have been considered as "cages" for the delivery of drugs and sources of new biomaterials mimicking natural tissue.[5; 12] XC, in providing a vision of biological materials at the molecular level, in this manner aided their transformation into biomaterials in the more conventional sense of the term.

References

1. Abergel C. Spectacular improvement of X-ray diffraction through fast desiccation of protein crystals. Acta Crystallographica Section D: Structural Biology. 2004;60(8):1413–16.

2. Authier A. Early days of x-ray crystallography. Oxford: Oxford University Press; 2013.
3. Beevers CA, Lipson H. A brief history of Fourier methods in crystal-structure determination, Australian Journal of Physics. 1985;38(3):263–72.
4. Berol DN. Living materials and the structural ideal: the development of the protein crystallography community in the 20th century [PhD dissertation]. Princeton, NJ: Princeton University; 2001.
5. Bhaskar S, Lim S. Engineering protein nanocages as carriers for biomedical applications. NPG Asia Materials. 2017;9(4):e371.
6. Brown A. J. D. Bernal: the sage of science. Oxford: Oxford University Press; 2005.
7. Chadarevian S de. Designs for life: molecular biology after World War II. Cambridge: Cambridge University Press; 2002.
8. Chadarevian S de, Kamminga H, editors. Molecularizing biology and medicine: new practices and alliances, 1910s–1970s. Amsterdam: Harwood Academic Publishers; 1998.
9. Cooper DR, Porebski PJ, Chruszcz M, Minor W. X-ray crystallography: Assessment and validation of protein-small molecule complexes for drug discovery. Expert Opinion on Drug Discovery. 2011 Aug 1;6(8):771–82.
10. Creager ANH, Morgan GJ. After the double helix: Rosalind Franklin's research on tobacco mosaic virus. Isis. 2008;99(2):239–72.
11. Dodson G. Dorothy Mary Crowfoot Hodgkin, O.M. 12 May 1910–29 July 1994. Biographical Memoirs of Fellows of the Royal Society. 2002;48: 179–219.
12. Gagner JE, Kim W, Chaikof EL. Designing protein-based biomaterials for medical applications. Acta Biomaterialia. 2014;10(4):1542–57.
13. Hall KT. The man in the monkeynut coat: William Astbury and the forgotten road to the double-helix. Oxford: Oxford University Press; 2014.
14. Law J. The development of specialties in science: the case of x-ray protein Crystallography. Social Studies of Science. 1973;3(3):275–303.
15. Maddox B. Rosalind Franklin: The dark lady of DNA. New York: HarperCollins; 2002.
16. Myers N. Rendering life molecular: models, modelers, and excitable matter. Durham, NC: Duke University Press; 2015.
17. Olby RC. The path to the double helix. London: Macmillan; 1974.
18. Park JB, Lakes RS. Biomaterials: an introduction, 3rd ed. New York: Springer; 2007.
19. Patterson AL. A direct method for the determination of the components of interatomic distances in crystals. Zeitschrift für Kristallographie — Crystalline Materials. 1935:1;90(1–6):517–42.
20. Scheffler RW. Interests and instrument: a micro-history of object Wh.3469 (X-ray powder diffraction camera, ca. 1940). Studies in History and Philosophy of Science Part A. 2009;40(4):396–404.

21. Shi Y. A glimpse of structural biology through x-ray crystallography. Cell. 2014 Nov 20;159(5):995–1014.
22. Schlichting I. Serial femtosecond crystallography: the first five years. IUCrJ. 2015;2(2):246–55.
23. Zheng H, Hou J, Zimmerman MD, Wlodawer A, Minor W. The future of crystallography in drug discovery. Expert Opinion on Drug Discovery. 2014;9(2):125–37.

CHAPTER 4.2

RAMAN SPECTROSCOPY

INDIANARA SILVA

In the twentieth century, Raman spectroscopy revolutionized our way of analyzing molecules and the structure of materials. It led to applications in areas from arts and museums — where it has been used as a non-destructive technique to identify pigments, dates, and even to detect falsifications — to the life sciences, where it has become an effective tool for studying human skin, bacteria cells, DNA, animal brains, and for intracellular imaging. The Raman method has been used to examine food and beverages, identifying their compositions and any contaminants. It has helped to detect controlled drugs and narcotics and even forged documents. In geology, solid state physics, nanomaterials, and pharmaceuticals, it has been of immeasurable value in analyzing and characterizing different types of material simply and accurately.[4; 6; 11] Raman spectroscopy was already a powerful and fruitful analytical tool in the late 1920s, when it was first developed, but just a few of its many applications were in sight then. It would reach its full potential with the advent of the laser in the 1960s.[5; 11; 12; 19]

In 1928, the Indian physicist Chandrasekhara Venkata Raman published his work on a new type of spectroscopy. Its impact was so striking that two years later he was awarded the Nobel Prize in Physics for "his work on the scattering of light and for the discovery of the effect named after him," becoming the first Asian scientist to win the most prestigious and coveted prize in science.[19] Raman was born in Trichinopoly, South India, and then moved to Visakhapatnam where he studied at St. Aloysius Anglo-Indian High School. At university, what is now the Presidency College of the University of Madras, Raman concluded his BA in 1904 receiving a Gold Medal in physics. Three years later he completed his MA with high distinction. During his university studies, he was primarily interested in acoustics and optics.[1; 2]

Starting a scientific career in India was not an easy task at that time. Talented and dedicated people like Raman found little encouragement. Despite his enthusiasm for science, in 1907 he joined the Indian Finance Department as an Assistant Accountant-General at Calcutta, where his work involved administrative duties, not physics. However, even while working for the government, Raman found opportunities to do experimental research at the Indian Association for the Cultivation of Science, the oldest Indian private research institute. His results would contribute later to the discovery of the Raman effect.[1; 2; 9; 10; 12; 15]

It was a productive time for Raman, who published approximately thirty original papers in several renowned journals.[12] The international scientific recognition he won did not go unnoticed by the Indian scientific community. In 1917 he left government service to accept the Palit Professorship of Physics at Calcutta University. The reader may wonder where and in which field he did his doctoral studies. Atypically, for a scientist of his stature, he never even began study for a PhD. Raman is among those very few physicists who won the Nobel Prize without having pursued formal education beyond a master's degree. He would nevertheless go on to receive innumerable honorary doctorate degrees from both Indian and international universities, countless awards and honors, and honorary membership in several local, national, and international societies.[1; 2; 9; 10; 12; 15]

Those honors stemmed from work Raman began in the late 1920s on the scattering of light in liquids, vapors, and gases. His first genuine concern was to explain why the Mediterranean Sea, which he had seen on a trip to Europe in 1921, reflected an opalescent blue so beautifully. Lord Rayleigh had answered this by saying that the blue color of the sea was a result of the reflection of the blue of the sky and was not related to the color of the water, a point Raman strongly disagreed with. His assumption was that the blue color of the sea was due to the scattering of the sunlight by water molecules. During his boat trip back to India, Raman decided to test his assumption with an experiment to analyze what would happen when the sunlight, captured by a heliostat on the roof, and then polarized through a Nicol prism, passed through water — isolated from the sky's reflection. He found that the color of the sea remained brilliantly blue, an effect of the scattering of light by water.[9; 10; 12; 15]

In order to investigate his experimental finding further, Raman initiated a research program in Calcutta, now using a mercury arc lamp instead of sunlight as a light source. Using this light source, a variety of samples, and a small, direct-vision pocket spectroscope, he and his team observed an interesting and unexpected effect.[9; 10; 12] When light scattered off molecules, a slight fraction of the scattered radiation changed its wavelength and consequently had a different frequency from the incident light — a process known as inelastic scattering. Raman observed a new kind of scattered light — "a feeble fluorescence" — whose new lines were close to the original lines of the incident light in the spectrum.[15; 18] This phenomenon is widely known as the Raman effect, which was discovered almost simultaneously by the Russian physicists Grigory S. Landsberg and Leonid I. Mandelstam and earlier predicted theoretically by the Austrian physicist Adolf Smekal.[15; 16] It is an optical analogue to the Compton effect.[17; 18]

What would cause a shift in the wavelength of a portion of the radiation scattered by molecules? Or, to put it differently, what would cause a change in the color of the light after being scattered by a sample? The surprising — and useful — answer to this question is that the shift in the frequency of the scattered light is a signature of the chemical and physical properties of the molecules. Because energy and momentum are transferred between the incident light and the molecule off which it scatters, the properties of that molecule determine the characteristics of the Raman shift. That is why Raman spectroscopy has become an indispensable technique for gaining knowledge about the nature and the structure of a material medium at the microscopic level. It provides a kind of "fingerprint" of the molecule from which one obtains quantitative and qualitative information. The Raman spectrometer generally consists of a monochromatic light source, a sample chamber, and a detector, allowing the examination of Raman shifts, which are related to the vibrational frequencies of the scattering substance and measured in wavenumbers.[4; 6; 12; 18]

The field of Raman spectroscopy advanced significantly when the laser began to replace the mercury arc lamp as a light source. Because they emit a completely monochromatic and coherent light, lasers are ideal light sources for Raman-type experiments because it becomes possible to obtain spectra samples with high quality and resolution.[12] Adopting laser sources solved the problem of low signal intensity when using an ordinary light source, which generates significant noise and therefore gives a poor-quality spectrum, producing better resolution of the Raman peaks.[11; 20] Successful use of the laser, and later the technical development of high-sensitivity and high-resolution detectors, was the turning point in the field.

In 1961 the Brazilian physicist Sergio Porto, in collaboration with his colleague Darwin Wood, working at the Bell Telephone Laboratories in Murray Hill, New Jersey, pioneered the key techniques of laser Raman spectroscopy.[19] Porto received his undergraduate qualifications, in 1946 and 1947 respectively, in chemistry and in teaching of chemistry from the Federal University of Rio de Janeiro, and his PhD in 1954 in physics from the Johns Hopkins University, Baltimore. He also worked as a physics professor at the Instituto Tecnológico da Aeronáutica in Brazil and at the University of Southern California in Los Angeles. Porto returned to his home country in 1973 and within three years founded the Department of Quantum Electronics at the University of Campinas in São Paulo. Porto is widely recognized as the pioneer of the laser Raman spectroscopy field.[13; 14]

Before 1963, the Raman effect only provided information on vibrational, rotational and anisotropic spectra of gaseous and liquid samples and of mostly transparent crystals. Nevertheless, using lasers of higher and lower power, more sensitive photomultipliers, electronic detection recording systems, and computer interfaces to analyze the Raman data, it became possible to identify the structure of many compounds and crystals more precisely and accurately, even those of colored and opaque substances such as metals. With gradual improvements in the technique, new types of Raman excitations came to the fore during the laser era.[4; 12]

The basic principle is as follows: when monochromatic light is scattered by a molecule, two kinds of light scattering can occur, elastic and inelastic. Whereas in elastic scattering there is no modification in the frequency or wavelength of the scattered light, in inelastic scattering there is a modification in its frequency as a result of the excitation or deactivation of molecular vibrations. In this case, three types of phenomena can occur: 1) Rayleigh scattering, in which there is no exchange of energy and the frequency of the scattered light is the same as that of the incident light; 2) anti-Stokes Raman scattering, in which the frequency of the scattered light is higher than that of the incident light because the photon gains vibration energy from the molecule; 3) Stokes Raman scattering, in which the frequency of the scattered radiation is lower compared to that of the incident light because the molecule gains energy from the photon.[4]

Raman spectrometers have since evolved into several types. Dispersive Raman spectroscopy and Fourier transform (FT) Raman spectroscopy are among the most common technologies used to obtain the Raman spectra. FT-Raman spectrometers with near-infrared or red excitation, developed in recent years, have shown promise for avoiding fluorescence from the sample and sample impurities, which interfere in the Raman signals. Likewise, the substantial progress in highly sensitive detectors used with coupling optical fibers and microscopes has had a great impact on the analysis process. The difference between these techniques is generally associated with the laser sources in use and how the Raman scattering is detected and analyzed. Both of them have specific advantages, and depending on the research problem, one can be more suitable than other.[4] For example, Dispersive Raman spectroscopy is well suited to the analysis of many types of micro-level samples, and is widely applied in pharmaceuticals, life sciences, geology, gemology, and crystal studies. Its apparatus usually contains a visible laser source, a small sample, a diffraction grating and a silicon charge-couple detector, from which it is possible to extract information of the Raman scattered light wavelengths.[7]

By contrast, FT-Raman spectroscopy is ideal for bulk sample analysis of materials such as bags, plastic, solids, films, tubes, and so on. It generally uses a near-infrared laser source, a sample, an interferometer, and a near-infrared detector to obtain the Raman spectrum through the Fourier transform. Its applications include the analysis of illicit drugs and clandestine laboratory samples, fibers, explosives, and the control and quality of pharmaceutical products.[7]

Raman spectroscopy has taken a long path to become one of the most successful and flexible analytical techniques for revealing the molecular structure of materials. It is the result of a gradual improvement in our experimental and theoretical knowledge, as well as instrumentation in science and technology.

References

1. Banerjee S. C. V. Raman and colonial physics: acoustics and the quantum. Physics in Perspective. 2014;16(2):146–78.
2. Bhagavantam S. Chandrasekhara Venkata Raman (1888–1970). Biographical Memories of Fellow of the Royal Society. 1971;17:564–92.
3. Barbieri J. Dissertação detalha a trajetória do físico Sergio Porto, introdutor do laser no país. Available from: http://www.unicamp.br/unicamp/unicamp_hoje/ju/outubro2007/ju375pag02.html [Accessed November 21, 2017].
4. Das RS, Agrawal YK. Raman spectroscopy: recent advancements, techniques and applications. Vibrational Spectroscopy. 2011;57(2):163–76.
5. Guimaraes WON, Lin CT, Mooradian A. Lasers and applications: proceedings of the Sergio Porto Memorial Symposium, Rio de Janeiro, Brasil, June 29–July 3, 1980. Berlin: Springer-Verlag; 1981.
6. Grasselli JG, Snavely MK, Bulkin BJ. Applications of Raman spectroscopy. Physics Reports. 1980;65(4):231–344.
7. Introduction to Raman Spectroscopy. Available from: http://www.biotechprofiles.com/companyfiles/madisonnetwork/5bdd0a9f37694d6c9fb6e62db5049477.pdf [Accessed October 10, 2017].
8. Jha SS. Some recent trends in Raman spectroscopy. Available from: http://journal.iisc.ernet.in/index.php/iisc/article/viewFile/3220/3274 [Accessed October 10, 2017].
9. Jayaraman A. Chandrasekhara Venkata Raman — A Memoir. Indian Academy of Science. Bengaluru: Lotus Printers Pvt. Ltd.; 2017.
10. Krishnan RS, Shankar RK. Raman effect: history of the discovery. Journal of Raman Spectroscopy. 1980;10(1):1–8.

11. Lewis IR, Edwards HGM. Handbook of Raman Spectroscopy: From the Research Laboratory to the Process Line. New York: Marcel Dekker; 2001.
12. Miller FA, Kauffman GB, C. V. Raman and the discovery of the Raman effect. Journal of Chemical Education 1989;66(10):795–801.
13. Mooradian A, Stoicheff BP. Obituaries: Sergio Pereira da Silva Porto. Physics Today. 1979;32(10):81–82.
14. Santana WA, Freire Junior O. Contribuição do físico brasileiro Sergio Porto para as aplicações do laser e sua introdução no Brasil. Revista Brasileira de Ensino de Física. 2010;32(3):3601.
15. Singh R, C. V. Raman and the discovery of the Raman effect. Physics in Perspective. 2002;4(4):399–420.
16. Singh R, Riess F. The 1930 Nobel Prize for Physics: a close decision? Notes and Records of the Royal Society of London. 2001;55(2):267–83.
17. Stuewer RH. The Compton Effect: turning point in physics. New York: Science History Publications; 1975.
18. Raman CV. Nobel lecture: the molecular scattering of light. 1930. Available from: http://www.nobelprize.org/nobel_prizes/physics/laureates/1930/raman-lecture.html [Accessed February 2, 2017].
19. Porto SPS, Wood DL. Ruby optical maser as a Raman source. Journal of the Optical Society of America. 1962;52(3):251–52.
20. Wise DL. Handbook of pharmaceutical controlled release technology. New York: Marcel Dekker; 2000.

CHAPTER 4.3

INFRARED SPECTROSCOPY

LOÏC PETITGIRARD

Infrared (IR) spectroscopy has become a widespread tool for investigating materials because an infrared spectrum offers kind of a material fingerprint, revealing qualitative and quantitative information about a substance's structure and components. The absorption of a specific infrared frequency by a molecule corresponds to what is called a "vibrational mode" — a regular, quantized molecular oscillation. Understanding the vibrational modes of a material sample reveals, among other things, the bond structure of the molecules within it. Infrared spectroscopy can thereby help identify the components that make up a material and their relative abundance in a sample. The size of the absorption peaks in the spectrum gives an indication of the amount of a component in a given mixture. Almost every type of sample (liquids, powders, films, surfaces, gases, and so on) can be analyzed using IR spectroscopy.

The development of infrared spectroscopy through the twentieth century was nonlinear. Some principles, such as interferometry for spectroscopy in the IR, were forgotten for some time before being (re)discovered and developed after 1950. The field was characterized for a long time by competition between technologies, before it was transformed by the advent of digital computers.[5]

The first infrared spectroscopes were dispersive instruments. Infrared frequencies were separated by a prism, much the same as with visible light, and later by diffraction gratings, which better separate the frequencies.[5] The detectors in those instruments measured the energy passing through the sample for each frequency, rendering its spectrum as a plot of each measured intensity/frequency. Until the 1960s, these instruments were the only tools for rendering spectra from the near-infrared (close to the visible, wavelengths from 0.8 to 2 μm, corresponding to vibrations of molecules) to the far-infrared (25 to 1,000 μm, revealing the properties of metals, semiconductors, and other solids).

These early instruments were nevertheless slow, mechanically complex, fragile, and clumsy to handle, and had poor sensitivity when compared with later tools. Fourier transform infrared spectroscopy (FTIR), conceived around 1950 and developed over the course of fifteen years, considerably improved performance in all these areas. FTIR was first applied to far-IR spectroscopy, then demonstrated in the near-IR. Since the 1970s it has been the essential tool for investigating mid-IR spectra (2.5 to 25 μm).

Thirty years after this pioneering period, FTIR has become a "black box" instrument, making it an efficient analytical instrument for material research as well as a quality-control tool in industry.[5; 10] FTIR techniques allow investigation of chemical components and structures, biological materials (cells, bacteria), impurities in materials, and various physical properties such as thickness (important in the semiconductor industry for wafer control, for example), crystallinity, phase changes, and other phenomena. It remains among the most widely used and popular analytic techniques. The following recounts the development of FTIR in the second half of the twentieth century, but it also touches upon some other dispersive techniques, which developed in parallel and remained in use for niche applications in the far or near-infrared ranges.

Interferometry

An interferometer is made of a beam splitter, BS, and two mirrors (a fixed one, M1, and a moving one, M2) in a precise configuration (see Figure 4.3.1). A light beam enters the interferometer, is divided in two parts by the beam splitter. After reflecting off the respective mirrors, the two beams are recombined. Their path

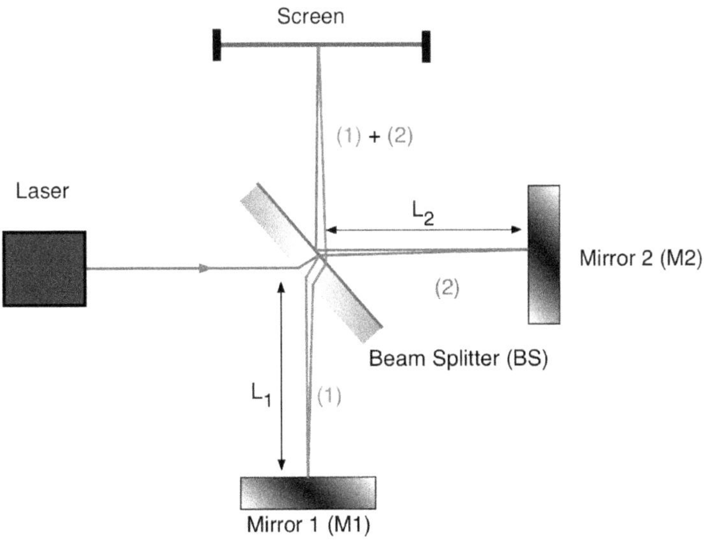

Fig. 4.3.1. The basic principle of the interferometer. *Credit*: Wikimedia commons.

difference can be discerned by analyzing the constructive or destructive interference produced by interfering the beams. Interferometers output light beams with a set of intricated wavelengths, modulated by the moving mirror. Each path difference renders different interference fringes, and thus produces an output beam with a specific set of wavelengths, whereas prisms and gratings select one specific wavelength at a time.

Fourier Transform

A Fourier transform is a mathematical transformation of a function f (mathematical function or derived from a signal) into another function $F = FT(f)$ that corresponds to the spectrum of the frequencies or wavelengths contained in the initial function. Both functions ($f(x)$ and $F(v)$) contain the same information but represented in two different manners (spatial or spectral domains). In FTIR spectroscopy the interferogram (output signal produced with the interferometer) is Fourier-transformed into the spectrum (intensity versus frequency).

1950–1965 Birth of FTIR

The idea of using an interferometer for spectroscopy, replacing a dispersive prism, was suggested and tested early in the twentieth century by Albert Michelson, Heinrich Rubens, and Herbert Hollnagel.[5]

But the relationship between the resulting interferograms and the spectrum was misunderstood. On one hand, Michelson explained and emphasized the importance of interferometers for spectroscopic measurement through his own experiments (mainly with visible light). But, on the other hand, he used his eye as a detector and could then only measure light contrasts, and not absolute intensities of light.

The complete spectrum an interferometer provides can be only determined with knowledge of the intensities of the fringes, obtained through the progressive increase of the path difference in the interferometer. Recording contrasts, Michelson could only get a partial interferogram, not a proper complete one. From 1910 onward, Rubens introduced the thermocouple as a detector, allowing him to measure complete interferograms. But they were not yet understood as Fourier transforms of the spectrum. Only Hollnagel, a student of Rubens, realized this. He published his PhD thesis on the topic in 1920, but subsequently disappeared from the scene — Rubens also moved to other instrument developments.[2; 5]

A complete analysis of the technique's advantages and a path to a true FTIR spectroscope would have to wait until 1948–1950 and the work of two physicists and spectroscopists: Peter Fellgett and Pierre Jacquinot.[2; 5] In his Cambridge University PhD thesis, Fellgett pointed out the wasted radiation in dispersive system: prisms and slits, as selecting device for wavelengths, lowered resolution and slowed measurements. He looked for a system to measure all frequencies simultaneously, and found in the Michelson interferometer a proper instrument to produce such a "multiplex."[3] Simultaneous measurement of different frequency ranges gave an advantage in signal-to-noise ratio, but the resulting signal, the interferogram, "encoded" the spectrum: the interferogram, in other words, is the Fourier transform of the spectrum. By 1950, Fellgett had rendered his first primitive spectra (with manual calculations) and the multiplex aspect of the system was subsequently called the "Fellgett advantage."

Pierre Jacquinot independently experimented with spectroscopes and looked into the luminosity of instruments.[4] He showed that interferometers, compared with dispersive instruments, did not strictly limit the amount of radiation passing through them. Interferometers had larger optical apertures, providing a "throughput advantage" (later called "Jacquinot advantage"). The advantages stemming from the work of each of these physicists flowed from reasoning about the physical characteristics of the instruments, applying lessons from quantum theory to (low light level) spectrometry.

A small community of instrumentalists, motivated by these theoretical advantages, came together to foster research and development of the technique.[9] H. Alistair Gebbie and George A. Vanasse produced the first complete spectra (more extensive than the first Fellgett's spectrum) using an IBM 650 computer in 1956. Janine Connes wrote the most complete theory of the FTIR instruments of her time in her PhD thesis in 1960, which covered methods for sampling the interferogram, improving signal-to-noise ratio, and calculating the Fourier transform. Her thesis was conducted at Bellevue (France), with Jacquinot as her supervisor. Physicists, especially spectroscopists, were the key players at that time because these instruments were high-precision spectroscopy in low optical infrared radiation (and first in the far-infrared domain). They were designed from scratch with the unique expertise of experimental physicists in a domain of IR where commercial dispersive instruments were impractical (not to say useless).

FTIR development is also the result of a transnational effort, based on a few small groups of physicists: mainly Jacquinot and the Connes in France, Fellgett, and Gebbie in Great Britain, Vanasse and Mertz in the United States. They gathered in dedicated conferences, in search of visibility among the vast community of physicists and legitimacy for their concepts and way to practice their "instrument science."[3; 9]

The "Infracord"

FTIR developed in parallel with dispersive instruments, which could make slow spectrum measurements; which, although less precise, proved faster and more efficient than early FTIR procedures.[5] High precision interferograms were easy to produce thanks to the instrumentalists' optical expertise, but conveyed no useful information until their Fourier transform was calculated, which could take hours. Although dispersive instruments were often slower to generate spectra, they provided output in a more readily interpreted format.

Introduced in 1957, the Model 137 "Infracord" of Perkin-Elmer epitomized the dispersive spectrometer of the 1960s.[5;8] It was a double-beam spectrometer in which two beams from the infrared source were projected on the sample area. Wavelengths (from 2.5 to 15 μm) could be selected sequentially with a rotating mirror and passed through the prism and detector (see Figures 4.3.2, 4.3.3,

Fig. 4.3.2. Perkin-Elmer Corporation. "Perkin-Elmer Model 137 Infracord Spectrophotometer," 1960–1969. Photographs from the Perkin-Elmer-Applera Collection, Box 3, Science History Institute. Philadelphia. *Credit*: Courtesy Science History Institute Collections.

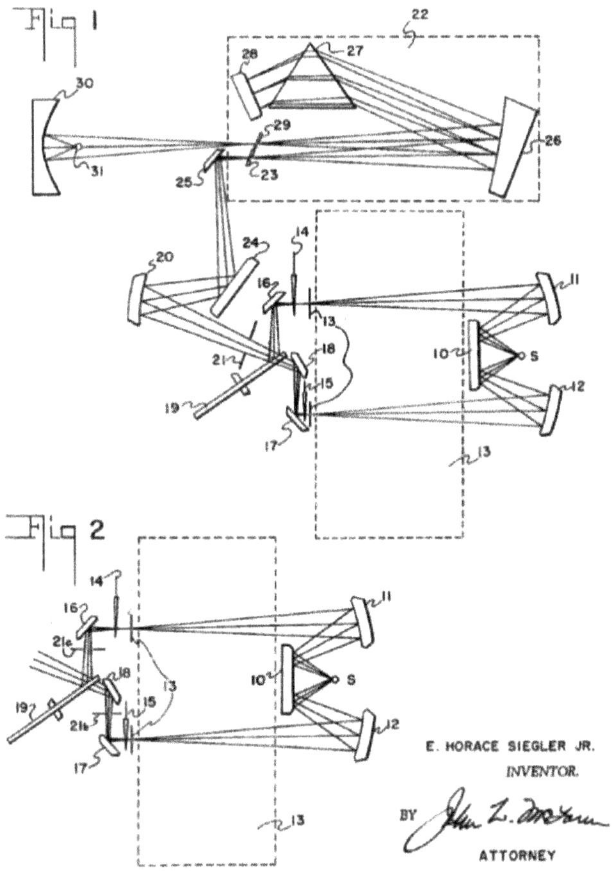

Fig. 4.3.3. Spatial Diagram of the Optical System of Perkin-Elmer Model 137. *Source*: United States Patent number 3011389.

and 4.3.4). Gratings would later replace the prism leading to a next-generation line of spectrophotometers. Perkin-Elmer had sold more than 10,000 instruments by the 1980s, benefiting from improvements in electronics and making them a major manufacturer in the domain (they also extended into instruments for UV spectroscopy).

Demonstrating FTIR's Superiority

The team led by Pierre and Janine Connes developed an instrument between 1964 and 1966 that demonstrated the superiority of

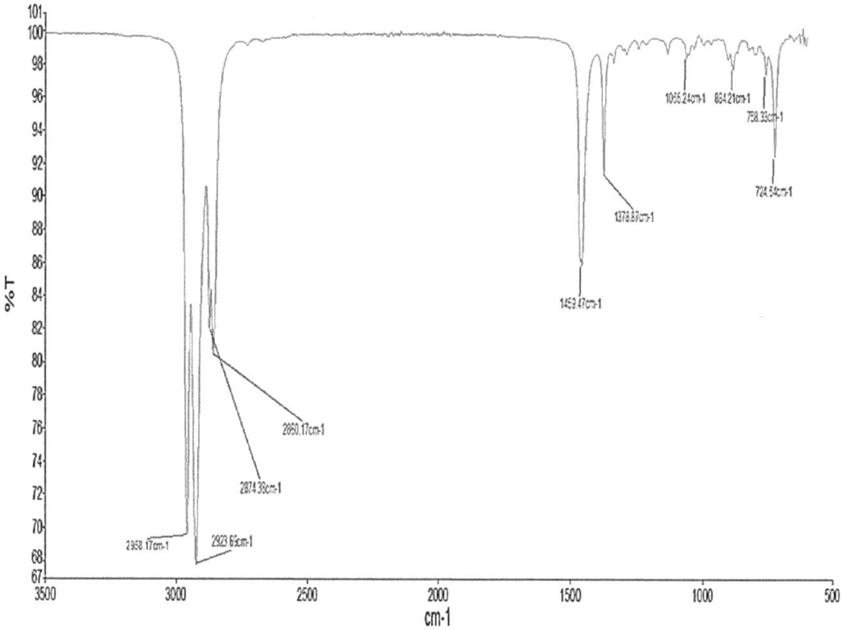

Fig. 4.3.4. Spectra for Hexane recorded with a modern Spectrometer (showing Transmittance for each wavenumber (cm^{-1})). *Credit*: Wikimedia Commons.

FTIR to dispersive techniques and gave many scientists the decisive incentive they needed to pursue FT spectroscopy.[2; 5; 9] The husband-and-wife team measured and calculated the spectra of Venus and Mars with unprecedented precision, in difficult conditions: taking near-infrared measurements through the atmosphere meant accounting for turbulence, fluctuations, and low-light conditions. They surmounted these challenges and were able to produce high-resolution results and a much better signal-to-noise ratio than that offered by any other instrument at that time. They ran Fourier transforms on an IBM 7040 mainframe, which took several hours to calculate each spectrum. Although this technological performance had a high symbolic impact, it nevertheless had no evident commercial applications.

A technical inflection point in the development of FTIR, apart from the optical developments the Connes team pioneered, was the capacity to calculate the Fourier transform of an interferogram faster and more completely. Until the advent of the Fast-Fourier Transform (FFT) algorithm, calculation was a chronophagous and expensive task, consuming costly hours on mainframe computers.[5] From 1966 on, the Cooley-Tukey algorithm for FFT, named for the mathematicians James Cooley and John Tukey, reduced calculation times from hours to minutes, and then to seconds. This was decisive. Hardware progress combined with software development made FTIR accessible outside specialized physics laboratories. The microprocessor, invented in the 1970s, was then used to control the instrument and made it easy to use. In the 1980s, FTIR spectrometers became powerful, cheap, reliable, and automated.

Commercial Development of FTIR

The first commercial FTIR spectrometers were produced in the 1960s. As lab experiments, FTIR spectrometers had begun, like most similar instruments, as custom-made devices. The first complete commercial system was FTS-14, manufactured by Digilab in 1969 (see Figures 4.3.5 and 4.3.6). It incorporated the heritage of this pioneering period alongside a number of new ideas, many of them promoted by Lawrence Mertz (then at Block engineering, which would become Digilab).[7; 8] In the FTS-14, as in many infrared spectrometers, the light beam was modulated: the displacement of the moving mirror (with constant speed) of the interferometer, allowing both what became known as fast-scanning technology for FTIR and repeated scanning to improve signal-to-noise ratio. The FTS-14 also included a computing facility and plotting device, and a laser reference system for internal calibration (inherited from the Connes's instruments). Beamsplitters, sources, and detectors were devised to explore a large range of wavelengths. Altogether, the FTS-14 set the basis for commercial FTIR spectrometers for the years to come because it encompassed all the advantages drawn from research in FTIR: ease of use, fast results, precision, and an even more user-friendly experience than dispersive techniques (which always required delicate manipulations).

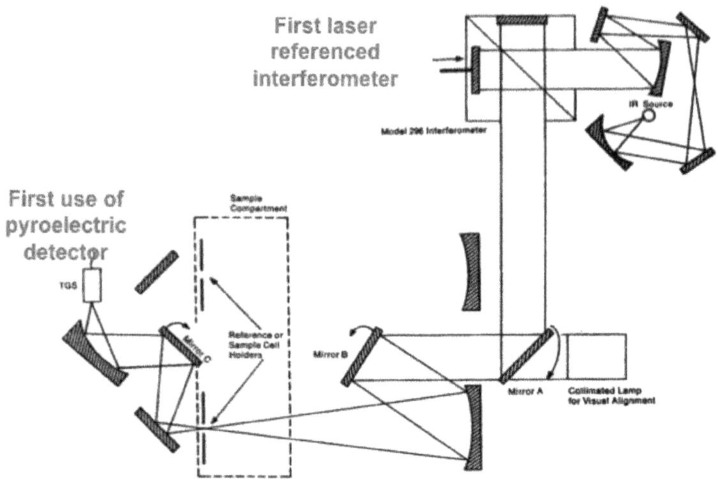

Fig. 4.3.5. Digilab FTS-14 — Schematic principle. *Source*: [8]. *Credit*: Courtesy of SAGE Publications.

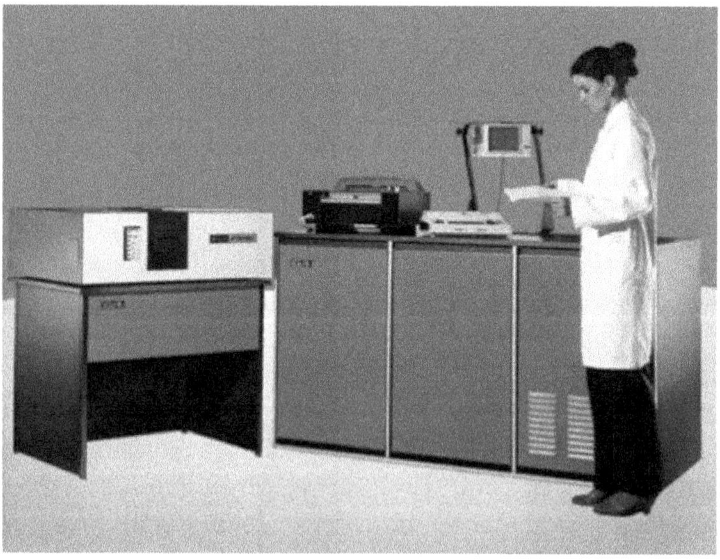

Fig. 4.3.6. Digilab FTS14 — 1972. *Source*: [8]. *Credit*: Courtesy of SAGE Publications.

Digilab turned out to be a leader as a manufacturer and it was acquired by Bio-Rad Laboratories in 1978: from a specialized manufacturer, it became part of a much larger catalogue of instruments, with clients in ever-wider domains. That was also a factor in the expansion of a mature technology outside the range of its first developments.

FTIR Expands its Uses

Starting in the hands of physicists, FTIR spectroscopy has been progressively adopted by chemists and industry at large.[1; 2; 5] On the one hand, such instruments were increasingly accessible and efficient. On the other hand, demands for IR spectroscopy rose in the 1970s coming from diverse horizons.[1] FTIR could answer questions from basic research in atomic and molecular physics, chemistry (producing data on wavelength molecular absorption), and astronomy (from planet spectra to deep space measurements). In solid-state studies, applications were numerous: superconductor energy-gap measurements, infrared detector and source-development projects, infrared-filter development, and metallic-surface property studies, to name a few. FTIR could also characterize compounds: corroded surfaces, paintings, oils and pigments on paper, optical opaque substances. The largest commercial applications were process and quality control: monitoring multicomponent mixtures, with high-speed results, meant a quick retroaction on production lines.[1] As spectra were becoming routine measurements, quality control was easier, detecting even low concentrations of impurities and additives for chemical, plastics, coatings, petroleum, and pharmaceutical industries.

Altogether FTIR offered non-destructive analysis, as well as easy and fast data collection on the order of a second per scan. FTIR allowed high-precision measurements without any complex manipulation for calibration of the instrument (it is self-calibrating thanks to the embedded laser beam reference passing through the interferometer), with high optical throughput. The device is mechanically simple with a single moving part: the mobile mirror for scanning.

A series of technical innovations and improvements made FTIR progressively faster and more accurate in the 1980s and afterwards. Different materials for the beam splitters in the interferometer and detectors (pyroelectric, or photovoltaic) have made it possible to investigate a range of wavelengths from 1 to 1,000 μm. Sampling techniques also improved: specialized sampling procedures such as attenuated total reflectance (ATR), diffuse reflectance (DR), and photoacoustic spectroscopy (PAS) increased efficiency greatly. ATR allows analysis of thick or very absorbent samples, directly in their liquid or solid state. Like DR, it is based on reflectance analysis of IR in the sample, not the transmission through the sample. That means a different (but easy) preparation of samples for FTIR. PAS is particularly useful for *in situ* state analysis of biological samples (without crushing such materials).

The control and complete analysis of the results by microcomputers has made the device user-friendly and rendered its operation routine. From veteran experimenters to inexperienced scientists and technicians, just about anybody can use FTIR.

Legacy of FTIR

In addition to its many industrial and scientific applications, FTIR has exerted a lasting influence over materials research by proving the feasibility and utility of Fourier transform spectroscopic techniques. First demonstrated with infrared, these techniques have since combined different technologies and unrelated ideas: among them spectroscopy, interferometry, digital computing, the mathematics of Fourier transform, and the algorithms for FFT. Since the 1950s, Fourier transform spectroscopy techniques have moved from the specialized domain of the physics laboratory and are now standard diagnostic tools in many branches of science and industry. These techniques have also extended to other domains: Fourier versions of nuclear magnetic resonance (NMR) spectroscopy and mass spectroscopy (MS) became reality after 1972 and showed a considerable gain in sensitivity and resolving power.[2; 9] FTIR had paved the way to Fourier techniques in spectroscopy, then pushed the limits of so-called classical instruments, such as spectroscopes and interferometers, and gave birth to major analytical tools.

References

1. Bell RJ. Introductory Fourier transform spectroscopy. New York: Academic Press; 1972.
2. Connes P. Early history of Fourier transform spectroscopy. Infrared Physics. 1984;24(2):69–93.
3. Fellgett PB. Three concepts make a million points. Infrared Physics. 1984;24(2):95–98.
4. Jacquinot P. How the search for a throughput advantage led to Fourier transform spectroscopy. Infrared Physics. 1984;24(2):99–101.
5. Johnston SF. Fourier transform infrared: a constantly evolving technology. New York: Ellis Horwood; 1992.
6. Kendall DN, editor. Applied infrared spectroscopy. New York: Van Nostrand Reinhold; 1966.
7. Mertz L. A perspective on Fourier spectroscopy. Infrared Physics. 1984;24(2):111–12.
8. Griffiths PR. The early days of commercial FT-IR spectrometry: a personal perspective. Applied Spectroscopy. 2017;71(3):329–40.
9. Johnston SF. In search of space: Fourier spectroscopy 1950–1970. In: Joerges B, Shinn T, editors. Instrumentation between science, state and industry. Dordrecht: Springer; 2001, pp. 121–41.
10. Johnston SF. The evolution of FTIR technology. Chemistry in Britain. 1990;26:573–77.

CHAPTER 4.4

ULTRAVIOLET SPECTROSCOPY

GERALD GALLWAS

"We shape our tools, and thereafter our tools shape us."

— Marshall McLuhan

The ultraviolet-visible (UV-vis) spectrophotometer was a major development in the electronic revolution of the chemical sciences that blossomed during the mid-twentieth century. The instrument enabled scientists to record the molecular (spectral) fingerprint of organic and biologically active compounds such as hormones and vitamins (Figure 4.4.1). Development of the first commercial product by Arnold O. Beckman (Figure 4.4.2) in 1941 was a seminal event in what is referred to as the "second revolution" or "electronic revolution" in chemistry. Although the study of electromagnetic radiation and its interaction with matter dates at least to Sir Isaac Newton's *Opticks* of 1704, lack of commercially available tools left the investigation of optical phenomena in the hands of those very few capable of building suitable instruments themselves.[44] The electronic revolution enabled the commercialization of instrumentation and fundamentally changed the sciences. Materials research shifted from the observation of experimental outcomes after the fact, to the measurement of physical properties before, during, and after an experiment.

Spectrophotometry as used in chemistry is primarily the measurement of absorbance of electromagnetic radiation by the sample as a function of wavelength. The spectrophotometer separates

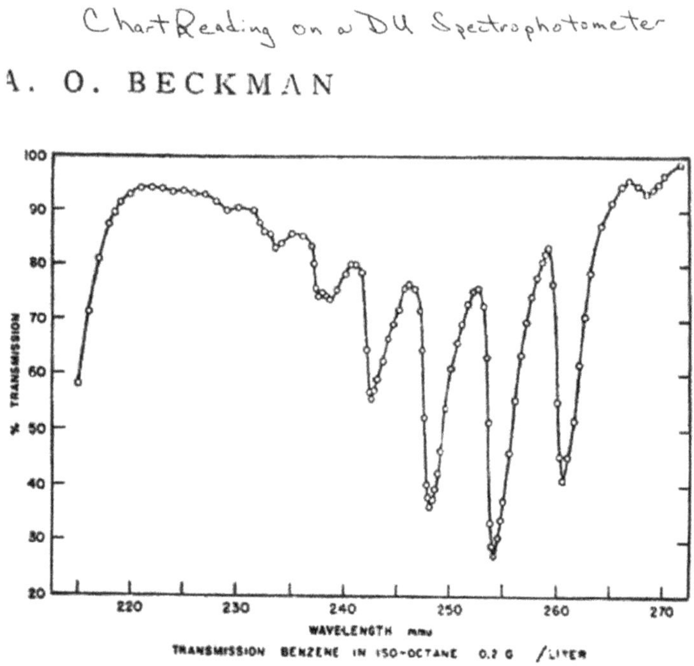

Fig. 4.4.1. Beckman Benzene Spectrum, 1941. Courtesy of the Arnold and Mabel Beckman Foundation.

Fig. 4.4.2. Arnold O. Beckman. Courtesy of the Arnold and Mabel Beckman Foundation.

individual wavelengths isolated by a dispersion device such as a prism or diffraction grating in the UV-vis region from 200 nm to 800 nm. The absorbance is based on electronic changes in energy levels within the molecule and accounts for the molecular (spectral) fingerprint.

Beckman was a faculty member at the California Institute of Technology until he left in 1939 to become the President of National Technical Laboratories, the company he had founded four years earlier. He is credited with the introduction of the first commercially successful glass electrode, electronic pH meter and ultraviolet spectrophotometer.

The conjunction of four forces fostered the introduction of Beckman's spectrophotometer and the emergence of the electronic chemical-instrumentation industry:

- The development of quantitative measurements based on light absorption by a sample;
- The quest for chemical structure using molecular fingerprints in both the ultraviolet and infrared regions of the spectrum;
- Development and commercialization of the spectrograph;
- Use of the vacuum tube for electronic amplification of an experimental signal.

Development of Quantitative Measurements Based on Light Absorption by the Sample

Although visual color comparison methods had been used for centuries, photometric comparison based on a commercial product dates to 1854, with introduction of the Duboscq colorimeter.[14; 56] It found early use in the measurement of caramel in syrup, dyes, liquors, and later for biological chemical determinations of creatinine and creatine in urine, as described in a classic paper by Otto Folin in 1904.[16] The basic Duboscq design was used well into the twentieth century for a variety of quantitative analyses, including that of hemoglobin, with instruments manufactured by Klett, Bausch & Lomb, Carl Zeiss, and others (Figure 4.4.3).

But the real power of photometric analytical measurements embodied in the Bouguer, Lambert, Beer absorption law — commonly

Fig. 4.4.3. Duboscq colorimeter circa 1920. Bosch & Lomb. *Credit*: Courtesy of the Arnold and Mabel Beckman Foundation.

known as Beer's law, which states that the absorption is proportional to the concentration of the absorbing component in solution — only began to be realized in the early 1930s with development of the first practical commercial colorimeter by scientists at Mayo Clinic. The Cenco-Sheard-Sanford Photelometer was a photoelectric filter instrument designed to improve hemoglobin measurement relative to the optical comparison methods then in common use (Figure 4.4.4).[5; 6; 34; 49; 53; 54] R. H. Muller summarized the state of the art in 1939 and of the instrumentation industry in 1939, 1940, and 1941.[39–42]

In 1942, the Coleman Electric Company introduced the Junior spectrophotometer, a direct-reading visible range unit that began to popularize photometric quantitative methods. By 1954, a low-cost visible region diffraction grating instrument by Bausch and Lomb,

Fig. 4.4.4. Cenco-Sheard-Sanford Photelometer circa 1933. *Credit*: Courtesy: Arnold and Mabel Beckman Foundation.

the SPECTRONIC 20, became the "go-to lab spectro" that put photometric measurement in thousands of laboratories.[32] By the 1960s, instrumental methods of analysis dominated the literature; in addition, the nature of journal advertisements shifted from laboratory apparatus to electronic instrumentation.

The Quest for Chemical Structure Based on Molecular (Spectral) Fingerprints

Exploration of the ultraviolet region of the spectrum had its roots in the development of the spectroscope by Bunsen and Kirchhoff in 1859. With it, they discovered cesium and rubidium shortly thereafter and went on to show the relationship of chemical elements to their unique emission spectra.[7] Using the basic Bunsen Kirchhoff spectroscope design but adding a camera to record the spectra,

W. A. Miller reported on the absorbance spectra of numerous liquids and gases as early as 1862. Miller's work inspired Walter Noel Hartley, who modified Miller's spectrograph and published extensively on his UV absorption band studies of metals and organic compounds later in the nineteenth century. He was the first to explore the perplexing issue of tautomerism with spectroscopy and succeeded in differentiating between the enol and keto structures.[26; 27; 36]

In the early 1900s, the French chemist Victor Henri did fundamental research in enzymology and showed that ultraviolet spectra were a function of certain chemical bonds rather than the basic molecule.[29] His library of more than 200 UV spectra appeared in the 1929 edition of the International Critical Tables.[30] Beginning in the 1940s, UV spectroscopy provided a powerful tool in the structural characterization of complex natural products based on the strong absorbance of conjugated double bonds such as those found in α, β-unsaturated ketones and 1, 3-dienes. The studies of molecular structure by R. B. Woodward are milestones in organic chemistry. Then in 1953, O. C. Rudolph & Sons introduced the first commercial spectropolarimeter, which Carl Djerassi described as "the prince who awoke the sleeping beauty of rotatory dispersion." It enabled the study of optically active, complex molecular structures found in steroids and other natural products including proteins, and nucleic acids. Djerassi later followed up on two newer spectroscopic techniques, optical rotatory dispersion/circular dichroism (ORD/CD) and magnetic circular dichroism (MCD). Although ORD/CD experiments are generally performed in the UV region, most but certainly not all MCD experiments are performed in the visible to near infrared regions, though they too can be performed in the UV.[12; 45; 48; 51; 52]

Development and Commercialization of the Spectrograph

The first commercial line of spectrographs was introduced in 1906 by Adam Hilger Ltd. Hilger products found wide use in research and the metallurgical industry (Figure 4.4.5). They offered the first commercially available instruments that expanded exploration of

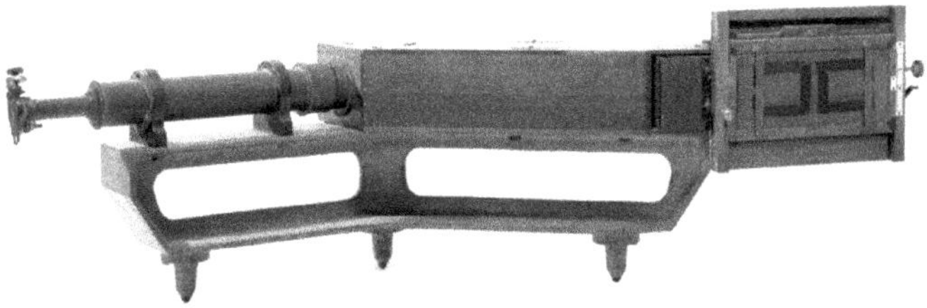

Fig. 4.4.5. Spectrograph ca. 1924. Adam Hilger Ltd. *Credit*: Courtesy of the Science History Institute Collections.

spectrophotometric methods in general and of the ultraviolet region in particular.[36] However at that time, spectroscopy was the purview of physicists and astronomers with little attention by chemists. Frank Twyman, the general manager of Hilger, recognized the opportunity to simplify the design from that of a research instrument to one for routine analytical procedures. He established a strong customer support program and published extensive procedures for a variety of chemical applications targeting chemists.

In the mid-1920s, Carl Zeiss and Bausch & Lomb entered the field as the focus of research began to include studies of the spectral absorption of food supplements such as vitamins. Vitamin A was of high interest at that time and it was found that selective UV absorption occurs in the region of 320 nm to 328 nm, which later became the basis of its quantitative measurement.[28; 38; 50] In 1938 UV spectra were used as evidence to support the equivalence of the natural and synthetic products of vitamin B (thiamine).[57] Spectrographs with photographic detection were used well into the mid-twentieth century.

During this period, few publications focused on ultraviolet spectroscopy, although serious research in infrared spectroscopy dates to the work of William Coblentz in 1903 to 1905 with subsequent studies in the 1930s.[1–3; 9; 17; 18] The first photoelectric recording spectrophotometer was designed by Arthur C. Hardy at MIT and commercialized by General Electric in 1933. A large and expensive visible range instrument intended for color measurements, it was used at the US National Bureau of Standards to establish visible color standards but did not extend into the ultraviolet region.[24; 25]

Fig. 4.4.6. Research UV-vis Spectrophotometer, University of Chicago, 1937. Hogness TR, Zscheile FP, Sidwell AE. Photoelectric spectrophotometry: an apparatus for the ultra-violet and visible spectral regions: its construction, calibration, and application to chemical problems. *Credit*: Reprinted with permission from Journal of Physical Chemistry 41(1937):380. © 1937 American Chemical Society.

Then in 1937, Thorfin R. Hogness, Frederick P. Zscheile Jr., and Albert E. Sidwell Jr. published the first extensive article on the design, construction, and application of a *Photoelectric Spectrophotometer in the Ultra-Violet and Visible Spectral Regions* (Figure 4.4.6). Occupying a small laboratory, the apparatus used a hydrogen lamp as the source of UV light with quartz optics to permit its transmittance, a cesium — cesium oxide photoelectric UV sensitive detector, and a Pliotron triode amplifier with a galvanometer readout. The publication included hand plotted spectra of benzene and potassium chromate.[31]

Application of the Vacuum Tube with the Electronic Amplification it Provided

The enabling technology of the electronic revolution was the Audion vacuum tube (triode) invented by Lee de Forest in 1906. It came to widespread use with the rise of AM radio in the 1920s. The earliest well-documented application of electronic amplification in chemical measurement occurred in 1922. Kenneth H. Goode published on a continuous-reading electro-titration apparatus with readings in Sorensen (pH) units. Goode went on to develop an electronic pH meter as the subject of his doctoral thesis at Pennsylvania State College and filed a US patent for an "electric meter" that was issued on February 2, 1932, number 1,843,320.[19–22] He commercialized his pH meter, but without the benefit of the glass electrode he was unable to make a practical product for routine use. Even as late as 1930, Henry C. Parker of the National Bureau of Standards stated in a review of instrumentation for use in teaching electrochemistry that the measurement of pH was more an experiment than a measurement.[46]

Researchers at New York University, University of Illinois, and Iowa State College all pursued an electronic pH meter with a glass electrode, but it was not until 1935 that Arnold Beckman achieved a commercially successful version.[15; 23; 47] The launch of his first product in September 1935 was closely followed by the Coleman Electric Company's announcement in December of a pH Electrometer with sealed glass electrodes.[10] The electronic revolution in the chemical sciences had sprung to life. By 1940, there were eleven competitors selling electronic pH meters.[13]

Beckman's first commercially successful ultraviolet spectrophotometer was inspired by engineers at Coleman, who had adapted the electronics of the Coleman pH Electrometer as the amplifier

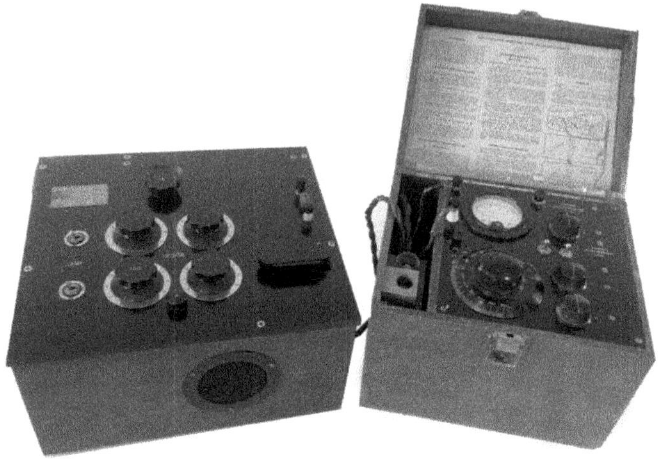

Fig. 4.4.7. Coleman Model 10 Spectrophotometer with pH Electrometer for Amplifier and Readout, 1939. *Credit:* Courtesy of the Arnold and Mabel Beckman Foundation.

and readout for a dual beam visual spectrophotometer announced in September 1939.[11] The Coleman Model 10 Spectrophotometer (Figure 4.4.7) was a diffraction grating instrument with a tungsten source lamp that did not reach into the ultraviolet region of the spectrum. The Beckman engineering team immediately recognized the shortfall as an opportunity to provide access to the region from 350 nm to 200 nm. It was there that biologically active molecules such as vitamins and hormones as well as numerous organic compounds had absorption spectra. Beckman had the electronics, as did Coleman, but he would need to develop an ultraviolet source lamp, a blue sensitive photodetector tube, and quartz optics to bring an ultraviolet spectrophotometer to reality.

Beckman first approached Bausch and Lomb, a leading optics firm and manufacturer of spectrographs, to collaborate on the project. He was advised that there was no market for such a product. Thus, Beckman decided to build it himself. He understood the value of rapid, accurate measurements and only had to reflect on the state of biological vitamin assays where, for instance, rat tail growth over a period of weeks was the basis for the measurement of vitamin A.

In contrast, he knew that an absorbance measurement of a diluted sample in the ultraviolet would require just minutes based on spectrophotometric studies in the 1920s.[38; 50]

Early in 1940, the Beckman team launched the project, despite a lack of experience in spectrophotometer design. Their two major assets were their understanding of the potential value of UV absorption band measurements and the available electronic amplifier, then a standard component in the Beckman pH meter. Four major design hurdles had to be overcome: 1) selection of the monochromator, 2) development of an emission source, 3) development of an ultraviolet sensitive photodetector, and 4) packaging all the components in a benchtop product.

The monochromator was designed with a folded light path to reduce the footprint to fit on a laboratory benchtop and used a quartz prism as it provided lower stray light in the ultraviolet than a diffraction grating. The hydrogen source lamp and blue sensitive photodetector had to be designed and manufactured by the Beckman team as there were no satisfactory commercial sources.[8]

In fourteen months, the Beckman team developed three prototypes, resulting in a design with quartz optics and a hydrogen source lamp which was introduced as the first commercial ultraviolet spectrophotometer at the Ninth Summer Conference on Spectroscopy at MIT in July 1941. Officially the Beckman Quartz Photoelectric Spectrophotometer, it came to be widely known as the Beckman DU — the D because Beckman named its sequential models alphabetically and the U for ultraviolet — with the addition of a blue sensitive photodetector a few months later (Figure 4.4.8). Due to wartime travel restrictions, Beckman and his chief engineer, Howard Cary, could not be in Cambridge for the introduction. Thus, Ed Patterson, the salesman at Arthur H. Thomas in Philadelphia who had originally encouraged Arnold Beckman to sell the pH meter, introduced the Beckman Spectrophotometer at the conference.

Originally priced at $723 in 1941 (around $12,800 in 2019 dollars), the instrument was an immediate success and continued in production until mid-1976 — thirty-five years later — with a run of more than 30,000 units. The first publication of results obtained with the

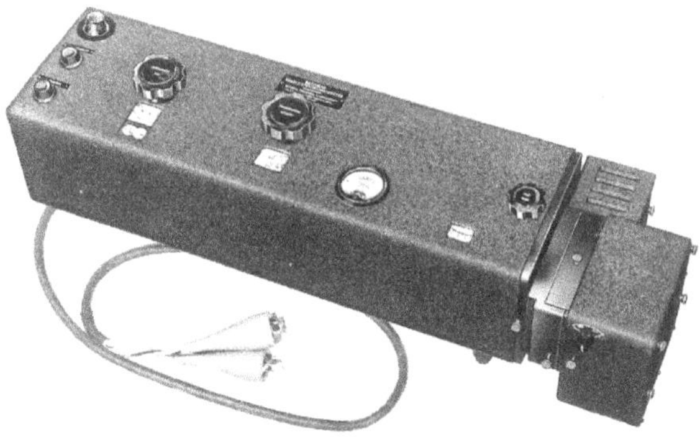

Fig. 4.4.8. Beckman Ultraviolet Spectrophotometer, July 1941. Courtesy: Arnold and Mabel Beckman Foundation.

Beckman DU appeared in the September 1942 issue of *Industrial and Engineering Chemistry, Analytical Edition* and described the analysis of vitamin A. By the end of production in 1976, the Model DU 2 sold for $7,885 (over $35,000 in 2019 dollars) and still used the original two-stage vacuum tube amplifier designed by Arnold Beckman in 1935 for his pH meter. Many accessories were developed over the years including flame emission for electrolyte measurements in laboratory medicine, fluorescence used in water quality studies, and total reflectance for color comparison to name a few. The Beckman DU changed the practice of the chemical sciences, replacing bioassays taking weeks with simple sample dilution and measurement requiring only minutes. Thousands of applications were developed over the years from process and quality control in manufacturing to food, drug, medical, and biochemical research.[4] The Beckman DU was used in the synthetic rubber program during World War II to monitor the polymerization of butadiene.[37] In 1945, Erwin Chargaff used the Beckman DU along with the newly developed paper chromatograph to carry out the first total biochemical analysis of DNA. His work resulted in Chargaff's rules and contributed directly to the later discovery of the structure of DNA.[55] That same year, Shimadzu introduced a UV-vis spectrophotometer, Model QF-60, that used a photographic detector and in 1949 Unicam Ltd. introduced a product, the SP500, like the Beckman DU.

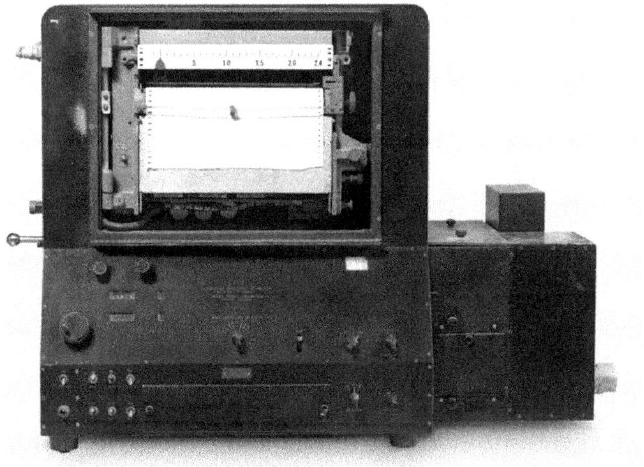

Fig. 4.4.9. Cary Model 11 UV-vis Spectrophotometer, 1947. *Credit*: Courtesy of the Science History Institute Collections.

Howard Cary left National Technical Laboratories (later Beckman Instruments) in January 1946 to found Applied Physics Corporation. A year later, he shipped the first double-beam recording UV-vis spectrophotometer, the Cary Model 11 (Figure 4.4.9).[32]

The proliferation of both ultraviolet and infrared spectrophotometers after World War II brought explosive growth and changed the chemical sciences profoundly. The 1965 US National Academy of Sciences publication, *Chemistry: Opportunity and Needs*, documents the growth in electronic instrumentation-based publications with ultraviolet and infrared being the most populous with thousands of journal citations.[43]

Beckman DU Spectrophotometer sales began to wane as interest in the near infrared and far ultraviolet developed. But the market also shifted due to changes in the technologies of data generation and storage. The need for automatic recording became a competitive advantage for the Cary product. In response, Beckman looked to develop a comparable system. Wilbur Kaye at Tennessee Eastman had adapted the Beckman DU monochromator to a double-beam recording instrument designated as DK-1. Beckman introduced the

DK-1 UV-vis recording spectrophotometer in 1952 and a lower-cost, flatbed recorder model, the DK-2, in 1955. Tennessee Eastman had no interest in chemical instrumentation patents, so Beckman manufactured both products royalty free.[33]

In 1977, Hitachi introduced the first microprocessor-controlled UV spectrophotometer, Model 340. Soon after, Hewlett-Packard introduced the first diode-array instrument, the Model HP8450A, which could scan from 200 nm to 800 nm in one second. By 1982, video displays offered menu selection, parameter setup, and applications software. Mouse operation was introduced in 1986 with the Unicam PU8700.[32]

Today, in the early decades of the twenty-first century, dozens of UV-vis spectrophotometer manufacturers operate throughout the world. They offer a vast range of products, from traditional bench top instruments to some that fit into the footprint of a cell phone. Sample volumes span from the one to three milliliters range down to the microliter level. Modern microprocessor operated products offer a menu of applications and are equipped with tungsten/deuterium source lamps and CCD (charged-coupled device) array detection with scanning rates of less than a second across the range of 200 nm to 1,000 nm.

The electronic revolution in the chemical sciences began with the pH meter and UV-vis spectrophotometer. What followed was an ever-escalating expansion of technologies that enable students to carry out experiments in a day that Nobel Prize–winning scientists of the past took years to accomplish. Lasers, fiber optics, microprocessors, carbon nanotubes, biochips, and the internet are notable examples of the impact. And the revolution continues.

References

1. Barnes RB, Bonner LG. A survey of infrared spectroscopy. Journal of Chemical Education. 1937;14:564–71.
2. Barnes RB, Brattain RR, Seitz F. On the structure and interpretation of the infrared absorption spectra of crystals. Physical Review. 1935;48:582–602.
3. Barnes RB. IR-Spectra and organic chemistry. Review of Scientific Instruments. 1936;7:265–71.

4. Beckman AO, Gallaway WS, Kaye W, Ulrich WF. History of Spectrophotometry at Beckman Instruments. Analytical Chemistry. 1977;49:280A–300A.
5. Beer A. Bestimmung der Absorption des rothen Lichts in farbigen Flüssigkeiten. Annalen der Physik und Chemie. 1852;86:78–88.
6. Bouguer P. Essai d' optique sur la gradation de la lumière. Paris: Claude Jombert; 1729.
7. Bunsen G, Kirchhoff R. Chemische analyse durch spectralbeobachtungen. Annalen der Physik und der Chemie (Poggendorff). 1860;110:161–89.
8. Cary HH, Beckman AO. A quartz photoelectric spectrophotometer. Journal of the Optical Society of America. 1941;31(11):682–89.
9. Coblentz W. Investigations of infrared spectra. Washington: Carnegie Institute of Washington, DC; 1905.
10. Coleman Electric Company. pH Electrometer Advertisement. Industrial and Engineering Chemistry. 1935;13(23):452.
11. Coleman Electric Company. Regional spectrophotometer advertisement. Industrial and Engineering Chemistry. 1939;17(17).
12. Djerassi C. Optical rotatory dispersion: applications to organic chemistry. New York; McGraw-Hill: 1960.
13. Dole M. The glass electrode: methods, applications, and theory. New York: Wiley; 1941.
14. Duboscq J, Mène C. Nouveau colorimèter pour l'analyse des matieres tinctoriales au point de vue commercial. Comptes Rendus Hebdomadaires des Seances de l'Acaldemie des Sciences. 1868;67:1330–31.
15. Elder LW, Wright WH. pH Measurement with the glass electrode and vacuum tube potentiometer. Proceedings of the National Academy of Sciences. 1928;14:936–39.
16. Folin O. Beitrag zur Chemie des Kreatinins und Kreatins im Harne. Zeitschrift fur Physiologishche Chemie. 1904;41:223–42.
17. Gershinowitz H, Wilson EB. Infra-red and Raman spectra of polyatomic molecules I. An automatic prism spectrometer for the infra-red. Journal of Chemical Physics. 1938;6:197.
18. Gershinowitz H, Wilson EB. Infra-red and Raman spectra of polyatomic molecules II. Cis-and transbutene-2. J. Chemical Physics. 1938;6:247.
19. Goode KH. A Continuous-reading electrotitration Apparatus. Journal of the American Chemical Society. 1922;44:26–29.
20. Goode KH. An improved, continuous-reading hydrogen-ion meter. Journal of the American. Chemical Society. 1926;47:2483–88.
21. Goode KH. Electric meter. United States Patent 1,843,320. February 2, 1932.
22. Goode KH. The 'acid-alkalimeter': a direct-reading pH meter. Journal of the Optical Society of America. 1928;17(1):59–71.
23. Goodhue LD, Schwarte LH, Fulmer EI. A simple, inexpensive electron tube potentiometer for use with the glass electrode. Iowa State College Journal of Science. 1933;2:111–18.
24. Hardy AC. A new recording spectrophotometer. Journal of the Optical Society of America. 1935;25(9):305–10.

25. Hardy AC. History of the Design of the Recording Spectrophotometer. Journal of the Optical Society of America. 1938;28:360–64.
26. Hartley WN, Huntington AK. Researches on the action of organic substances on the ultra-violet rays of the spectrum. Philosophical Transactions of the Royal Society of London. 1879;170:257–74.
27. Hartley WN. The absorption spectra of the vapours of benzene and its homologues, at different temperatures and pressures, and of solutions of benzene. Philosophical Transactions of the Royal Society of London. 1908;208: 475–528.
28. Heilbron IM, Kamm ED, Morton RA. The absorption spectra of oils and oil constituents with special reference to pro-vitamin D. Biochemical Journal. 1927;21;6:1279–83.
29. Henri V. Etudes de Photochimie, 1919; Structure des molecules. Editions E.D.K: 1934.
30. Henri V. Quantitative determination of ultra-violet absorption spectra in solutions of organic substances. International Critical Tables. 1929:359–79.
31. Hogness TR, Zscheile FP, Sidwell AE. Photoelectric spectrophotometry, an apparatus for the ultra-violet and visible spectral regions: its construction, calibration, and application to chemical problems. Journal of Physical Chemistry. 1937;41:379–415.
32. Jarnutowski R, Ferraro JR, Lankin DC. Fifty years of commercial instrumentation in absorption spectroscopy, part II, landmark instruments in UV/Vis. Spectroscopy. 1992;7:22–35.
33. Kaye WI. Transcript of an interview, February 11 and 27, 2002. Philadelphia; Chemical Heritage Foundation: 2002.
34. Lambert HJ. Photometria, sive de Mensura et Gradibus Luminis, Colorum et Umbrae. Augsburg: 1760.
35. Meehan EJ, Kolthoff IM. Report XIV: effect of soap composition on GR-S polymerization. CR 1943;117(13).
36. Morris PJT, editor. From classical to modern chemistry the instrument revolution, chapters: Morris, Peter JT, Travis SS, The role of physical instrumentation in structural organic chemistry in the twentieth century(57–84); Bigg, Charlotte, Adam Hilger, Ltd and the Development of Spectrochemical Analysis(111–128): The Royal Society of Chemistry with the Science Museum, London and the Chemical Heritage Foundation; 2002.
37. Morris PJT. The American synthetic rubber research program. Philadelphia: University of Pennsylvania Press; 1989.
38. Morton RA, Heilbron IM. The absorption spectrum of vitamin A. Journal of Biological Chemistry. 1928;22;4:987–96.
39. Muller RH. American apparatus, instruments, and instrumentation. Industrial and Engineering Chemistry, Analytical Edition. 1939;11:563–82;
40. Muller RH. American apparatus, instruments, and instrumentation. Industrial and Engineering Chemistry, Analytical Edition. 1940;12:571–630
41. Muller RH. American apparatus, instruments, and instrumentation. Industrial and Engineering Chemistry, Analytical Edition. 1940;13:667–754.

42. Muller RH. Photoelectric Methods in analytical chemistry. Industrial and Engineering Chemistry, Analytical Edition. 1939;11:1–17.
43. National Research Council. Chemistry: opportunity and needs. Washington: National Academy of Sciences; 1965.
44. Newton I. Opticks: Or, a Treatise of the Reflexions, Refractions, Inflexions and Colours of Light. London; 1704.
45. O. C. Rudolph & Sons. Advertisement. Analytical Chemistry. 1953;25(1): 17A.
46. Parker HC. Electrical measuring instruments for use in teaching electrochemistry. Journal of Chemical Education. 1930;7(1):69–80.
47. Partridge HM. A Vacuum Tube Potentiometer for Rapid E.M.F Measurements. Journal of the American Chemical Society. 1929;51(1):1–7.
48. Rudolph H. Photoelectric polarimeter attachment. Journal of the Optical Society of America. 1955;45(1):50–59.
49. Sanford AH, Sheard C. The determination of hemoglobin with the photoelectrometer. Journal of Laboratory and Clinical Medicine. 1930;15:483–89.
50. Schlutz FW, Ziegler MR. Some spectroscopic observations on cod liver oil: II. The absorption bands of cholesterol. Journal of Biological Chemistry. 1926;69:415–19.
51. Seeman JI. Angewandte Chemie International Edition-Wiley Online Library, Carl Djerassi in his own words, Chem. Int. Ed. 2014;53:3268–79.
52. Seeman JI. R. B. Woodward, a great physical organic chemist. Journal of Physical Organic Chemistry. 2014;27(9):708–21.
53. Sheard C, Rogers DA. Photelometer. United States Patent 2,051,317. August 8, 1936.
54. States MN. Apparatus for using photoelectric cells quantitatively. United States Patent 2,051,320. August 18, 1936.
55. Vischer E, Chargaff E. The separation and quantitative estimation of purines and pyrimidines in minute amounts. Journal of Biological Chemistry. 1948;176(2):703–14.
56. Warner DJ. The Duboscq colorimeter. Bulletin of the Scientific Instrument Society. 2006;88:68–70.
57. Williams RR. The chemistry of thiamine (Vitamin B1). Journal of the American Medical Association. 1938;110(10):727–32.

CHAPTER 4.5

MÖSSBAUER SPECTROSCOPY

CATHERINE WESTFALL

From PhD Project to Nobel Prize

Mössbauer spectroscopy, a spectroscopic technique based on the Mössbauer effect, provides information on a wide variety of materials by probing tiny energy-level changes in an atomic nucleus resulting from its environment. This valuable tool, which is used in many fields of science and engineering — a measure of the wide applicability of the Mössbauer effect can be taken from the existence of an International Board on the Applications of the Mössbauer Effect and the Mössbauer Effect Data Center — began in a surprising way: it was first developed as a graduate student project.

It all started in the mid-1950s, when Rudolf Mössbauer (Figure 4.5.1), then a German graduate student working at the Max Planck Institute for Medical Research in Heidelberg and the Technical University of Munich, began a series of experiments using techniques pioneered by Philip Moon and others to measure nuclear resonance absorption when osmium-191 (^{191}Os) beta decays to iridium-191 (^{191}Ir). During the course of his experiments, Mössbauer made the unexpected discovery of the effect that bears his name: an atomic nucleus in a crystal, because it is held firmly in place by the crystal lattice, undergoes near-zero recoil when it emits a low-energy gamma ray, and hence provides its entire recoil energy to the gamma ray.

Fig. 4.5.1. Rudolf Mössbauer. *Credit*: AIP Emilio Segrè Visual Archives, Physics Today Collection, W. F. Meggers Gallery of Nobel Laureates.

Mössbauer would often explain this newly discovered phenomenon with the example of a person throwing a stone from a boat: "The majority of the energy is submitted to the stone, but a small amount goes into the kinetic energy of the recoiling boat. During summer time, the boat will simply pick up this recoil energy. If, however, the person throws the stone during winter time, with the boat frozen into the lake, then practically all the energy is going into the stone thrown and only a negligible amount is submitted to the boat. The entire lake will, thus, take up the recoil and this procedure occurs as [a] recoilless process."[12 p8]

Mössbauer quickly realized that the effect he discovered could be used for a new kind of spectroscopy with an exceedingly high energy resolution: He was the first to use Mössbauer spectroscopy, measuring the width of the first excited state of ^{191}Ir. In a move that would awe and intimidate subsequent generations of physics graduate students, Mössbauer submitted papers documenting his work to receive his PhD degree as well as the 1961 Nobel Prize in Physics.[13–15]

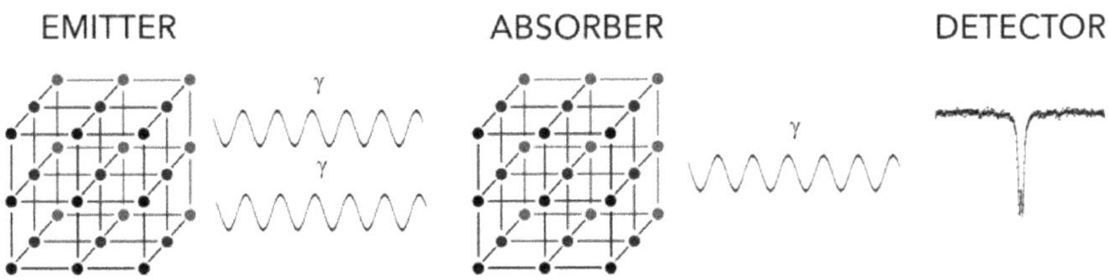

Fig. 4.5.2. A schematic representation of a Mössbauer experiment. The gamma source emits a spectrum of frequencies, some of which are absorbed by nuclei in the target. The detector records a dip at the frequency of the absorbed rays.

The basic elements of Mössbauer's first spectroscopic experiments were common to nuclear resonance measurements; these same elements would also be used by the many physicists who would subsequently perform Mössbauer spectroscopy. His experimental setup (Figure 4.5.2) consisted of a source of gamma radiation mounted on a rotating disk, a stationary absorber, and a detector to measure the transmitted radiation. This equipment allowed measurement of the recoilless nuclear-resonant absorption of gamma rays using Doppler shifts of the absorbing relative to the emitting nuclei.

Recoilless Emission

Typically, when a gamma ray is emitted by an atom in a solid, the recoil energy is carried away by a phonon — a normal mode of vibration in a crystal lattice. The emitted gamma ray therefore carries the original emission energy minus the energy of the recoil phonon. The reduction in the energy of the emitted gamma ray is small in this process, on the order of one part in a million. For most purposes, that reduction can be ignored, but it is enough to prevent the gamma ray from being reabsorbed by another nucleus of the same kind. The loss of a miniscule amount of recoil energy makes the gamma-ray energy fall short of the amount required, which, according to quantum mechanics, is given by the difference between the initial and final energy states of the absorbing nucleus. Pre-Mössbauer experiments allowed nuclei to reabsorb emitted gamma rays by supplying them with additional push to compensate for the lost recoil energy. This could be done by heating the source or by spinning it in a fast centrifuge. Mössbauer's discovery showed that this is not necessary when the emitting atom is "frozen" in the lattice of a crystal. In this case, the gamma ray "resonates" with the energy in the absorbing nucleus, and the structure of the resonance can be scanned by varying the relative speed of the source and the absorber by tiny amounts, a few millimeters per second.

Mössbauer Spectroscopy Spreads

Mössbauer was able to carry out his experiments using the minimal equipment available in post–World War II Germany; for example, his first spectroscopic experiment employed a turntable with gears from Heidelberg's toy shops. The confirmation of the Mössbauer effect, however, was made not by Mössbauer but by scientists at Argonne and Los Alamos National Laboratory in the United States.[6; 11] Subsequent innovations — including the discovery that iron-57 (^{57}Fe) sources provided a much larger effect and therefore created a more powerful spectroscopic tool — were made mostly at large laboratories. In addition to Argonne and Los Alamos, these included Brookhaven National Laboratory and the British Atomic Energy Research Establishment at Harwell.

The first conference on Mössbauer spectroscopy was organized by Hans Frauenfelder in 1960 and the affiliations of the participants show the dominance of large laboratories in this early work.[8] However, as Mössbauer noted: "This field is a fine example of a branch within experimental physics which does not require the application of much equipment, a fact which allows even the smaller research institutions to compete with large, well-equipped laboratories."[8 p55–56] As Frauenfelder pointed out in a 1962 review, which includes reprints of the most important early papers, the advent of ^{57}Fe sources helped in the spread of the field beyond a few large laboratories.[7 p12–13]

Another factor in the spread of Mössbauer spectroscopy was the discovery in the early 1960s that the unprecedentedly high resolution of the Mössbauer resonance had the ability to resolve the so-called hyperfine interactions that connect nuclei with their surrounding electrons. Since these interactions connect the nuclear moments (nuclear-monopole moment, nuclear-magnetic moment, nuclear-quadrupole moment) and corresponding solid-state properties (electron-charge distribution at the nucleus, magnetic field at the nucleus, electric-field gradient at the nucleus), understanding and exploiting the Mössbauer effect would require work at the intersection of nuclear and solid state physics and the skills and knowledge of physicists in both specialties. Indeed, Mössbauer spectroscopy would become an important tool in determining the properties of matter for physicists working in a wide variety of specialties.

The rapid expansion of Mössbauer spectroscopy can be gauged by comparing the first conference on Mössbauer spectroscopy in 1960 with the third in 1963. The 1960 conference recorded fifty-two papers on the subject; the bibliography in 1963 had over six hundred entries. By 1963 numerous small institutions were represented as well as large laboratories, and, as was explained in the proceedings, work was being done in "physics, chemistry, and biology."[5 p333]

Although smaller institutions became increasingly involved in Mössbauer spectroscopy after 1960, large institutions that facilitated cutting-edge research retained advantages through the mid-1960s. Historian Klaus Hentschel made this point in his study of a series of landmark experiments that were conducted between 1959 and 1965, and which provided the first terrestrial evidence of the equivalence principle in Albert Einstein's general theory of relativity. The competition to measure the predicted gravitational redshift was joined by two teams, Robert V. Pound and Glen A. Rebka at Harvard University and a team that included John Schiffer (an Argonne physicist on leave), Ted Cranshaw, and A. Bruce Whitehead at Harwell.[16; 17] To explain why others did not "enter the race," Hentschel observed that "to reduce systematic errors sufficiently it was indispensable to have strong and chemically pure sources of γ radiation: in the case of Schiffer and his team it involved exposing iron plates in the Birmingham cyclotron and subsequent chemical treatment by the Radiochemical Centre in Amersham. Pound and Rebka were provided with similar services by Oak Ridge National Laboratory, the Nuclear Science and Engineering Corporation and by Nuclear Metals Inc."[9 p273]

Having access to such resources was an important advantage during this period for all work in Mössbauer spectroscopy. In particular, "the presence of a cyclotron or of a nuclear reactor was most convenient," because such equipment was needed for "the preparation of strong sources."[9 p274]

The Turn to Materials

Starting in the 1970s, Mössbauer spectroscopy moved from an era of discovery, during which it was of interest to nuclear physics, into an era of application, in which it was of interest to materials scientists.[1–4] Instead of using a relatively small number of resonances to explore nuclear properties, these scientists employed a large Mössbauer periodic table. In the process, increased variety of measurement became more valuable than increased scale: Mössbauer spectroscopy was sometimes performed using particle accelerators and at other times was performed using small-scale devices little different from the ones used by Mössbauer himself.

In this endeavor, the unprecedented ultra-high resolution of Mössbauer spectroscopy gave scientists the opportunity to make a broad range of hyperfine-structure measurements. They could add this hyperfine-structure information to the mix of other data, such as measurements of electronic and magnetic properties, the measurements of thermal and electrical conductivities, and imaging, to piece together the vastly complicated picture of the magnetic, thermal, electrical, mechanical, and biological properties of a large variety of materials. These new data were crucial because materials scientists depended upon a large quantity of high-quality information to achieve the key goals of their field: delineating the basic properties of matter and finding materials suitable for researchers in other fields as well as a rich variety of medical and industrial applications.

References

1. Argonne National Laboratory. Physics division status report, April 1, 1967–March 31, 1968. Argonne, IL: Argonne National Laboratory; 1968 [ANL-7481].

2. Argonne National Laboratory. Solid state division research status, October 1, 1971–September 30, 1972. Argonne, IL: Argonne National Laboratory; 1972 [no ANL number assigned].
3. Argonne National Laboratory. Solid state division research status, October 1, 1974-September 30, 1975. Argonne, IL: Argonne National Laboratory; 1975 [ANL-SSS-75-10].
4. Argonne National Laboratory. Solid state division research status, October 1, 1977–September 30, 1987. Argonne, IL: Argonne National Laboratory; 1987 [ANL-SSS-78-10].
5. Bearden A. Third international conference on the Mössbauer effect. Reviews of Modern Physics. 1964;36(1):333–34.
6. Craig PP, Dash JG, McGuire AD, Nagle D, Reiswig RR. Nuclear resonance absorption of gamma rays in ^{191}Ir. Physical Review Letters. 1959;3(5): 221–23.
7. Frauenfelder H. The Mössbauer effect. New York: W. A. Benjamin; 1962.
8. Frauenfelder H, Lustig H. Mössbauer effect: recoilless emission and absorption of gamma rays. University of Illinois: Allerton House; 1960.
9. Hentschel K. Measurements of gravitational redshift between 1959 and 1971. Annals of Science. 1996;52(5):273–95.
10. Hentschel K. Zum zusammenspiel von instrument, experiment und theorie: Rotverschiebung im sonnenspektrum und verwandte spektrale verschiebungseffekte von 1880–1960. Hamburg: Verlag Dr. Kovac; 1995.
11. Lee LL, Meyer-Schutzmeister L, Schiffer JP, Reiszig RR. Nuclear resonance absorption of gamma rays at low temperatures. Physical Review Letters. 1959;3(5):223–25.
12. Mössbauer RL. The discovery of the Mössbauer effect. Hyperfine Interactions. 2000;126(1):1–12.
13. Mössbauer RL. Kernresonanzabsorption von Gammastrahlung in ^{191}Ir. Die Naturwissenschaften. 1958;45(22):538–39.
14. Mössbauer RL. Kernresonanzfluororeszenz von Gammastrahlung in ^{191}Ir. Zeitscrift für Physik. 1958;151(2):124–43.
15. Mössbauer RL. Kernresonanzabsorption von γ-Strahlung in ^{191}Ir. Zeitschrift für Naturforschung A. 1959;14(3):211–16.
16. Pound RV. Weighing photons, I. Physics in Perspective. 2000;2(3):224–68.
17. Pound RV. Weighing photons, II. Physics in Perspective. 2001;3(1):4–51.
18. Sanderson T. Charles Johnson and the Liverpool Mössbauer group [master's thesis]. Liverpool, UK: University of Liverpool; 2002.
19. Westfall C. A different laboratory tale: fifty years of Mössbauer spectroscopy. Physics in Perspective. 2006;8(2):189–213.

CHAPTER 4.6

LASERS

BENJAMIN WILSON

Laser light is coherent, monochromatic, highly directional, and capable of high intensities. It can probe matter within narrow frequency bands, at tiny scales, and with fantastically large energies. After the demonstration of the first working model in 1960, laser researchers quickly grasped the device's potential as a tool for the study of matter, light, and their interaction.

What follows is a brief look at the emergence of two closely related fields of laser science — nonlinear optics and laser spectroscopy — each of which took shape during a period of dramatic growth in laser research in the 1960s and early 1970s.[4] These fields were nourished by government funding, especially from the military services, whose interest in technologies of stimulated emission was longstanding.[8; 19] More than contracts lay beneath the rise of laser science, however, especially in the case of nonlinear optics. Specific Cold War military challenges, and the efforts of a highly placed group of defense consultants and laser experts, were critical to the development of the field.[22]

Optical Harmonic Generation

Only a year after the announcement of the first working ruby laser at Hughes Research Laboratories, scientists were using laser–matter interactions to unlock a completely new regime of optics. When the University of Michigan physicist Peter Franken attended a meeting of the Optical Society of America in early 1961, he began to wonder how much energy laser light could deliver to a solid. Scribbling a quick calculation, he realized that even the modest power output of an early model laser, if focused into a small spot several micrometers in width, would create huge optical electric field amplitudes at the center of the spot. Comparable in strength to the electric fields binding electrons and nuclei within atoms, these optical fields would tug the electrons from their normal orbits.[22]

Franken was particularly interested in the possibility of generating optical harmonics. Harmonics of an incident wave oscillate at multiples of the frequency of the original wave, like the overtones of a musical note. Physicists and engineers had long known how to generate harmonics at microwave and radiofrequencies by passing radiation through appropriate materials. The phenomenon had never been observed in the visible region of the spectrum, however, because there had never been a source of light intense enough to excite a sufficiently strong response. Franken's calculation suggested that a laser might well have sufficient intensity to cause second harmonic generation, in which the frequency of an input wave is doubled as it passes through the medium. In the case of the ruby laser, its red light would be converted to blue.

During the spring and summer of 1961, Franken and three colleagues tried an experiment using a ruby laser supplied by Trion Instruments, an Ann Arbor–based company founded by one of Franken's students. They fired three-joule laser pulses of one-millisecond duration at a sample of crystalline quartz, split the output into its frequency components using a spectrometer, and recorded the results on a photographic plate. True to Franken's guess, a weak second harmonic beam emerged from the quartz at exactly twice the frequency of the laser.[9]

Nicolaas Bloembergen, a Dutch-born American physicist working at Harvard University, saw a preprint of Franken's paper and decided to attempt an observation of second harmonic generation himself. A skilled experimentalist and theorist, Bloembergen had conducted his doctoral research on nuclear magnetic resonance. In 1956 he had invented a new scheme for the maser (the microwave-frequency precursor of the laser) using a solid-state crystal as the amplifying medium, rather than the atomic beam of Charles Townes's original design. But in 1961 Bloembergen and his younger colleague Peter Pershan had difficulty getting their homebuilt ruby laser to work. In the meantime, they turned to the quantum mechanical theory of nonlinear optics. After months of painstaking calculations, they published an article showing how a variety of nonlinear optical processes (second harmonic generation, the frequency mixing of two beams of unequal frequency, and higher order effects) could be described as the coupling of electromagnetic fields in nonlinear media. The paper laid the theoretical groundwork for the new field of nonlinear optics.[1]

Stimulated Scattering

Bloembergen later said that the laser itself deserved much of the credit for the advent of nonlinear optics. In his Nobel Prize lecture in 1981, he recounted that the famous Dutch physicist Hendrik Lorentz had shown, by modeling electrons bound within atoms as harmonic oscillators, that the laws of classical optics arose from Maxwell's equations of electrodynamics. Lorentz, Bloembergen insisted, might have discovered nonlinear optics in the late nineteenth century had he simply modeled the electrons as anharmonic oscillators instead. The missed opportunity had a straightforward explanation, said Bloembergen: "Lorentz lacked the stimulation from stimulated emission of radiation." The laser, still several decades away, had not yet produced the laboratory observations that would demand a nonlinear theory.[3 p13]

Yet there is a much richer story to be told about the context and motivation behind the development of nonlinear optics and related fields in the United States, which opened up new avenues for characterizing the basic properties of materials. At the center of this story is one of the Cold War's greatest technical challenges — ballistic missile defense — and a close-knit community of defense consultants.[22]

In December 1961, just as he began his new research program in nonlinear optics, Bloembergen was invited to Washington, DC, to participate in a top-secret meeting at the Institute for Defense Analyses (IDA), a nonprofit think tank consulting for the Department of Defense. Keith Brueckner, a physicist on leave from the University of California who served as IDA's director of research, had called together a special laser committee for IDA, including Bloembergen, Townes, the physicist Normal Kroll, and a handful of others. The group's task was to assess the military potential of high-power lasers — especially the possibility that lasers might be capable of shooting down a ballistic missile.

At the committee's first meeting in 1961, Bloembergen made a remarkable proposal. Estimating that a million joules of energy would be necessary to destroy a ballistic missile in flight, he calculated that a cubic meter of neodymium-doped glass, pumped with an enormous amount of energy (perhaps to the point of melting), would deliver the required missile-killing pulse. Referred to thereafter by the consultants as "the Bloembergen Cube," the idea stoked the imaginations of IDA's management and military sponsors. Industrial contracts totaling several million dollars soon issued from a new classified program to develop high-power lasers, administered by the Office of Naval Research and Project DEFENDER, the missile defense effort of the Advanced Research Projects Agency (ARPA).

Generating and propagating a high-intensity beam for missile defense would prove no easy feat, however. As the consultants began to elaborate in a series of classified papers and briefings, the laser light was certain to suffer distortion and power loss through scattering in the atmosphere. And yet, the problem had a welcome side benefit for the consultants, motivating important early research in a new area of nonlinear optics known as stimulated scattering. In 1962, scientists at Hughes had noticed that their pulsed lasers were producing output at narrow frequency lines lower than the dominant laser frequency. The culprit, it soon became clear, had been the nitrobenzene cell placed within the laser cavity to create the pulses. As the laser light passed through this cell, it excited internal quantum states within the nitrobenzene molecules, then scattered inelastically from them. In analogy to a scattering process familiar from classical optics, the effect came to be called stimulated Raman scattering.[7]

The IDA consultants strove to understand stimulated Raman scattering in great depth, concerned on one hand that such scattering would degrade the power and quality of high-power beams, and hopeful on the other that intentionally Raman-shifting the laser frequency might locate better propagation channels through the atmosphere. Townes returned to his lab at the Massachusetts Institute of Technology after attending a classified IDA summer study on lasers in 1963 and, with two colleagues, developed a semi-classical model of the effect. Thanks to a special agreement between IDA and its ARPA sponsor, the consultants were able to publish the unclassified results of their classified work without the delay of an extensive declassification review. Townes's paper quickly made its way into the physics discipline's premier journal, *Physical Review Letters*.[10]

In similar fashion, other challenges led to new advances in nonlinear optics. Researchers working on high-power laser prototypes noticed mysterious cracks and streaks in the optical materials of their prototypes during testing. The IDA consultants began to consider possible explanations after being made aware of the defects by John Atwood, the director of research at the laser defense contractor Perkin-Elmer. Norman Kroll believed the cause might be an effect known as electrostriction, in which the strong electric fields of the laser light pulled the surrounding atoms out of their normal positions, creating hypersonic vibrations in the optical materials. In a subsequent experiment at MIT to measure the coupling of laser light and sound waves, Townes and a colleague focused intense laser pulses in samples of quartz and sapphire and detected scattered light at slightly lower frequencies emitted in the reverse direction. This new effect, which Townes called stimulated Brillouin scattering, resulted from the interaction between the laser light and induced acoustic phonons in the sample.[6]

At the same time, Townes realized that some of the filament-like streaks of damage might also be the result of a related nonlinear optical effect: the intensity-dependent index of refraction. At high enough light intensity, a material's index of refraction (which determines the propagation speed and bending of light passing through it) varies with the intensity of the light. Moreover, the cross-sectional intensity of a laser beam is not uniform, but strongest in the middle and decreasing with radial distance. The intensity-dependent refraction combined with this uneven intensity profile means that, from the point of view of the beam, the index of refraction is highest at its center. As Townes and collaborators described in a theoretical paper in 1964, above a certain power threshold, the beam will undergo total internal reflection at its edges, overcoming ordinary dispersion and trapping itself inside a narrow waveguide. All of the beam's energy bores through this tunnel, producing a thin streak of damage.[5]

Nicolaas Bloembergen, who had used his generous Project DEFENDER grant to outfit his lab at Harvard with state-of-the-art lasers, made the first laboratory confirmation of the self-trapping effect in 1965.[14] And in another foundational theory paper that same year, he bundled the new discoveries together within the framework of his coupled-wave theory of light–matter interactions.[20]

Laser Spectroscopy

Nonlinear optics found numerous applications, not least in the study of spectroscopy, one of physicists' and chemists' oldest techniques for investigating the structure and properties of matter. The laser had always been recognized as a promising instrument for precision spectroscopy. But spectral resolution — the ability to distinguish separate atomic or molecular emission and absorption lines from each other — was limited by a phenomenon known as Doppler broadening. As with sound, light emitted by a moving source is Doppler-shifted up or down depending on whether the source moves toward or away from the observer. Thus, for a thermal sample of randomly moving molecules, their closely spaced spectral lines are smeared together by the Doppler shifts. The molecular motion washes out the subtle spectral details, especially the so-called hyperfine structure, which arises from interactions between the spins of electrons and nuclei. "Many spectroscopists," wrote the laser pioneer Arthur Schawlow in 1978, "must have wished that they could somehow hold those molecules still, and stop them from moving around."[18 p142]

Saturated absorption spectroscopy, a new nonlinear-optical technique developed in the late 1960s and 1970s, effectively allowed laser scientists to hold the molecules still, all but eliminating Doppler broadening. An early breakthrough came from work by two pairs of researchers, one at the defense contractor Perkin-Elmer, and another at the Joint Institute for Laboratory Astrophysics (JILA) in Boulder, Colorado. (JILA, operated by the National Bureau of Standards and the University of Colorado, had been established in 1962 with the support of ARPA, in part to conduct spectroscopic research relevant to missile defense.) These researchers found that by placing an absorbing cell inside the resonant cavities of their helium–neon gas lasers, a sharp spike was observed in the laser output at the absorption frequency of the active molecules. At this frequency, the molecules in the cavity were no longer able to absorb light — they were *saturated* — yielding the spike in output. The spike was narrow in frequency width because only one particular group of molecules in the cavity could become saturated at resonance — namely, those able to interact with light waves traveling in either direction as the waves bounced back and forth between the cavity mirrors. This group consisted of those molecules at rest with respect to the cavity mirrors — precisely the molecules for which Doppler broadening was absent.[2; 15]

In 1970 a young German researcher named Theodor Hänsch realized that saturated absorption could be used to do spectroscopy of extraordinary precision on samples outside the laser cavity. Hänsch was soon to arrive in the United States on a NATO-sponsored postdoctoral fellowship to work with Arthur Schawlow at Stanford University. In the clever arrangement Hänsch and Schawlow designed (see Figure 4.6.1), a single laser beam was split into stronger "saturating" and weaker "probe" beams, each beam then directed with mirrors to cross one another in nearly opposite directions through a sample under investigation. When the saturating beam was blocked before entering the sample, more of the probe beam would interact with the sample molecules, so less of it would reach a detector on the far side. When the saturating beam was allowed to pass, the

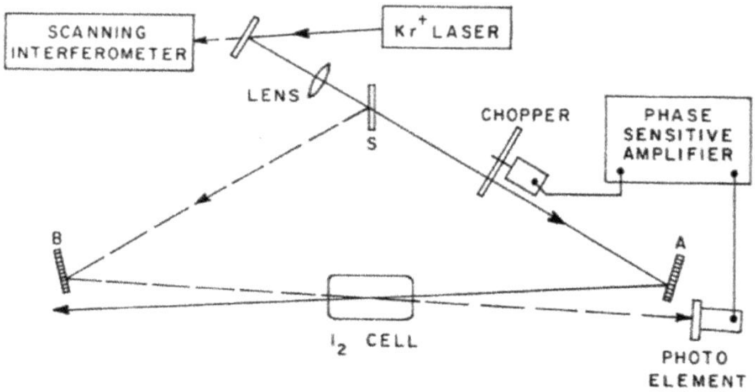

Fig. 4.6.1. The experimental arrangement for laser saturation spectroscopy. First a laser beam is split, then each of the split beams directed with mirrors to pass through a sample in nearly opposite directions. (In this diagram, the sample is molecular iodine.) A chopper alternately blocks and admits a "saturating beam" to the sample. When the saturating beam is admitted, more of the opposite "probe beam" passes through the sample; when it is blocked, less of the probe passes through the sample, since more of it is absorbed. However, this is true only for sample molecules that are motionless along the beam axis. These molecules suffer no Doppler broadening in their absorption and emission spectrum. The probe beam can then be tuned across a tiny frequency range, mapping out an extremely high-resolution spectrum for the sample.[11]

molecules of the sample became saturated and absorbed less energy from the probe beam, allowing more of the probe to reach the detector. Because the two beams were identical in frequency, however, this modulation of the probe beam by the saturating beam only occurred for molecules capable of absorbing *both* beams identically. Once again, because the beams traveled in opposite directions, this was true only of molecules at rest. Doppler broadening was erased, and the probe beam could now be tuned across several lines to reveal a spectrum of extraordinary resolution.

In 1971 Schawlow, Hänsch, and their colleagues used saturated absorption to resolve hyperfine splitting in the Balmer series of the hydrogen spectrum. They also achieved the first-ever direct optical observation of the Lamb shift (a tiny quantum-electrodynamical correction to hydrogen's energy levels).[13] Hänsch and his student Carl Wieman later made a further refinement to the method, using polarizers to greatly improve the signal-to-noise ratio of the probe beam. Other researchers developed an even more efficient procedure for eliminating Doppler broadening, known as two-photon absorption, extending precision spectroscopy into the ultraviolet. Together, these experiments offered not only a proof of concept for laser spectroscopy, but an exacting test of the quantum theory of atoms, the foundation for scientists' basic understanding of the structure of matter.[12; 21]

Nonlinear optics and laser spectroscopy continued to grow in tandem during the 1970s. Higher-order harmonic generation, stimulated Raman emission, and other nonlinear effects were harnessed to greatly expand the spectrum of laser wavelengths available for spectroscopy.[16] Researchers also began to use nonlinear processes to shorten laser pulses to astoundingly brief timescales, from picoseconds (10^{-12} s) down to femtoseconds (10^{-15} s). With this ultimate high-speed camera, phase transitions, chemical reactions, and other transient, near-instantaneous phenomena could be watched as they unfolded.

And progress continues. In a recent experiment at Stanford, an infrared laser was fired at a semiconductor film of only a few atoms' thickness to produce attosecond-scale (10^{-18} s) pulses and harmonic generation at thirteenth order and higher. Such techniques not only promise to push the boundaries of semiconductor and nanostructure fabrication, but also to make possible delicate measurements of electron dynamics in solids.[17] Laser science has come a long way since Peter Franken's 1961 experiment in second harmonic generation, and the top-secret laser meetings of the Institute for Defense Analyses. Yet the field maintains its dual pursuit of basic knowledge and engineering application.

References

1. Armstrong JA, Bloembergen N, Ducuing J, Pershan PS. Interactions between light waves in a nonlinear dielectric. Physical Review. 1962;127(4): 1918–39.

2. Barger RL, Hall JL. Pressure shift and broadening of methane line at 3.39 μ studied by laser-saturated molecular absorption. Physical Review Letters. 1969;22(1):4–8.
3. Bloembergen N. Nonlinear optics and spectroscopy. In Frängsmyr T, Ekspong G, editors. Nobel lectures, physics 1981–1990. Singapore: World Scientific Publishing Co.; 1993.
4. Bromberg JL. The laser in America, 1950–1970. Cambridge, MA: MIT Press; 1991.
5. Chiao RY, Garmire E, Townes CH. Self-trapping of optical beams. Physical Review Letters. 1964;13(15):479–82.
6. Chiao RY, Townes CH. Stimulated Brillouin scattering and coherent generation of intense hypersonic waves. Physical Review Letters. 1964;12(21): 592–95.
7. Eckhardt G, Hellwart RW, McClung FJ, Schwarz SE, Weiner D, Woodbury EJ. Stimulated Raman scattering from organic liquids. Physical Review Letters. 1962;9(11):455–57.
8. Forman P. Behind quantum electronics: national security as basis for physical research in the United States, 1940–1960. Historical Studies in the Physical and Biological Sciences. 1987;18(1):149–229.
9. Franken PA, Hill AE, Peters CW, Weinreich G. Generation of optical harmonics. Physical Review Letters. 1961;7(4):118–20.
10. Garmire E, Pandarese F, Townes CH. Coherently driven molecular vibrations and light modulation. Physical Review Letters. 1963;11(4):160–63.
11. Hänsch TW, Levenson MD, Schawlow AL. Complete hyperfine structure of a molecular iodine line. Physical Review Letters. 1971;26(16):946–49.
12. Hänsch TW, Schawlow AL, Series GW. The spectrum of atomic hydrogen. Scientific American. 1979;240(3):94–110.
13. Hänsch TW, Shahin IS, Schawlow AL. Optical resolution of the Lamb shift in atomic hydrogen by laser saturation spectroscopy. Nature Physical Science. 1972;235:63–65.
14. Lallemand P, Bloembergen N. Self-focusing of laser beams and stimulated Raman gain in liquids. Physical Review Letters. 1965;15(26):1010–13.
15. Lee PH, Skolnick ML. Saturated neon absorption inside a 6238-Å laser. Applied Physics Letters. 1967;10(11);303–5.
16. Levenson MD. Introduction to nonlinear laser spectroscopy. New York: Academic Press; 1982.
17. Liu H, Li Y, You YS, Ghimire S, Heinz TF, Reis DA. High-harmonic generation from an atomically thin semiconductor. Nature Physics. 2017;13:262–66.
18. Schawlow AL. Laser spectroscopy of atoms and molecules. Science. 1978;202(4364):141–47.
19. Seidel RW. From glow to flow: a history of military laser research and development. Historical Studies in the Physical and Biological Sciences. 1987;18(1):111–47.

20. Shen YR, Bloembergen N. Theory of stimulated Brillouin and Raman scattering. Physical Review. 1965;137(6A):A1787–805.
21. Wieman C, Hänsch TW. Doppler-free laser polarization spectroscopy. Physical Review Letters. 1976;36(20):1170–73.
22. Wilson B. The consultants: nonlinear optics and the social world of Cold War science. Historical Studies in the Natural Sciences. 2015;45(5): 758–804.

CHAPTER 4.7

PROBE AND OTHER MICROSCOPIES

CYRUS C. M. MODY

The history of twentieth-century microscopy is often told as a transition from imaging with light (focused using glass lenses) to imaging with electrons (focused using magnetic lenses). In that story, electrons are simply particle–waves, which act like photons but have a much higher frequency and therefore better theoretical resolution (see chapters on transmission electron microscopy, scanning electron microscopy, and lithography). Even in the 1930s, though, researchers saw that electrons could image surfaces in ways not analogous to light. Such techniques developed and diffused slowly, despite achievements such as the first images of individual atoms. In the 1980s, though, one variant — the scanning tunneling microscope — caught on and led to the invention of the atomic force microscope and its relatives. These "probe" microscopes are now as common as electron microscopes. Their success has catalyzed the union of materials research and the life sciences that accelerated at the turn of the twentieth century.

Pre-1980

All electron microscopies descend from the kinds of cathode-ray-tube experiments profiled in the introduction to part 4 of this volume. In scanning and transmission electron microscopy, the cathode is pictured as the source of a ray of electrons, which illuminates a sample in the same way that light illuminates the sample in optical microscopy. In the 1930s, though, Erwin Müller — then working at Siemens in Germany — realized that the same filament (usually of tungsten) could serve as both sample *and* electron source. Placing a fluorescing screen near the filament would cause an image to form on the screen corresponding to the places on the filament from which electrons were emitted. In 1951 he invented the field ion microscope (FIM), which did the same thing but with ions, thereby allowing a radical jump in resolution. Indeed, in 1955 one of Müller's graduate students at Pennsylvania State University, Kanwar Bahadur, showed — somewhat over Müller's opposition — that the FIM could reveal the locations of individual *atoms* in the filament under certain circumstances.[3]

Another of Müller's students, Russell Young, left Penn State in 1961 for the US National Bureau of Standards, where he joined a group working to improve the resolution of conventional electron microscopes. Starting in 1965, though, Young began to think about ways to adapt Müller's field emission microscope to a classic Bureau problem: calibrating the smoothness of a surface and/or the distance between two surfaces. In the FEM, the field emission current between the cathode and screen depends strongly on the distance between the two; Young therefore saw that if he substituted a sample for the screen, he could use the strength of the field-emission current to calibrate the distance between cathode and sample. (Young and his set-up are pictured in Figure 2.3.1 of the entry on vacuum.) Later, he realized that if he scanned the cathode across the sample and fed the field emission current into an oscilloscope, then he could build up a three-dimensional image of the surface of the sample. He dubbed this device the Topografiner.[6]

Not many FEMs and FIMs — and only one Topografiner — were ever built. Müller was a careful gatekeeper who did not want people to use his inventions "improperly," and the FEM and FIM were restricted in the samples they could image. Young was eager for others to adopt the Topografiner, but his managers at the Bureau viewed it as a distraction during a period of organizational turmoil.

Young did have a protégé, Clayton Teague, who showed that the Topografiner could operate not just with a field emission current (where electrons move ballistically between cathode and sample) but also with a higher-resolution tunneling current (where electrons quantum mechanically disappear from the cathode and magically reappear at the sample). Teague, however, failed to communicate his findings to the wider scientific community, possibly because of the same managerial opposition Young faced.

Tunneling to Victory

Meanwhile, at the IBM research lab near Zurich, Heinrich Rohrer hired a new PhD, Gerd Binnig, in 1978 to help solve a technological problem: "pinholes" in thin films used to make superconducting circuits. Binnig and Rohrer both had experience with electron tunneling, so they made the same connection as Young: the tunneling current can be used to measure very small distances between a probe and sample, and if the probe is scanned an image of the surface can be built up. By 1981 they had repeated Teague's experiment, and a year later they showed that the STM could produce atomic (or near-atomic) resolution images of gold. Few were impressed, though, until they published atomic-resolution images of silicon in 1983 (Figure 4.7.1). At that point, surface scientists (see the entry on vacuum in this volume) — particularly at Bell Laboratories and IBM's US research sites — frantically began building their own STMs.[4 Ch. 2–3]

Surface scientists were intrigued by the information STM revealed about "surface reconstructions" — the intricate pattern of atoms at the surface of a semiconductor (and some metals) that rearranges how atoms are organized in the bulk material. Surface reconstructions had been crucial "test objects" for training new surface scientists and developing new surface science techniques since the 1950s, but few reconstructions had actually been solved by the 1980s. The STM did not solve any either, but it did provide data which — in conjunction with other techniques — brought the field closer to unlocking one of its central mysteries. Largely for that reason (and

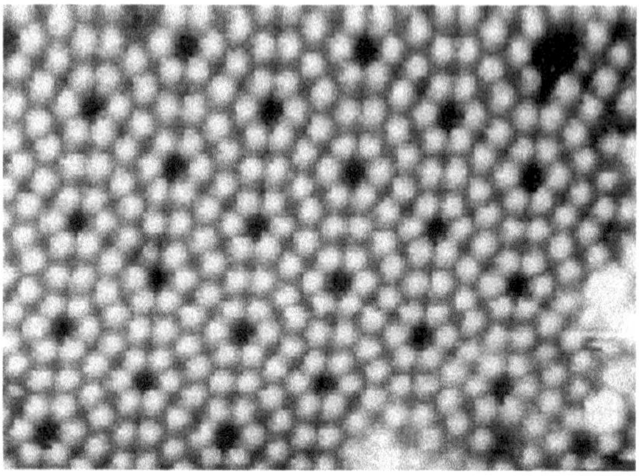

Fig. 4.7.1. Digitized STM image of the silicon (111) 7×7 surface reconstruction made at IBM's Yorktown, New York research lab. Reprinted from Demuth JE, *et al.* A scanning tunneling microscope for surface science studies. IBM Journal of Research and Development. 1986;30(4):396–402. *Credit*: Courtesy of International Business Machines Corporation, ©1986 International Business Machines Corporation.

with help from IBM's public relations machine) Binnig and Rohrer shared the 1986 Nobel Prize in Physics with Ernst Ruska, who had invented the first electron microscope *five decades* earlier!

Binnig and Rohrer were not themselves surface scientists, though, and they were sometimes frustrated with the field's conventions. They wanted the STM to find wider application, and so through their (and their employer's) networks of colleagues, they began recruiting non-surface scientists to build STMs. Groups in Madrid, Basel, Santa Barbara, Stanford, Caltech, the Bureau of Standards, and a few other places bravely attempted it, but the technology was still primitive and imaging surface reconstructions was still the only reliable means of ascertaining whether an STM was working properly. Most non-surface scientists struggled and/or had to learn surface science in order to build a *functioning* microscope, much less a microscope capable of contributing to new knowledge and discoveries.

Stepwise Innovation

By mid-1985, replication of the STM was becoming somewhat more routine (Figure 4.7.2). However, its popularity outside surface science

Fig. 4.7.2. Early STM built by John Baldeschieler's group at Caltech based on a design developed at IBM Zurich. *Credit*: Photograph by Gregory Tobias. Courtesy of Science History Institute Collections.

was still limited because it could only image conducting (or semiconducting) samples in ultrahigh vacuum — or so it was thought. A few groups more or less simultaneously discovered that images could be generated of insulating samples in vacuum, and also of insulating and conducting samples in air. Surface scientists were highly suspicious of these images, though, and controversy about their status has never fully been resolved.

Binnig — along with a Stanford faculty member, Calvin Quate, and an IBM Zurich technician, Christoph Gerber — therefore decided to try a different approach.[1] If you take a very small, flexible cantilever — a tiny rectangle of aluminum foil, for instance — and attach a small, pointed weight (the "tip") at one end, then you can drag the tip over a surface and the top of the cantilever will bend ("deflect") in correspondence with the topography of the surface. This was the basis for surface profilometry, an established technique used — for instance — to ensure the smoothness of silicon wafers. However, Binnig believed that if he positioned an STM on the back of the cantilever, he could achieve much higher resolution. He hoped to let the tip hover over the surface without contacting it, with the deflections caused by van der Waals and other forces between the atoms of the tip and sample — hence, an "atomic force microscope" (or AFM).

The AFM was taken up by a few non-surface scientists, but because it was even more unreliable than the early STMs its popularity was initially limited. AFMers such as Paul Hansma at the University of California–Santa Barbara developed innovations — such as optical sensing (where the cantilever's deflection is detected with a laser rather than an STM), which set the AFM on a path toward reliability. But air STM initially remained much more popular for two reasons: the first commercial manufacturers started offering low-cost versions to people who lacked the time or expertise to build their own; and beautiful images created with the STM started to appear on the cover of prestigious journals such as *Science* and *Nature* almost every month. However, in the early 1990s it became apparent that some of those images were unreliable; and because the first commercial AFMs started to appear on the market at exactly that time, almost all air STMers switched over the next few years. From that point on, AFM was used predominantly in air and liquid and STM in ultrahigh vacuum, with hardly anyone staying at the state of the art in both techniques at once.

Children of the AFM

With the invention of the AFM, it became clear that a variety of interactions between a small solid probe and a surface could be used to map a number of different properties of the surface: thermal, magnetic, electrostatic, and so on. New kinds of "probe microscopes" started to appear regularly, with several dozen named variants now in existence.[2] Most of these are only used by a handful of research labs but a few — particularly the magnetic force microscope — have achieved wider popularity. Specialized ways of using STM and AFM have also spread widely: for example, "tapping" a surface with the AFM tip to measure softness without damaging the sample; or "dragging" atoms into selected positions with an STM tip to form quantum showpieces, tiny logic gates, and words such as "IBM." The latter, in particular, have become iconic examples of the power of IBM and probe microscopy specifically, and of nanotechnology more generally.[5]

The challenge for all of these techniques — stretching back to the Topografiner — is that their inventors want them to be used to image samples about which the inventors know little. They therefore find it difficult to generate images at all; and when they do, people with expertise in those samples do not understand the images and do not regard the inventors as credible. Young never resolved that problem. Binnig and Rohrer drew on IBM's substantial surface science contingent to learn how to choose and prepare samples and establish credibility in the field. Once they won the Nobel, it became much easier for them to establish partnerships with representatives from a variety of fields who promoted the STM and AFM among their colleagues. The earliest AFMers, such as Quate and Hansma, adopted the technique because they believe it had greater potential than the STM for interdisciplinary use. Thus, they also adopted organizational techniques — such as a steady flow of postdocs from various disciplines — that complemented their technological innovations.

The spread of STM and especially AFM was also aided by the start-up companies founded in the 1980s and 1990s to sell these instruments. In particular, Digital Instruments — a Santa Barbara manufacturer somewhat associated with Hansma — developed relatively low-cost, user-friendly machines that practitioners of many disciplines could afford to use. It also developed techniques for incorporating others' innovations into its instruments. And it created more expensive versions of the AFM for use by industries manufacturing everything from microchips to contact lenses.

As a result, probe microscopy since 1990 has exemplified — and to some extent driven — the late-century shift in materials research from an exclusive focus on "classical" materials (metals, ceramics, semiconductors, and to a lesser extent polymers) toward a focus on biomaterials and other material systems inspired by the life sciences. For many probe microscopists, that shift was encapsulated by the sudden appearance of centers, journals, and other institutions bearing the prefix "nano" around the year 2000 — institutions which almost always tied together the life, physical, and engineering sciences and cultivated ties among academic labs, start-ups, and established firms. Indeed, because of their experience with a technology where interdisciplinarity and university-industry cooperation were at a premium, probe microscopists were often leaders in the establishment of nano institutions. That is, the probe microscopy example shows us that tools gain popularity for how they can be used in materials research; but materials research is often led in new directions not just by the novel technological capacities of new tools but also by the social experiences of the communities devoted to those tools.

References

1. Binnig G, Quate CF, Gerber C. Atomic force microscope. Physical Review Letters. 1986;56(9):930–33.
2. Friedbacher G, Fuchs H. Classification of scanning probe microscopies (technical report). Pure and Applied Chemistry. 1999;71(7):1337–57.
3. Jacoby M. Atomic imaging turns 50. Chemical & Engineering News. 2005;83(48):13–16.
4. Mody CCM. Instrumental community: probe microscopy and the path to nanotechnology. Cambridge, MA: MIT Press; 2011.
5. Toumey C. 35 atoms that changed the nanoworld. Nature Nanotechnology. 2010;5(4):239–41.
6. Young RD. Surface microtopography. Physics Today. 1971;24(11):42–49.

CHAPTER 4.8

SCANNING ELECTRON MICROSCOPY

PEDRO RUIZ-CASTELL

Electron microscopes produce a magnified image through the interaction of a beam of high-energy electrons with a specimen — the sample under study. Electron microscopes come in two main varieties: the transmission electron microscope (TEM) and the scanning electron microscope (SEM). This essay focuses on the latter, but it will first be necessary to review the history of early electron microscopy that the two techniques have in common. The TEM, in which an electron beam passes through the specimen, requires a very thin specimen that electrons can penetrate. After passing through the specimen, the electron beam is focused by electron lenses to form an image on a fluorescent screen or a photographic film (Figure 4.8.1). In the SEM, on the other hand, an electron beam passes over the specimen in a regular raster pattern, generating various signals from X-rays, backscattered, and secondarily released electrons that are registered by a detector placed appropriately to reconstruct a picture of the specimen surface (Figure 4.8.2).[3; 7] Even though the SEM is a closer analog of the conventional optical microscope than the TEM — using high-energy electrons rather than photons — the TEM was developed before the SEM.

Originally conceived to work in such fields as electron optics, solid state physics, and metallographic studies, the electron microscope soon captured the imaginations of a heterogeneous group of scholars interested in the use of new physical techniques for the microscopical

Table 4.8.1. TEM and SEM compared.

TEM	SEM
Electron beam passes through the specimen	Electron beam passes over the specimen
Specially prepared thin samples	Samples of any thickness
After passing through the specimen, electron lenses focus the beam to form an image	After interacting with the specimen, a detector registers the signals generated
Provides information on the specimen inner structure	Reconstructs the specimen surface

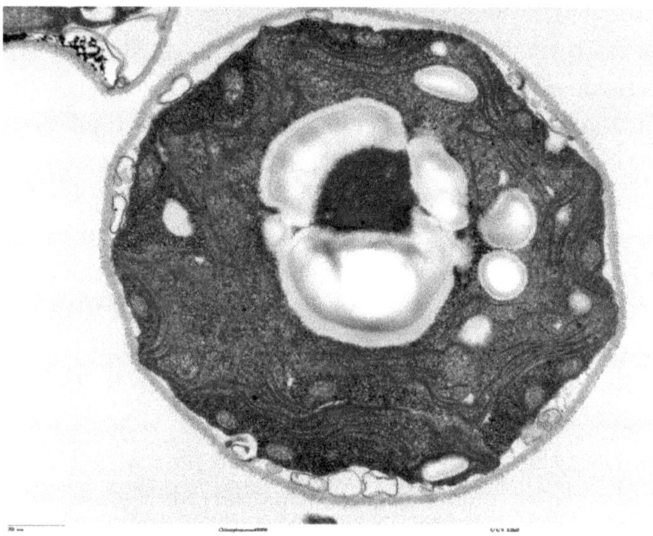

Fig. 4.8.1. TEM image, showing an example of green algae (Chlorophyta). *Credit*: Dartmouth Electron Microscope Facility, Dartmouth College.

study of different material structures, both organic and inorganic. This group included experimental physicists, life scientists, engineers, material scientists, and industrial scientists involved in work such as metallurgy.[6; 8] In particular, during its early years of development, the electron microscope offered an exceptional tool to complement research on X-ray analysis of complex structures.

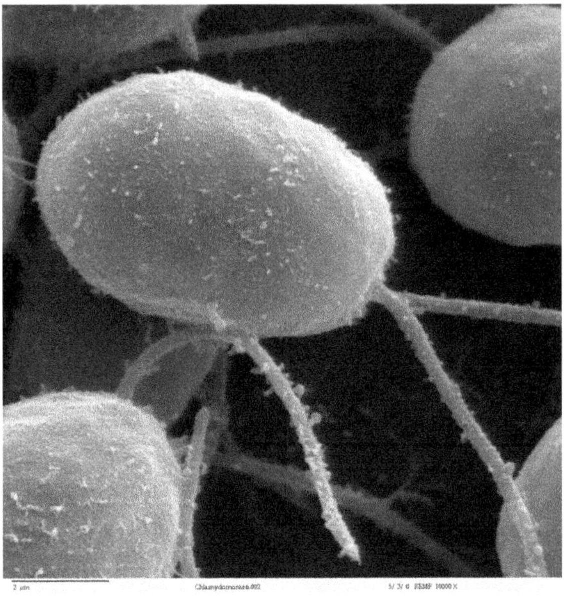

Fig. 4.8.2. SEM image, showing an example of green algae (Chlorophyta). *Credit*: Dartmouth Electron Microscope Facility, Dartmouth College.

The Early Days of Electron Microscopy

The electron microscope was developed in the 1930s thanks to an improved understanding of cathode-ray tubes and progress in techniques for achieving high vacuums. A cathode-ray tube is a device in which a beam of electrons produces a luminous image on a fluorescent screen. Therefore, specimen images could be displayed on such tubes, which required high vacuums in order to minimize scattering of the electron beam by air particles and maintain its focus before reaching the specimen, securing the stable operation of these instruments.

Soon after Hans Busch (1884–1973) demonstrated in the late 1920s how coils generating a magnetic field of rotational symmetry could be used to focus electron beams, work on the development of electron microscopy began in Germany. Two engineers from the Technische Hochschule of Berlin, Max Knoll (1897–1969) and

Ernst Ruska (1906–1988), using some of the methods and apparatus already developed for the purpose of cathode-ray oscillography, pioneered the construction of the electron microscope in the early 1930s.

Knoll and Ruska's results opened the doors for the development of electron microscope projects in countries such as the United States, Canada, Belgium, and Great Britain. The Radio Corporation of America (RCA) played a particularly crucial role, providing the first American commercial electron microscope in February 1940 — just a few months after the German firm Siemens had offered the world's first commercial model. RCA's electron microscopy research, based on the work of James Hillier (1915–2007), had been promoted by Vladimir Zworykin (1889–1982), then head of electronics research at RCA. Zworykin's interest was linked to the firm's investment in television and the desire to gain prestige for RCA's cathode-ray technique.[6]

With these pioneering forays into electron microscopy, Germany and the United States were well positioned to take the first steps toward development of the SEM. The first instrument based on the principles that characterize the SEM was proposed by Knoll in 1935, after he had moved to the Telefunken Company to work on the development of television, focusing on iconoscope-type camera tubes. Here, he built an electron beam scanner that had many of the features of a modern SEM. A few years later, Manfred von Ardenne (1907–1997) built the first instrument in which mechanical scanning was employed in his private laboratory at Berlin, thereby establishing the theoretical basis of the SEM. His instrument, however, operated as a scanning transmission electron microscope (STEM), but not as a SEM, because it had no suitable detector.[3–5] In America, a team led by Zworykin at RCA gave an account of the first working SEM in 1942 (Figure 4.8.3), which made use of the variation of secondary emission over the surface of the specimen and was employed chiefly for the examination of metals.[9]

These projects, however, were not developed further. The SEM was inherently a more complicated instrument than the TEM, since it involved the use of several additional pieces of apparatus.[9] Ironically,

Fig. 4.8.3. First SEM of high resolution from Manfred von Ardenne, 1937. *Credit*: Alexander von Ardenne via Wikimedia commons.

SEM was also delayed by innovations in "replica" techniques used in TEM. Replicas coat a non-conducting specimen with a conducting layer so that the electron beam does not charge (and potentially destroy) the specimen. In the early 1940s, Hans Mahl (1909–1988) reported replica techniques that provided satisfactory results in surface imaging. These replicas, thin enough to be imaged in a TEM, were membranes of plastic or oxide which bore the imprint of the surface being studied.[3] The leaps in TEM technology led to an increasing perception that only limited applications could be expected for an instrument with a lower resolution than a TEM.[1;5] As a result, efforts were mainly concentrated on the production of TEMs during these years.

Reconsidering the SEM

Charles Oatley (1904–1996) provided the boost needed for further developing a microscope that could image surfaces directly, which would lead to the modern type of SEM. Oatley had been engaged during World War II in the development of "radio location" or "radar" for detecting the approach of aircraft, which demanded the design and development of high-power radiofrequency transmitting tubes operating with a demountable vacuum system. In 1945 he was appointed as a lecturer at the Cambridge University engineering department, where he transformed the teaching of electrical subjects and formed a group of researchers that would play a central role in the development of the SEM. Oatley was well aware of the improvements in electronic technologies and components that had resulted from war efforts and that could be applied in such an instrument.[5; 10]

Based on the work of Zworykin and his team and under Oatley's leadership and guidance, several research students during the 1950s contributed to the establishment and improvement of the theoretical and instrumental foundations of the SEM.[1; 2; 4] One of these students was Dennis McMullan (b. 1923), who had had some experience in the radar and television industries, particularly in relation to the development and manufacture of cathode-ray tubes. After completing an electrostatically focused TEM, he converted it into an STEM and then — adding scan coils, an electron multiplier detector and a long persistence cathode-ray tube monitor — an SEM.[5] McMullan left Cambridge in 1953 and Kenneth C. A. Smith assumed responsibility for the instrument, making substantial improvements.[10] By the mid-1950s, many different specimens had been examined under the SEM. The main field of application was the examination of opaque objects when replica techniques could not be used — for instance, when it was desirable to examine the surface at a temperature at which the replica would be destroyed or to watch the changes in the structure as the surface was heated.[9]

Particularly relevant for showing the capabilities of the SEM were the imaging of a tungsten/germanium point-contact rectifier and the study of the forming processes used in its manufacture. Replicas were inapplicable, and the TEM inadequate, for these samples, so by the semiconductor industry enthusiastically supported Cambridge's SEM research on this topic.[10] Over the course of the 1950s, further industrial and academic materials arrived from outside the Cambridge engineering department to be studied under the SEM. Good examples are the collaborations established with the Ericsson Telephones factory in Nottingham to study the features observed on the etched surfaces of zone-grown germanium and with the Pulp and Paper Research Institute of Canada for the examination of wood and chemical fibers, woodchips, and newsprint.

Marketing the SEM

This kind of cooperation was not new. During World War II, several RCA electron microscopes had been sent to Great Britain under a lend-lease agreement. Among the research centers that benefited from such arrangements was the Cavendish Laboratory in Cambridge, where some interest had been shown in the acquisition of an electron microscope for studying the structure of metals. The Cavendish Laboratory was run by the Nobel Prize winner in Physics William Lawrence Bragg (1890–1971), who was engaged in X-ray investigations of different structures and keenly interested in the application of new forms of microscopy as a complement to his research. Once the RCA electron microscope was set up at Cambridge, several firms and corporations sent samples to assess the type of information revealed by the new instrument, such as the Petroleum Warfare Research Association, the Severn Valley Gas Corporation Ltd, the British Pottery Research Association, the British Aluminium Company Ltd, and the Iron and Steel Industrial Research Council. The electron microscope was expected to find widespread use in the chemical, metallurgical, textile, food, and petroleum industries, as well as in bacteriological laboratories.[8]

Furthermore, the addition of Vernon Ellis Cosslett (1908–1990) to the Cavendish Laboratory team in 1946 would prove particularly relevant for the development of both the TEM and the SEM. Following work by Raimond Castaing (1921–1998) setting out the principles for electron-probe microanalysis, Cosslett asked one of his students, Peter Duncumb, to build an instrument in which the sample could be imaged in selected X-ray radiation, much like how an SEM imaged the surface in secondary electrons — as had been shown by Oatley and McMullan. The result was a scanning electron microanalyser, which was able to illustrate the value of imaging to metallurgy. This instrument, further developed at the Tube Investments Research Laboratories near Cambridge, was marketed in 1960 by the Cambridge Instrument Company as the Microscan.[2]

The commercial success of the scanning electron probe X-ray microanalyser was a key factor in the fabrication and rapid success of the first commercial SEM (Figure 4.8.4), produced in 1965 by the Cambridge Instrument Company and based on the instruments developed by Oatley and his group.[11] Marketed under the trade name Stereoscan, several hundred were sold within a few years, setting the pattern for other SEMs marketed by other manufacturers.[3] In the early 1970s, the introduction of the solid-state X-ray detector and the microprocessor, together with the flexibility in scan speed and scan area selection provided by digital electronics, further improved the performance of the SEM.[2]

Research carried out during these years led to the first SEM capable of maintaining a relatively high pressure, removing the need to dry and coat the specimens. The so-called environmental SEM (ESEM) introduced gas — typically water vapor — into the specimen chamber, so that samples could be viewed under a deteriorated rather than high vacuum. A wide range of insulating materials — oils, liquids, and others — can be observed at equilibrium or during *in situ* evolutions with this form of electron microscopy.[1] This application enhances the many uses traditionally provided by the SEM both in advanced research and in analysis for science and industry,

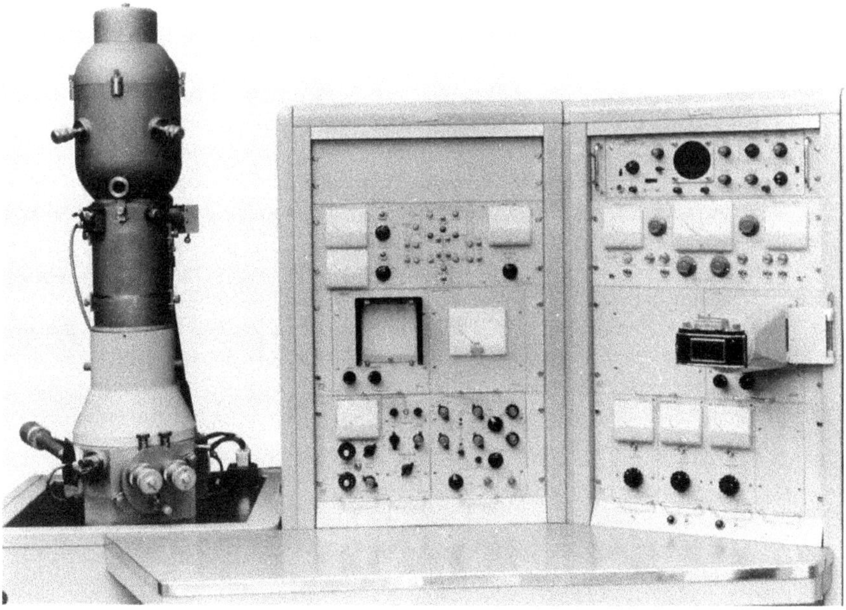

Fig. 4.8.4. Stereoscan, the first commercial SEM built by the Cambridge Instrument Company. *Credit*: Zeiss Microscopy via Wikimedia Commons.

including imaging of biological cells, rock specimens, microelectronic components, and metallurgical defects, all currently being carried out at nanometer scale.[3]

References

1. Bogner A, Jouneau P-H, Thollet G, Basset D, Gauthier C. A history of scanning electron microscopy developments: towards "wet-STEM" imaging. Micron. 2007;38:390–401.
2. Breton PJ. From microns to nanometres: early landmarks in the science of scanning electron microscope imaging. Scanning Microscopy. 1999;13:1–6.
3. McCullan D. Microscopy, electron scanning. In: Hempstead CA, editor. Encyclopedia of 20th-century technology, New York: Routledge; 2005.
4. McCullan D. Recollections of the early days of SEM in the Cambridge University engineering department, 1948–53. Journal of Microscopy. 1985;139: 129–38.
5. McCullan D. Scanning electron microscopy, 1928–1965. Scanning. 1995;17: 175–85.

6. Rasmussen N. Picture control: the electron microscope and the transformation of biology in America, 1940–1960. Stanford: Stanford University Press; 1997.
7. Rasmussen N, Hawkes P. Microscope, electron. In: Bud R, Johnston S, Warner D, editors. Instruments of science: an historical encyclopedia. New York: Garland; 1998.
8. Ruiz-Castell P. Seeing the invisible: the introduction and development of electron microscopy in Britain, 1935–1945. History of Science. 2013;51: 221–49.
9. Smith KCA, Oatley CW. The scanning electron microscope and its fields of application. British Journal of Applied Physics. 1955;6:391–99.
10. Smith KCA, Wells OC, McMullan D. The fiftieth anniversary of the first applications of the scanning electron microscope in materials research. Physics Procedia. 2008;1:3–12.
11. Stewart ADG. The origins and development of scanning electron microscopy. Journal of Microscopy. 1985;139:121–27.

CHAPTER 4.9

ABERRATION-CORRECTED ELECTRON MICROSCOPY

THOMAS VOGT

Microscopy allows us to observe objects we cannot see with our eyes alone. With a light microscope, we can distinguish objects at the scale of the wavelengths of visible light just under a micrometer. Around 1870 Ernst Abbe, who laid the foundation of modern optics, suggested that the resolution of a microscope would improve by using some yet-unknown radiation with shorter wavelengths than visible light, that is, below 390 nanometers ($1\,\text{nm} = 10^{-9}\,\text{m}$). Electrons can have wavelengths near 1 picometer ($1\ \text{pm} = 10^{-12}\,\text{m}$) and should therefore allow atoms to be distinguished, since they are typically at least a few hundred pm apart. In this oversimplified view, further decreasing the wavelength of the radiation should allow us to increase the resolution of an electron microscopy even more. However, this can only be done by increasing the energy of the radiation, which would ultimately destroy the samples. Other factors also affect the resolution of electron microscopes, among them non-ideal imaging properties of electromagnetic lenses, which result in false images. It took more than half a century to understand and control these aberrations and separate objects less than 1 Ångstrom ($1\,\text{Å} = 10^{-10}\,\text{m}$) apart.

At the Technische Hochschule of Berlin in 1931, Max Knoll and his doctoral student Ernst Ruska showed that magnetic coils could be used as lenses for electrons and thereby cleared the path to developing transmission electron microscopy (TEM).[19] In TEM, a broad electron beam impinges on a sample and the electrons exiting the sample are magnified using an electro-optical lens. These "transmitted" electrons carry information about the structure of the material in the sample. A year after their first publication, the resolution limit of the light microscope was surpassed. The wavelength of electrons can be controlled by their accelerating voltage. It is about 4 pm for a 100 keV electron, smaller than an atomic diameter. In comparison, green light has a wavelength of about 550 nm and good light microscopes can separate two objects about 300 nm apart, which corresponds to about 1,000 atomic diameters.

Subsequently, Max Knoll and Manfred von Ardenne, who headed his private research laboratory, the Forschungslaboratorium für Elektronenphysik in Berlin-Lichterfelde made early attempts to scan a specimen with a fine electron beam and thereby also create an image.[18; 33] Although conventional TEM continued to dominate electron microscopy for over fifty years; the scanning technique was further developed in the 1960s by Albert Crewe and his collaborators in Chicago,[9] leading to the imaging of heavy atoms in Scanning Transmission Electron Microscopy (STEM).[8] This led to a revolution in sub-Ångstrom imaging in the late 1990s, when correcting aberrations associated with electro-optical lenses (as outlined in the next section) finally became possible. Today, STEM imaging is ubiquitous and routinely used in materials science and engineering as well as in condensed matter physics and chemistry. Many of the entries in this volume examine tools that were introduced to the materials research community, rapidly gained widespread acceptance, and then continued to slowly evolve. Other tools, however, struggled to live up to their promises, sometimes over many decades, before finally achieving success. STEM is a particularly clear instance of the latter category.

Theoretical Reasons for Doubt and Hope

In 1926 Hans Busch from Jena University described the focusing of electrons in a manner similar to how glass lenses can focus visible light. No glass or electromagnetic lens is perfect, but in light optics aberrations can be corrected by a series of convex and concave lenses. In electron optics this is not possible because there are no concave lenses. Thus, new ways of correcting for the aberrations caused by round electron lenses had to be developed.

The resolution of a STEM depends on the size of the electron beam one scans across the sample. In turn, the size of the beam depends on three main electro-optical effects. The first is the diffraction limit, which relates the resolving power to distinguish two objects and is proportional to the wavelength and inversely proportional to the illumination angle. Maximizing resolution would thus aim for high energies and small wavelengths of electrons as mentioned above and large illumination angles. The second is the spherical aberration of lenses: in STEM imaging, the smallest achievable electron probe is proportional to a measure of the deviation from perfect lensing called the constant of spherical aberration and the third power of the illumination angle. In a simple ray-tracing picture, this can be understood as creating a disk of least confusion before the image plane (see Figure 4.9.1). One should attempt to minimize both the wavelength and spherical aberration using appropriate optics. In contrast to the diffraction limit, one would aim for the smallest possible illumination angle. Third, the chromatic aberration in the image plane is due to the presence of electrons with different energies coming from their source and creating a disk of confusion whose size is proportional to the chromatic aberration constant and the width of the energy distribution. At small illumination angles, the diffraction limit determines the resolution, while at larger ones the spherical aberration dominates, provided that the spherical and chromatic aberration constants are of similar orders of magnitude and the energy spread of the electrons is about 1 eV. STEM imaging using energies above 100 kV will have spherical and not chromatic aberration as the main limiting effect.

Early work by Otto Scherzer at the Technische Hochschule Darmstadt pointed to the unavoidable fact that round electron lenses have positive spherical and chromatic aberration effects, which will limit the resolution of an electron microscope ("Scherzer limit") to about

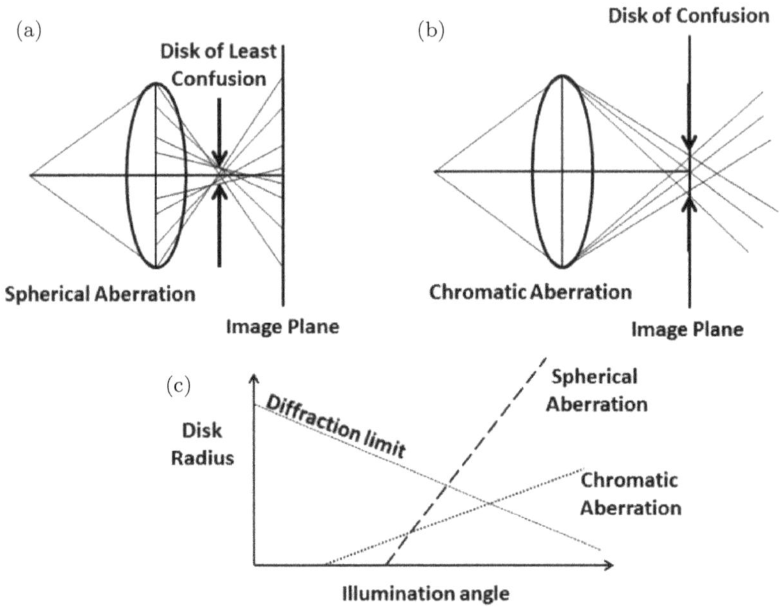

Fig. 4.9.1. Ray-tracing of spherical (a) and chromatic (b) aberration effect in cylindrical electro-optical lenses and their dependence on the illumination angle for acceleration voltages higher than 100 kV. An electrostatic hexapole sketched to the right of the ray tracing of the spherical aberration effect allows for its correction as it displaces the electrons with electric fields as indicated. Part (c) shows how diffraction limit, spherical aberration, and chromatic aberration vary with the radius of the disk of least confusion and the angle of the illuminating electron beam.

one hundred times the electron wavelength, which in a TEM is near 2–2.5 picometers.[30] In electron optics, round lenses will always display a positive aberration and no negative aberration is possible. But in a groundbreaking paper in 1947, Scherzer showed that both chromatic and spherical axial aberrations could be corrected by deviating from round lenses with a rotationally symmetric electromagnetic field using an electromagnetic multipole.[29] He predicted, based on a "gut-feeling" (German: *rein gefuehlsmaessig*) that this resulted in a negative correction effect to counteract the positive aberration. Think of a multipole as a geometrical figure approximating a sphere by placing points on the surface of the sphere and connecting them. Rotating a quadrupole, hexapole, or octupole, made of electrical fields pointing in different directions, only approximates the rotational symmetry of the sphere and it is this small deviation from the sphere's shape that is needed for aberration correction. A sketch is given in Figure 4.9.1.

The First Arduous Steps toward Aberration Correction

Between 1949 and 1954, Robert Seeliger, a student of Scherzer, was the first who attempted to build a corrector for a TEM comprised of two round lenses and three octupoles; however, he failed to overcome mechanical and electromagnetic instabilities.[31] Nevertheless, he proved Scherzer's conjecture that such a multipole corrector leads to a negative spherical aberration needed to compensate for the positive one of the round objective lens. Seeliger's corrector was subsequently moved to the University of Tübingen where Gottfried Möllenstedt intentionally enhanced the spherical aberration by using an unusually large illumination angle to degrade the resolution (see Figure 4.9.1).[24] Using octupoles, he then improved the resolution by about a factor of seven and significantly increased the contrast. Largely ignored work by Möllenstedt's PhD student Werner E. Meyer at the University of Tübingen showed that misalignment, static imperfections, and charging, and alternating external electromagnetic fields, as well as mechanical instabilities, were the main roadblocks for aberration correction.[23] This proved to be correct until the 1990s! Geoffrey Archard,[1] working at the Associated Electrical Industries (AEI) research laboratory at Aldermaston Court, built on this early work establishing that cylindrical lenses can be substituted by quadrupoles. He pointed out that an early attempt by Jack C. Burfoot at the University of Cambridge could be simplified by using a sequence of four optical elements: a quadrupole-octupole, a round lens-octupole, a second quadrupole-octupole, followed by a quadrupole. In 1964 Hans Deltrap built the first quadrupole-octupole corrector to eliminate spherical aberration.[5; 12] However, the resolution was limited not by the spherical resolution, but other qualities of the lenses at that time and his corrector was only tested on an electron-optical bench not in an electron microscope.

An important international workshop took place in 1966 at Argonne National Laboratory, at which the design of a high-voltage, high-resolution electron microscope with a quadrupole-octupole corrector was proposed but subsequently not funded. In 1971, Harald Rose at TU Darmstadt, after showing that all correctors to date suffered from off-axis coma, a "comet-like blur away from the optical axis," developed a corrector that could simultaneously correct for chromatic and spherical aberration.[27] In 1972 Crewe and Vernon Beck made new attempts to correct for spherical aberration of a STEM using a quadrupole-octupole corrector alleviating some of the difficulties encountered in the Darmstadt project.[7] However, they were unable to eliminate aberrations due to inhomogeneity of the metal used in the lens and the lack of fast diagnostic and control electronics to establish suitable optical settings for the many parameters. Without such electronics, no feedback control was possible. At the International Congress on Electron Microscopy in Toronto in 1978, Scherzer quipped that the resolution of the best uncorrected TEMs "is clearly limited by the unavailability of the necessary funds."[28] The Darmstadt project continued until Scherzer's death in 1982, when it was terminated despite significant progress.[4]

Up to this point, techniques using sextupole correctors had been ignored since Peter Hawkes had shown in 1965 that the large second-order aberrations prevent the exploitation of the favorable third-order aberrations to correct for spherical aberration of round lenses.[16] In an important breakthrough, Beck showed in 1979 that one can arrange sextupoles in a way to suppress this large second-order aberration.[3] This design was subsequently improved by Rose and later morphed into a prototype of the sextupole correctors which are now produced commercially by the CEOS Company in Heidelberg.[26]

The Darkest Hour . . . and a New Dawn

In the late 1980s "aberration fatigue" set in after the Darmstadt project ended with Scherzer's death. Crewe gave up, saying "unfortunately, we could never make the corrector work. . . . After many heartbreaking attempts, we were forced to admit defeat," and the US National Science Foundation decided to no longer fund efforts to build correctors.[15] The materials science community, which had always been seen as the main user of high-resolution imaging, set their hopes on high-voltage electron microscopy where resolution was simply enhanced by reducing the wavelength of the electron beam (see diffraction-limit above and Figure 4.9.1c) by going to acceleration voltages of 1 million volts and higher. These gigantic instruments reached Ångstrom resolution and the reduced electron inelastic scattering at such high energies allowed the use of thicker samples. However, atom displacement damage ("knock-on" damage) destroyed the samples, often within minutes.

There is a certain irony that funding agencies in Europe, Japan, and the United States were prepared to invest an order of magnitude more in high-voltage instruments (one microscope would cost tens of millions of US dollars) than was ever allocated for aberration correction of electron lenses. During these dark days for the aberration-correction community, it was realized that the required machining, manufacturing, and positioning of the many required optical elements was beyond what could reproducibly be achieved at that time. Furthermore, neither the stable hardware (that is, power supplies) nor the software existed to excite the various multipoles and other optical elements and integrate them with diagnostic tools based on fast detector readout to ensure feedback for auto-tuning algorithms. As Harold Rose was quoted by Knut Urban, "it took more time to adjust a specific electron optical state than this would have time to last."[32] Also, a quadrupole-octupole corrector is not very suitable for conventional TEM as one needs to correct over the whole region imaged. A much more suitable imaging system would be to correct a probe-forming SEM or STEM. Fortunately, the electron optics community proved to be resilient and was ultimately able to complete the revolution in high-resolution electron imaging after a long march of over four decades. Three main efforts in Europe funded internally by the European Molecular Biology Laboratory (EMBL), privately by the Volkswagen Foundation and by the Paul Instrument Fund of the Royal Society in the

United Kingdom provided the funds needed to realize imaging at the atomic scale.

Due to the intrinsic difficulty in applying quadrupole correctors to TEM, Max Haider convinced management that a SEM with spherical and chromatic corrections aligned with the mission of EMBL as it had the potential to be used for imaging larger biomolecules. EMBL had a tradition of instrument development and could support this effort by in-house funding. With Joachim Zach, Haider proved the working principle of a quadrupole-octupole corrector using a low-voltage SEM. Despite its intrinsic lower resolution compared to a TEM, the instrument's resolution was improved from 5.6 to 1.8 nm.[34; 35] Their design was a corrector based on Hardy's thesis from 1967 in Cambridge.

Parallel to these efforts Haider, Rose, and Urban lobbied for and received high-risk funding from the Volkswagen Foundation which allowed them to design, build, and test a hexapole-based corrector at the EMBL in Heidelberg. Rose's corrector used a telescope round lens doublet and two identical sextupoles, one centered at the front focal point of the first round lens and the other on the back focal point of the second round lens. The hexapole arrangement cancels out all second order path deviations, whereas the third order rotationally symmetric path deviations add up. The result is a negative spherical aberration proportional to the square of the hexapole strength compensating the positive one of the objective lens. The prolonged struggle to secure funding is described in detail in Urban's 2015 paper. The final decision by the Volkswagen foundation was made by a one-vote majority and only part of the requested funding was granted; the rest was contingent on achieving a critical milestone showing that aberration correction could be achieved on a conventional TEM, a worst-case test as mentioned above. First images were obtained on a Philips CM200 microscope with a Rose corrector in Heidelberg by Bernd Kabius on June 24, 1997, before the instrument was shipped to Jülich since EMBL management had decided to shut down in-house electron microscopy development. He showed that one could resolve dumbbells (two atoms in close proximity in a projection) 1.4 Å apart when viewing along the (110) direction in a GaAs sample. A paper was submitted first to *Nature* and then *Science* and initially rejected by both but finally accepted in *Nature*.[14]

During the same period, Ondrej Krivanek and Niklas Delby continued to work on adapting quadrupole-octupole correctors to a Vacuum Generator (VG) STEM at the Cavendish Laboratory in Cambridge. They were funded by the Royal Society (Paul Instrument Fund) and now had access to high speed diagnostic and control electronics to tune the corrector. Furthermore, they installed additional corrector coils to cancel parasitic aberrations. These improvements allowed them to present initial results in 1997 — one hundred years after the discovery of the electron by J. J. Thomson.[21] Impressive results using dark field (DF) imaging were published in 1998.[20] Subsequently Krivanek and Delby moved to Seattle and started the company NION (NIklas and ONdrei). In 2000 their MARK-2 corrector was working and installed in a VG STEM at a specially shielded site at IBM where it allowed DF images at a resolution of 1.4 Å to be recorded. An iconic image of silicon dumbbells (Figure 4.9.2) is shown in a 2001 paper.[11] The same instrument at IBM was used to record sub-Angstrom resolution images.[2]

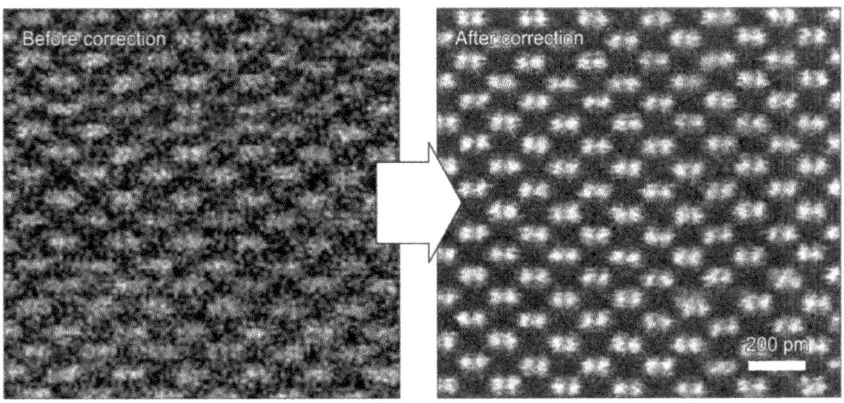

Fig. 4.9.2. Scanning TEM micrograph of silicon "dumbbells" before and after automatic aberration correction. *Source*: Inada H, *et al.* High-speed and sensitive x-ray analysis system with automated aberration correction scanning transmission electron microscope. Applied Microscopy. 2015;45(1):1–8. *Credit*: Creative Commons CC BY-NC 3.0.

As spherical aberration correction was being successfully implemented in commercially available electron microscopes, chromatic aberration was now the major hurdle to higher resolution. Many commercial electron microscopy manufacturers simply limited the energy spread of the incident beam by using a monochromator which minimized but did not correct for chromatic aberration.

After a workshop convened by Murray Gibson in July 2000, the TEAM (Transmission Electron Aberration-corrected Microscope) project was funded by the US Department of Energy involving the national laboratories at Argonne, Brookhaven, Oak Ridge, and Lawrence Berkeley as well as the Frederick Seitz Materials Research Laboratory at the University of Illinois. Harald Rose's new design replaced each sextupole in the original corrector by a telescopic quadrupole-octupole quintuplet. CEOS further optimized the design creating an achromatic aplanatic system with stabilities at the order of 2×10^{-8} for all electric and magnetic quadrupole fields. FEI and CEOS built two instruments, TEAM 0.5, a TEM and STEM with spherical aberration correction, and TEAM 1, a STEM with spherical aberration correction and TEM with chromatic and spherical correction, which were installed at the National Center for Electron Microscopy at the Lawrence Berkeley National Laboratory in 2008. These best-in-class instruments have resolutions of 65 and 55 pm at 200 and 300 kV,[17] respectively, and the STEM resolves a 47 pm dumbbell spacing along the 114 direction in a germanium crystal.[13]

Samples containing light elements such as carbon, oxygen, or lithium degrade rapidly in an electron beam due to "knock-on" damage, where atoms are displaced from their original positions by high-speed electrons. However, reducing the accelerating voltage increases the damage due to radiolysis and chromatic aberration becomes dominant. At the University of Ulm in collaboration with two companies, FEI and CEOS, the SALVE (Sub-Ångstrom Low-Voltage Electron microscopy) project initiated by Ute Kaiser has shown spherical and chromatic aberration correction for accelerating voltages between 20 and 80 kV.[22] Between 40 and 80 kV sub-Ångstrom resolution was achieved. More recently, a Japanese collaboration by the microscopy manufacturer JEOL and the National

Institute of Advanced Industrial Science and Technology built a TEM with a spatial resolution of 1.4 Å at 15 kV.[25] This allowed the observation of a monolayer of graphene at atomic resolution. This bodes well for the structural characterization of very beam-sensitive materials.

Over half a century of experimental struggles since the emergence of electron microscopes in 1932 allowed us to finally achieve the goal of sub-Ångstrom imaging. Although the key strategy, namely the deviation from round electro-optical elements by using multipoles, was recognized early on by Scherzer in 1947 and the first corrector was built by Robert Seeliger in the early 1950s, we had to wait until the 1990s until we could control the mechanical and electronic instabilities that were limiting the resolution. Only after these "secondary effects" were understood and controlled did aberration limit the resolution and could be corrected for. The length of such developments tests the patience of private and public funding agencies; however, transformative scientific advances are more and more becoming battles of attrition rather than singular strokes of geniuses.

References

1. Archard GD. An unconventional electron lens. Proceedings of the Physical Society. 1958;72(1):135–37.
2. Batson PE, Delby N, Krivanek OL. Sub-angstrom resolution using aberration-corrected electron optics. Nature. 2002;418:617–20.
3. Beck VD. A hexapole spherical aberration corrector. Optik. 1979;53:241–55.
4. Bernhard W. Erprobung eines sphaerisch korrigierten Elektronenmikroskops. Optik. 1980;57:73–94.
5. Burfoot J. Correction of electrostatic lenses by departing from rotational symmetry. Proceedings of the Physical Society. 1953;B66:775–92.
6. Crewe AV. Scanning electron microscopes: is high resolution possible? Science. 1966;154(3750):729–38.
7. Crewe AV. Some Chicago aberrations. Microscopy and Microanalysis. 2002;8(Suppl. 2):4–5.
8. Crewe AV, Wall J, Langmore J. Science. 1970;168:1338–40.
9. Crewe AV, Wall J, Welter LM. Journal of Applied Physics. 1968;39(13): 5861–68.
10. Dahmen U, Erni R, Kisielowki C, Radmilovic V, Ramasse Q, Schmid A, Duden T, Watanabe M, Minor A, Denes P. An update on the TEAM

project — first results from the TEAM 0.5 microscope, and its future development. In Luysberg M, Tillmann K, Weirich T, editors. Proceedings of the 14th European Microscopy Congress, Aachen, Germany, vol. 1. Berlin: Springer; 2008.
11. Delby N, Krivanek OL, Nellist PD, Batson PE, Lupini, AR. Progress in aberration-corrected scanning transmission electron microscopy. Journal of Electron Microscopy. 2001;50:177–85.
12. Deltrap JHM. Correction of spherical aberration with combined quadrupole-octupole units. In Titlbach M, editor. Proceedings of the Third European Regional Conference on Electron Microscopy, Prague, vol. A, Prague: Czechoslovakian Academy of Sciences; 1964. 45–46.
13. Erni S, Rossell MD, Kiselowski C, Dahmen U. Atomic-resolution imaging with sub-50 nm electron probe. Physical Review Letters. 2009;102:096101.
14. Haider M, Uhlemann S, Schwan E, Rose H, Kabius B. Electron microscopy image enhanced. Nature. 1998;392:768–69.
15. Hawkes PW. The correction of electron lens aberrations. Ultramicroscopy. 2015;156:A1–A64.
16. Hawkes PW. The geometrical aberration of general optical systems. Philosophical Transactions of the Royal Society of London. 1965;A257:479–662.
17. Kisielowski C, Freitag B, Bischoff M, van Lin H, Lazor S, Knippels G, Tiemeijer P, van der Starn M, von Harrach S, Stekelenburg M, Haider M, Uhlemann S, Müller H, Hartel P, Kabius B, Miller D, Petrov I, Olson EA, Donchev T, Kenik EA, Lupini AR, Bentley J, Pennycook SJ, Anderson IM, Minor AM, Schmid AK, Duden T, Radmilovic V, Ramasse QM, Watanabe M, Erni R, Stach EA, Denes P, Dahmen U. Detection of single atoms and buried defects in three dimensions by aberration-corrected electron microscope with 0.5 Å information limit. Microscopy Microanalysis. 2008;14:469–77.
18. Knoll M. Aufladepotentiel und Sekundäremission elektronenbestrahlter Körper. Zeitschrift für Technische Physik. 1935;16:467–75.
19. Knoll M, Ruska E. Das Elektronenmikroskop. Zeitschrift für Physik. 1932;78(5–6):318–39.
20. Krivanek OL, Delby N, Lupini AR. Towards sub-Å electron beams. Ultramicroscopy. 1999;78:1–11.
21. Krivanek OL, Delby N, Spence AJ, Camps RA, Brown LM. In Sturgess JM, editor. Proceedings of the Ninth International Congress on Electron Microscopy, Toronto, Canada, vol. 3. Toronto: Microscopical Society of Canada, 1978. 185–96.
22. Linck M, Hartel P, Uhlemann S, Kahl F, Müller H, Zach J, Haider M, Niestadt M, Bischoff M, Biskupek J, Lee Z, Lehnert T, Börrnert F, Rose H, Kaiser U. Chromatic aberration correction for atomic resolution TEM imaging from 20 to 80 kV. Physical Review Letters. 2016;117:076101.
23. Meyer W.E. Das praktische Aufloesungsvermoegen von Elektronenmikroskopen. Optik. 1961;18:101–14.
24. Moellenstedt G. Elektronenmikroskopische Bilder mit einem nach O. Scherzer sphaerisch korrigierten Objektiv. Optik. 1956;13:209–15.

25. Morishita S, Mukai M, Suenaga K, Sawada H. Atomic resolution imaging at an ultralow accelerating voltage by a monochromatic transmission electron microscope. Physical Review Letters. 2016;117:153004.
26. Rose H. Correction of aperture, aberrations in magnetic systems with threefold symmetry. Nuclear Instruments and Methods in Physics Research. 1981;187:187–299.
27. Rose H. Elektronenoptische Aplanate. Optik. 1971;34:285–311.
28. Scherzer O. Limitations for the resolving power of electron microscopes. In Sturgess JM, editor. Proceedings of the Ninth International Conference on Electron Microscopy, vol. 3. Toronto: Microscopical Society of Canada;1978.
29. Scherzer O. Sphärische und chromatische Korrektur von Elektronenlinsen. Optik. 1947;2:114–32.
30. Scherzer O. Über einige Fehler von Elektronenlinsen. Zeitschrift für Physik. 1936;101:593–693.
31. Seeliger R. Spherical-aberration correction of electron lenses by means of image-forming elements without rotational symmetry. Optik. 1951;8:311–17.
32. Urban KM. In quest of perfection in electron optics: a biographical sketch of Harald Rose on the occasion of his 80th birthday. Ultramicroscopy. 2015;151:2–10.
33. von Ardenne M. Das Elektronen-Rastermikroskop: Theoretische Grundlagen. Zeitschrift für Physik. 1938;109(9–10);553–72.
34. Zach J, Haider M. Aberration correction in a low voltage SEM by a multipole corrector. Nuclear Instruments and Methods in Physics Research Section A. 1995;363(1–2):316–25.
35. Zach J, Haider M. Correction of spherical aberration in a low voltage SEM. Optik. 1995;93:112–18.

CHAPTER 4.10

MASS SPECTROMETRY

KEITH A. NIER

Mass spectrometry encompasses an amazingly diverse group of devices and techniques and is widely employed throughout science and much of engineering. The common aspect that identifies these devices as a group is functional: each involves apparatus that, at atomic and molecular scales, can sort particles by their masses. This is done by ionizing particles from a sample or target of interest, applying some kind of electromagnetic field or fields, and detecting the particles according to their ratios of mass to charge.[22; 25] If the particles are molecular in nature, some of them may fragment during or after the ionization process and the masses of the fragment particles can also be determined. All mass spectrometers are able to characterize materials at least to the extent of identifying substances in the sample, so it could be said in the very loosest sense that all mass spectrometers are used in materials science and engineering, though this is only minimally informative. More useful understanding requires differentiation of device types and the worlds of practice in which they are found.

Every mass spectrometer, and every device that uses the methods of mass spectrometry even if it is incorporated into an apparently different kind of apparatus, has as its core the combination of a source where ions are produced from the material of interest, an analyzer where ions are sorted by their mass and charge characteristics, and a detector for the results of the analyzer's work (Figure 4.10.1).

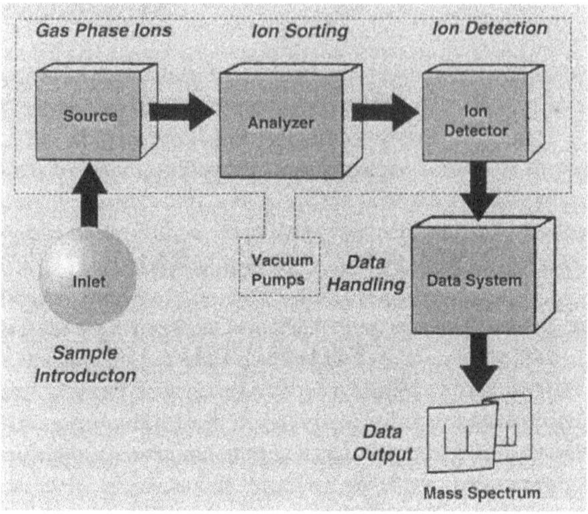

Fig. 4.10.1. Functional Elements of a Mass Spectrometer. *Source*: Anne B. Giordani *et al.*, *What is Mass Spectrometry?*, ASMS The American Society for Mass Spectrometry, 3rd ed., 1998, Figure 1, p. 4.

Controlling this core are arrangements to introduce material into the machine, record detected output, sustain the requisite vacuum for isolating the ions from the machine's surroundings, and stabilize and regulate all operations. The last aspect is now commonly done by computer.

Mass spectrometers are commonly described by mentioning which of the myriads of ways of ionizing, modes of application of fields, and methods of detection a given device employs. The major types of mass spectrometer are broadly specified by the type of analyzer, though in more recent years very many combine more than one type in a single, complex device.

Mass analyzers of the earliest type are generally called deflection, or sector-field instruments. These instruments work by deflecting the paths of charged particles as they move through a region with a strong uniform magnetic or electrostatic field. The greater the mass of the ions, the less they are deflected. Ions of different mass can be detected at different locations (and displayed in a manner analogous to an optical line spectrum) or they can be collected sequentially

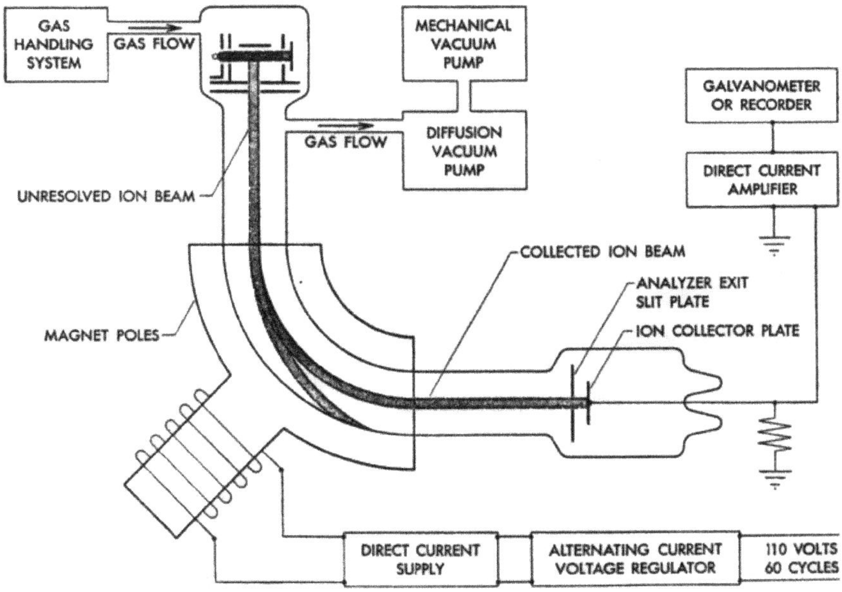

Fig. 4.10.2. Schematic Diagram of a Magnetic Deflection Mass Spectrometer. *Source*: "The Mass Spectrometer," Alfred O. Nier, Scientific American, 188 (March 1953), 68–74, p. 68.

at a single point by controlled changes in either the strength of the deflecting field or of the velocity of the ions as they enter that field (Figure 4.10.2).

Almost all mass spectrometers before World War II and for many years thereafter were deflection instruments. However, significant steps toward the development of three other major types of mass spectrometer analyzers took place in the decade following the war. One of these types, time-of-flight (or TOF) machines, analyzes a batch of ions by momentarily applying an electric field to accelerate them into a field-free region. Ions of lesser mass are accelerated more and reach the detector sooner than those of greater mass. Several decades of development, including some major innovations, have made this into a prominent category of instrumentation. A second type, ion-cyclotron-resonance (ICR) mass spectrometers, uses combinations of fields to constrain ions in cyclotron motion, with ions of different mass having different resonance frequencies. In Fourier

Transform ICR (FTICR), developed decades later and now the most common version, the cycling ions induce signals in part of the containing cell, with the signals analyzed by computer to indicate what ions are involved. In the third major type of analyzer stemming from mid-century beginnings, quadrupole instruments, ions are projected into a region surrounded by four parallel electrodes that subject the ions to both a constant electrostatic and a rapidly alternating field. For specific field values, only ions of a particular mass can make it all the way along the passage, whereas others impact the electrodes or the inside surfaces of the instrument. When the field values are changed, ions of different masses can be collected. Quadrupole instruments became much more prevalent over the last third of the twentieth century and they now are probably the type of mass spectrometer in most widespread use. One more type of instrument, ion traps, began as a considerably later transformation of the quadrupole type and were commonly named quadrupole ion traps (QITs), despite not having four poles (Figure 4.10.3).

Several other ways of sorting ions by mass have been tried, the most dramatic being the use of linear accelerators as mass analyzers, but none were extensively adopted before the end of the twentieth century.

The applicability of any of these types of instruments to the characterization of the materials generally considered under the purview of materials science and engineering has been governed far more by the way ions can be produced from a sample than by any capabilities or limitations of analyzers or detectors. If one can manage to ionize a material, some kind of mass analyzer can handle the ions. The most common ionization method for mass spectrometry in the first half of the twentieth century was electron ionization, in which an already vaporized sample was hit with an electron beam. This could be used readily only for gaseous substances and those that are easily vaporizable without significant molecular change, neither of which describes the sorts of substances around which the field of materials science and engineering emerged.

Several other notable approaches to ionization developed by mid-century, however, and many more since then. Each made it

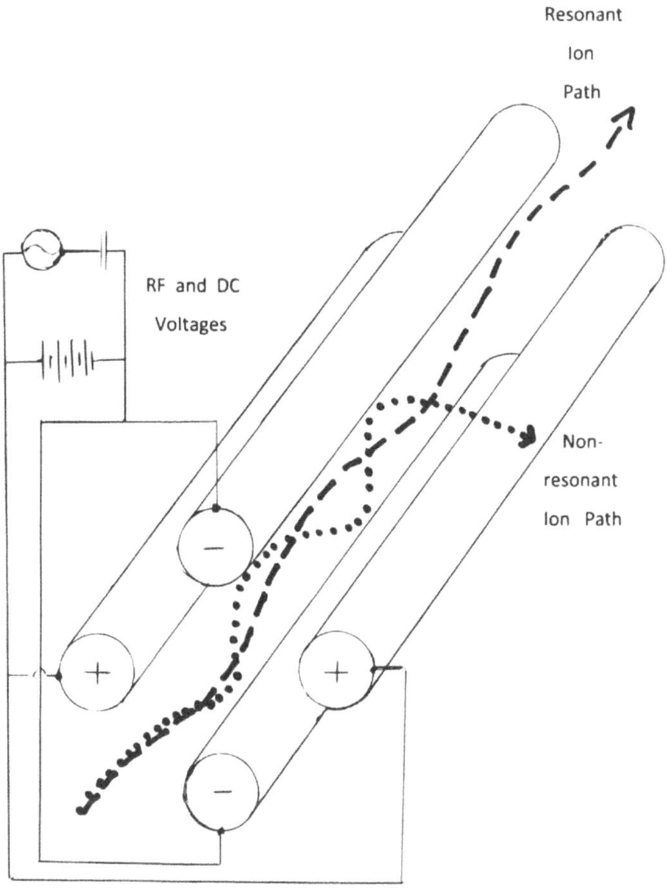

Fig. 4.10.3. Schematic of a quadrupole mass analyzer. Drawing by the author.

possible to apply mass spectrometry to more and more materials. Ionization by thermal, spark source, glow discharge, field desorption, chemical, particle beam, radioactive disintegration, laser, electrospray, matrix-assisted laser desorption, inductively coupled plasma, and other methods have all been extensively employed. Furthermore, linking multiple analyzers, known as tandem mass spectrometry, enables not only ever more precise analysis but also the use of the output of one analyzer as the highly selected input for experiments from which the products pass to further analyzers. The overall result of such developments is the growth of immense bodies of information about all sorts of substances.[26–28]

Standard introductory texts old and new and major reference works review the principle types of mass spectrometers, methods of ionization, and more, but most have tended to place little emphasis on the varieties of instrumentation most used in materials research during the decades in which materials science and engineering became an established field.[4–6; 10; 32] Instead, standard references have concentrated on the approaches that dominate the broad area in which mass spectrometry has had its greatest and most obvious growth, that is, analytic organic and biological chemistry.[17]

A more comprehensive approach would regard mass spectrometry as roughly divided among four fairly distinct worlds of practice which we can briefly label as nuclear, cosmochemical, molecular, and industrial. The extent to which any of these is connected to the commonly understood scope of materials science and engineering is quite uneven.

In the nuclear domain, highly specialized, large, and unique machines are presently on line in various particle accelerator facilities to determine the masses of products of sub-atomic particle interactions, of nuclear reactions, and of efforts to generate further transuranic elements. Ever since World War II, fissionable substances have been a special category of material of great importance, and thus the analysis of their production, possible interactions, and products has been a notable but quite distinct aspect of materials research and engineering. Apart from that, however, mass spectrometry of the sort involved in nuclear and subatomic physics has had little to do with the general development of materials science and engineering.[6; 16]

In the cosmochemical domain, mass spectrometers have become fundamental for the whole vast area of the earth and planetary sciences. Identification of the composition of geological and astronomical materials, including nanoscale variations, is the most obvious use here. To the extent that the scope of materials science runs from mineralogy through meteoritics, mass spectrometry is involved. Most of the materials at issue are inorganic, though the presence of organics and even biological materials in geological and cosmic samples received growing attention as the twentieth century progressed. This naturally affected what types of mass spectrometry were used. For these cosmochemical fields, however, mass spectrometry has become even more important as the foundation of chronology and thus all large-scale temporal analysis for the earth and other bodies — including micro-bodies — in the solar system. Furthermore, since essentially all physical and chemical processes can result in some degree not only of elemental but of isotopic differentiation, mass spectrometry is crucial in identifying relevant evidence for all sorts of research questions. These even include characterizing, via products of extinct radioactivities, materials that no longer exist.[3; 23]

The world of mass spectrometry focused on organic, bio-, and biological chemistry, including environmental topics, includes the largest number of known practitioners and mass spectrometers. This area includes forensic and regulatory applications. And although elemental substances are significant in many cases, and inorganic substances can be compounds, it is understandable that this area has been called molecular mass spectrometry. Most of the development of new ionization methods and of various analyzers has been oriented to organic and biological substances. Although the initial application of mass spectrometers in this area was the analysis of hydrocarbon mixtures in oil refineries, they soon were turned to all aspects of analytical chemistry and increasingly to the identification of and determination of the structure of natural products.[11] From this it was an inevitable step to address synthesized substances. And to just the extent that materials research embraces biological materials, the importance of mass spectrometric characterization will grow.

It is worth noting that mass spectrometers can be component parts of other sorts of apparatus, playing the role of a selective input device or of a detector to identify experimental products.[31] This is, for example, how a mass spectrometer figured in the discovery of fullerenes.[30]

The fourth general world of mass spectrometry, the industrial one, is more diverse and disjointed than the others. For various reasons, it is also the least well known, even among practitioners. The use of mass spectrometry in industrial contexts can effectively disappear from view when it is taken for granted and thus not publicized. When it is important, but for proprietary topics or applications, it may not be publicly discussed. Furthermore, industrial mass spectrometry is sometimes subsumed in a more generic category such as elemental analysis, leak detection, or residual gas analysis, so that descriptions of processes and plants may be uninformative as to just what instruments are involved in any given situation. And just as there are more historians of science than of technology, there has been more notice in historical studies of mass spectrometry in scientific research than in its industrial roles.

It also should be emphasized that, in this area, mass spectrometry is not only employed in research characterizing materials but also has significant roles as part of infrastructure and production.[6; 16; 33; 34] First, a major aspect of characterizing materials is itself a matter of both production and infrastructure rather than research, namely monitoring, whether it be analysis of inputs and outputs for identity and purity, analysis of byproducts, or analysis of internal or external environments to identify possible contaminants or leaks. Second, ion implantation to modify a material generally employs a kind of mass spectrometer where the target material replaces the detector. This shows again how the technology of mass spectrometry can be present but remain unnamed. Third, mass spectrometry has been and remains crucial in the preparation and certification of various reference materials.

It is clear that most applications of mass spectrometry in materials research have come in industrial or technological contexts, though some cases could also be considered part of molecular mass spectrometry. The clearest early example of mass spectrometry in materials research is just such a case of overlapping categories. This was work on producing synthetic rubber during World War II. Although the electron ionization deflection instruments of the time were irrelevant for handling macromolecules, they were crucial for the analysis of the petroleum derivatives that were the feedstock for experiments and production.[20]

More extensive and varied mass spectrometry involvement in materials research came during and after the 1950s and relied heavily on developing ionization methods. Thermal, spark-source, and glow-discharge ionization were soon employed in the analysis of semiconductors and in trace analysis more generally.[1; 2; 21; 36] Research on many properties of diverse sorts of materials was reported in hundreds of research studies by the middle of the 1960s.[33] By the 1980s, the rate and range of studies had increased many times over. Metals, alloys, glasses, ceramics, semiconductors, superconductors, composites, and catalysts were being investigated with mass spectrometry. Topics of research on these materials included diffusion, phase change, corrosion resistance, grain boundaries and micro-structure, thin films, isotope effects, and channeling in single crystals, among others.[12; 34]

Perhaps the greatest influence in these developments was the creation, diversification, and refinement of secondary-ionization mass spectrometry (SIMS), in which a beam of ions of known type is directed onto a target and thus produces new, secondary ions from the impacted material.[18] Although the roots of these instruments and techniques also go back at least to World War II, the practical development came slowly in the 1950s and more rapidly thereafter.[7–9; 13; 14; 24] With SIMS came the widening application of mass spectrometry to the study of surfaces. SIMS proved to be capable of much greater sensitivity than other methods in identifying impurities on surfaces; in fact, it was at times considered too sensitive to be very useful.[37] Still further development led to the use of SIMS to provide depth analysis in addition to information about the outermost surface of things and to producing images showing the distribution of various components or impurities across both artificial and biological materials.[35]

Late in the twentieth century, the development of analyzer and ionization techniques, particularly matrix-assisted laser desorption/ionization (MALDI), allowed increasing use of mass spectrometry in the characterization of polymers. Although there had been notable earlier work, the quantity and range of such studies greatly increased in the 1990s.[15; 19] The last decades of the century also brought a major development in elemental and isotopic analysis with the introduction of inductively coupled plasma (ICP) mass spectrometry, most notably the sector field multi-collector version (MC-ICP) in the 1990s.[29] Thus, the range of substances that can be analyzed by one or another type of mass spectrometry and the types of information obtainable continue to increase. And here again materials characterization includes cases that in some way bridge the disparate worlds of mass spectrometry, the industrial and biochemical molecular on one hand, and the industrial and geo- and cosmochemical on the other.

References

1. Ahearn AJ, editor. Mass spectrometric analysis of solids. Amsterdam: Elsevier; 1966.
2. Ahearn AJ, editor. Trace analysis by mass spectrometry. New York: Academic Press; 1972.
3. Allègre CJ. Isotope geology. Cambridge: Cambridge University Press; 2008 [original French edition 2005].
4. Bensaude-Vincent B. The construction of a discipline: materials science in the United States. Historical Studies in the Physical and Biological Sciences. 2001;31(2):223–48.
5. Duckworth HE. Mass spectroscopy. Cambridge: Cambridge University Press; 1958.
6. Duckworth HE, Barber RC, Venkatasubramanian VS. Mass spectroscopy, 2nd ed. Cambridge: Cambridge University Press; 1986.
7. Gale PJ. Alfred Benninghoven. In Nier KA, Yergey AL, Gale PJ, editors. Historical perspectives: notable people in mass spectrometry. Amsterdam: Elsevier; 2015. Gross ML, Caprioli RM, editors. The encyclopedia of mass spectrometry, vol. 9, pt. B.
8. Gale PJ, Bentz BL. The early history of secondary ion mass spectrometry (SIMS). In Nier KA, Yergey AL, Gale PJ, editors. Historical perspectives: the development of mass spectrometry. Amsterdam: Elsevier; 2015. Gross ML, Caprioli RM, editors. The encyclopedia of mass spectrometry, vol. 9, pt. A.
9. Gale PJ, Bentz BL. Richard Edward Honig. In Nier KA, Yergey AL, Gale PJ, editors. Historical perspectives: notable people in mass spectrometry. Amsterdam: Elsevier; 2015. Gross ML, Caprioli RM, editors. The encyclopedia of mass spectrometry, vol. 9, pt. B.
10. Gross ML, Caprioli RM, editors. The encyclopedia of mass spectrometry. Amsterdam: Elsevier; 2015.
11. Gross ML, Grayson MA. The beginnings of organic mass spectrometry. In Nier KA, Yergey AL, Gale PJ, editors. Historical perspectives: the development of mass spectrometry. Amsterdam: Elsevier; 2015. Gross ML, Caprioli RM, editors. The encyclopedia of mass spectrometry, vol. 9, pt. A.

12. Grasserbauer M. Solid state mass spectrometry for materials science. Pure and Applied Chemistry. 1992;64(4):485–95.
13. Honig RE. The development of secondary ion mass spectrometry (SIMS): a retrospective. International Journal of Mass Spectrometry and Ion Processes. 1985;66:31–54.
14. Honig RE. The growth of secondary ion mass spectrometry (SIMS): a personal view of its development. In Benninghoven A, Colton RJ, Simons DS, Werner HW, editors. Secondary ion mass spectrometry: SIMS. Berlin: V. Springer-Verlag; 1985.
15. Hummel DO, Schüddemage H-D, Rübenacker K. Mass spectrometry. In Hummel DO, editor. Polymer spectroscopy. Verlag Chemie: Weinheim; 1974. Chapter 5.
16. Laeter JR de. Applications of inorganic mass spectrometry. New York: John Wiley & Sons; 2001.
17. McLafferty FW. A century of progress in molecular mass spectrometry. Annual Review of Analytical Chemistry. 2011;4:1–22.
18. McPhail DS. Applications of secondary ion mass spectrometry (SIMS) in materials science. Journal of Materials Science. 2006;41(3):873–903.
19. Montaudo G, Lattimer RP, editors. Mass spectrometry of polymers. Boca Raton: CRC Press; 2002.
20. Morris PJT. The American synthetic rubber research program. Philadelphia: University of Pennsylvania Press; 1989.
21. Nier KA. Arthur John Ahearn. In Nier KA, Yergey AL, Gale PJ, editors. Historical perspectives: notable people in mass spectrometry. Amsterdam: Elsevier; 2015. Gross ML, Caprioli RM, editors. The encyclopedia of mass spectrometry, vol. 9, pt. B.
22. Nier KA. Beyond platonic prejudice: a laboratory perspective on science as a subset of technology. In McClellan JE III, editor, The applied-science problem: papers from a workshop at Stevens Institute of Technology. Jersey City, NJ: Jansen/Daniels; 2008.
23. Nier KA. The development of mass spectrometry in the earth and planetary sciences. In Nier KA, Yergey AL, Gale PJ, editors. Historical perspectives: the development of mass spectrometry. Amsterdam: Elsevier; 2015. Gross ML, Caprioli RM, editors. The encyclopedia of mass spectrometry, vol. 9, pt. A.
24. Nier KA. Richard Franz Karl Herzog. In Nier KA, Yergey AL, Gale PJ, editors. Historical perspectives: notable people in mass spectrometry. Amsterdam: Elsevier; 2015. Gross ML, Caprioli RM, editors. The encyclopedia of mass spectrometry, vol. 9, pt. B.
25. Nier KA. Spectrometer, mass. In Bud R, Warner DJ, editors. Instruments of science: an historical encyclopedia. New York & London: The Science Museum, London, The National Museum of American History, Smithsonian Institution, Garland Publishing; 1998.
26. Nier KA, Yergey AL, Gale PJ. A general chronicle of mass spectrometry and guide to the scope of the volume. In Nier KA, Yergey AL, Gale PJ, editors.

Historical perspectives: the development of mass spectrometry. Amsterdam: Elsevier; 2015. Gross ML, Caprioli RM, editors. The encyclopedia of mass spectrometry, vol. 9, pt. A.
27. Nier KA, Yergey AL, Gale PJ, editors. Historical perspectives: the development of mass spectrometry. Amsterdam: Elsevier; 2015. Gross ML, Caprioli RM, editors. The encyclopedia of mass spectrometry, vol. 9, pt. A.
28. Nier KA, Yergey AL, Gale PJ, editors. Historical perspectives: notable people in mass spectrometry. Amsterdam: Elsevier; 2015. Gross ML, Caprioli RM, editors. The encyclopedia of mass spectrometry, vol. 9, pt. B.
29. Rettberg T. The development of inductively-coupled plasma mass spectrometry (ICP-MS). In Nier KA, Yergey AL, Gale PJ, editors. Historical perspectives: the development of mass spectrometry. Amsterdam: Elsevier; 2015. Gross ML, Caprioli RM, editors. The encyclopedia of mass spectrometry, vol. 9, pt. A.
30. Smalley RE. Discovering the fullerenes. In Grenthe I, editor. Nobel lectures, chemistry 1996–2000. Singapore: World Scientific Publishing Co.; 2003.
31. Tsong TT. Microscope, field ion. In Bud R, Warner, DJ, editors. Instruments of science: an historical encyclopedia. New York & London: The Science Museum, London, The National Museum of American History, Smithsonian Institution, Garland Publishing; 1998.
32. Watson JT, Sparkman DO. Introduction to mass spectrometry: instrumentation, applications, and strategies for data interpretation, 4th ed. Chichester, UK: John Wiley & Sons; 2007.
33. White FA. Mass spectrometry in science and technology. New York: John Wiley & Sons; 1968.
34. White FA, Wood GM. Mass spectrometry: applications in science and engineering. New York: John Wiley & Sons; 1986.
35. Winograd N. The development of secondary ion mass spectrometry (SIMS) for imaging. In Nier KA, Yergey AL, Gale PJ, editors. Historical perspectives: the development of mass spectrometry. Amsterdam: Elsevier; 2015. Gross ML, Caprioli RM, editors. The encyclopedia of mass spectrometry, vol. 9, pt. A.
36. Yergey AL, Bentz BL, Gale PJ. The development of thermal, spark source, and glow discharge ionization. In Nier KA, Yergey AL, Gale PJ, editors. Historical perspectives: the development of mass spectrometry. Amsterdam: Elsevier; 2015. Gross ML, Caprioli RM, editors. The encyclopedia of mass spectrometry, vol. 9, pt. A.
37. Zangwill A. Physics at surfaces. Cambridge: Cambridge University Press; 1988.

CHAPTER 4.11

THERMOGRAPHY

ANDREAS T. HAKA

Thermography, a noncontact procedure using an infrared detector to image electromagnetic radiation emitted from surfaces, was not an invention.[24; 25; 34] Rather, it was the product of a combination of technological and scientific developments and findings, which eventually led to this technology and its fields of application. These fields include, for example: building construction; medicine; emergency medical services; electrical engineering; geography; materials science; the testing, processing, and working of materials; architectural engineering; manufacturing and production technologies; the military; and property surveillance.

Scientific Foundations and Stages in the Development of Thermography

Thermography emerged from a number of scientific findings and established itself as a technology during the twentieth century. Its foundations can be traced to the Enlightenment.

The Pomeranian pharmacist Carl Wilhelm Scheele (1742–1786) was probably the first to focus on "radiant heat" (*strahlende Hitze*) and he proposed a scientific interpretation of the phenomenon around 1775. William Herschel (1738–1822) pursued this topic further shortly afterwards in the course of investigating light, and he discovered thermal radiation.[11; 12; 23] The experiment that Herschel performed, in which sunlight is cast through a prism, allowed him to detect heat radiation beyond the red range of the visual spectrum with a thermometer. The word "infrared" was later also derived from this (with the Latin prefix *infra*) for "under red." But in Europe it was referred to as "ultrared" (*ultra* for beyond) up to the early 1950s before "infrared" established itself as the internationally accepted term, mainly at the instigation of British and American scientists.[6 p1]

The Evolution of Thermography

The kaleidoscope of discoveries, developmental processes, and contributions by pertinent persons that set the foundations of thermography cannot be presented here in full.[11; 15; 21; 32] Below are just a few representative contributors to the evolution of thermography:

- Thomas Johann Seebeck (1770–1831): 1822 discovery of the thermoelectric effect (Seebeck effect);
- Leopoldo Nobili (1784–1835) and Macedonio Melloni (1798–1854): 1822/1830 the first thermocouple, the first thermopile;
- Macedonio Melloni: 1839 discovery of the solar absorption spectrum;
- Jean Charles Peltier (1785–1845): 1834 discovery of the Peltier effect;
- André-Marie Ampère (1775–1836): 1835 assumption that the properties of light and electromagnetic radiation are the same;
- Gustav Kirchhoff (1824–1887): 1859 Kirchhoff's law;
- James Clerk Maxwell (1831–1879): 1864 theory of electromagnetic radiation;
- Samuel Pierpont Langley (1834–1906): 1880 analysis of the atmosphere by means of a bolometer;
- Josef Stefan (1835–1893), Ludwig Boltzmann (1844–1906): 1884 total radiation law;
- Wilhelm Wien (1864–1928), John William Strutt, 3rd Baron Rayleigh (1842–1919): dependence of blackbody radiation on wavelength;
- Heinrich Hertz (1857–1894): 1886 photoelectric effect;
- Heinrich Rubens (1865–1922): 1898 description of the reflectivity of infrared radiation.

The path toward thermography as a technology was not yet foreseeable when Scheele and Herschel were conducting their investigations in the late eighteenth century. Discoveries of other natural phenomena were required to ground the technical developments that interlinked over a 150-year period and led to the appearance of thermography as a technical science for the imaging of surfaces.

The development of the bolometer in the mid-nineteenth century by the American astrophysicist and aviation pioneer Samuel Pierpont Langley (1834–1906) constituted one of the most important building blocks of present-day infrared thermography.[1] A bolometer is a sensitive thermal-radiation detector that uses the temperature dependence of electrical resistance. The sensitive element is composed of a resistance that alters its value with the absorption of heat. The change in resistance elicits a change in the signal's decreasing tension as it travels through the bolometer resistance and is thus measurable.

Further Development of Infrared Technology in the Civilian and Military Sectors

In 1913 the British entrepreneur and instrument maker Leonhard Bellingham (1888–1953) patented an infrared-measuring instrument (Figure 4.11.1) for locating icebergs, steamships, or other objects that differed in temperature from their surroundings.[4]

Soon afterwards, infrared technology was developed for military purposes during World War I — in particular, range finding, secure communication, and guidance systems for "flying torpedoes." These efforts had no broader impact, however, because they were virtually all prototype trials.[10 p247] From the 1920s on, infrared research became an lively and expanding specialty within physics, particularly in Germany and Great Britain.

One of the most important researchers in the field of infrared research during that decade was the Austrian physicist Gerda Laski (1893–1928).[29] She was employed as a teaching assistant to Peter Debye (1884–1966) — who would go on to win the Nobel Prize in Chemistry — at the University of Göttingen between 1918 and 1920 and subsequently worked with Heinrich Rubens (1865–1922)

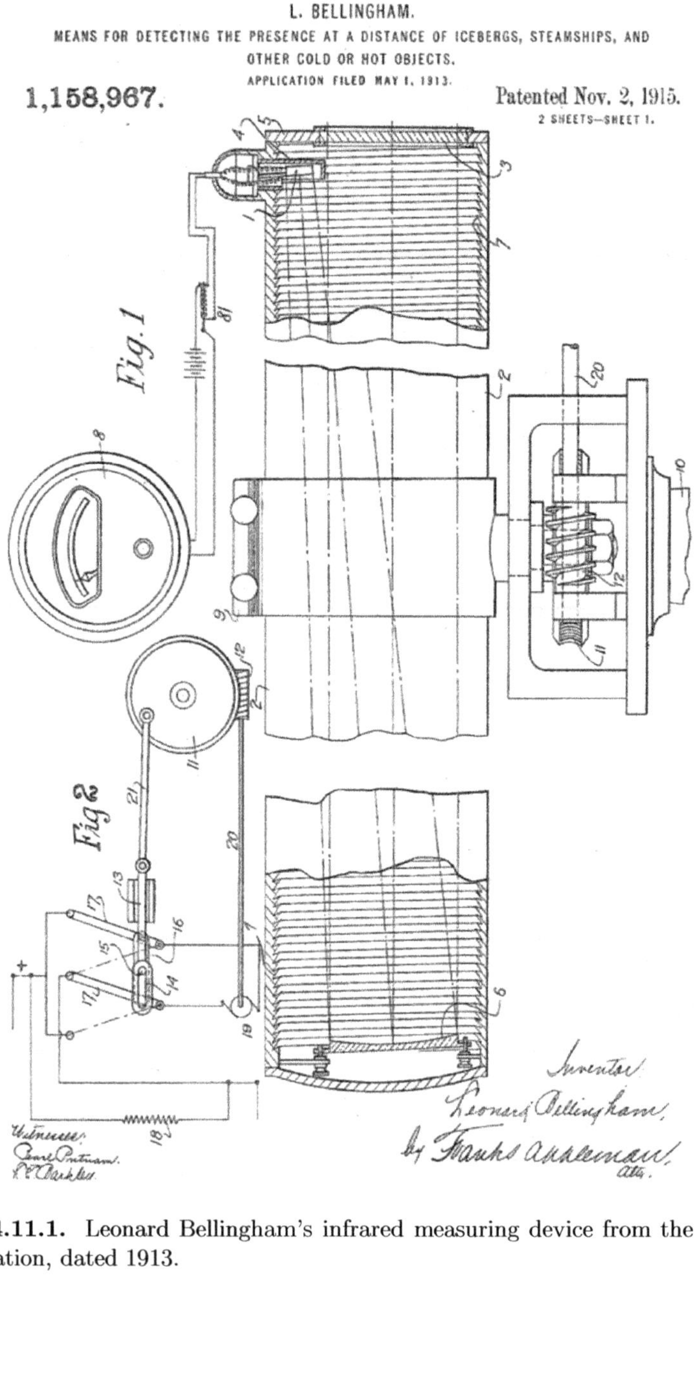

Fig. 4.11.1. Leonard Bellingham's infrared measuring device from the patent illustration, dated 1913.

in the physics department of the University of Berlin. In 1924 she became head of the section for ultrared research at the Kaiser Wilhelm Institute for Fibrous Materials Chemistry in Berlin. In 1928, shortly before her untimely death, she began setting up an ultrared research laboratory at the Berlin bureau of standards, the *Physikalisch-Technische Reichsanstalt*. Laski's research interests were focused on the infrared range of the spectrum itself as well as on its applications in materials analysis, specifically synthetic fibers.[16]

Two other influential scientists in infrared research at that time were the German physicist Clemens Schaefer (1878–1968) at the University of Breslau (now Wrocław) as well as his former doctoral student Frank Matossi (1902–1968), who later accepted appointments as a professor of physics at Graz (Austria), at White Oak (Maryland, USA), and at Freiburg in Breisgau (Germany), also serving as an expert for the Berlin army ordnance, the *Heereswaffenamt*. They jointly presented their results in a book published in 1930: *Das Ultrarote Spektrum*.[22]

Infrared research became institutionalized in Germany in 1942 in response to pressure from interested industrial firms and the military. The Reich Research Council (*Reichsforschungsrat*) coordinated this research and situated it with Johannes Plendl (1900–1991), the "plenipotentiary for high-frequency engineering." Abraham Esau (1884–1955) then took over this position in 1943.[31] The aim was to develop new technical systems for enemy assessment, target control of missiles, and active optical support for night combat.

In this context, the work by the German physicist Edgar Kutzscher (1906–1996) must be mentioned. His dissertation, submitted to the physics department of the University of Berlin, already treated long-distance measurement of infrared radiation.[14] He pursued this research further at the company *Electroacustik* in Kiel, where he developed infrared photocells for the flight control of guided missiles.[27] These infrared cells figured, for instance, as a basic component of an automatic target-finding device that found implementation in the glide bomb designated "Blohm & Voss BV 143." This bomb was specially designed to be dropped from a plane some distance away from the target. The bomb could then continue to glide on its own directly above the water surface in search of its target. Infrared cells permitted the bomb to be deployed at dusk or in the dark.

Research on infrared technology in Great Britain also mostly concentrated on military applications. One of the most active researchers in this area was the physicist and son of German émigrés Frederick Alexander Lindemann (1886–1957) (from 1941, Baron Cherwell, later 1st Viscount); he also served as science advisor to the British prime minister Winston Churchill (1874–1965).[33] As professor and head of Clarendon Laboratory at Oxford University, he and his collaborators worked on infrared sensors in target-finding systems for the Royal Air Force. Lindemann attempted to introduce his research on infrared technology to the government's board on air defense during World War II but became embroiled in an argument with the chairman Sir Henry Tizard (1885–1995), who advocated the development of radar technology. This quarrel ended with the dissolution of the board.[18 p284]

The First Thermocamera and Developments after World War II

Ernst Brüche (1900–1985), the founder of the specialty electron optics, made a decisive advance in the area of infrared detectors. Brüche developed the image sensor in the early 1930s in the research laboratories of the electrical combine *Allgemeine Elektricitäts-Gesellschaft* (AEG) in Berlin.[30; 20 p23,485; 5 p448–450] The bolometer had been the basis of systems developed up to that point. For military applications, the image sensor had a disadvantage, however: the sensitivity of the system only worked smoothly for nearby targets. Remote targets had to be "illuminated" by infrared search lights, which were detectable by enemy combatants with similar systems technology, revealing the attacker's position.[10 p250] This weakness was redressed by the sensitive photon detector, which had been developed almost in tandem and did not have this disadvantage. However, this new development only began to be deployed after World War II.

The first infrared camera for imaging thermal signals in the form of infrared radiation was developed by a Hungarian exile living in Britain — the physicist Kálmán Tihanyi (1897–1947).[26 p401] His invention, which he called a "radioscope," had already contributed significantly to television technology. Thus he must be regarded as a television pioneer together with the German inventor Manfred von Ardenne (1907–1997). Not long afterwards, in 1929, Tihanyi also developed a special infrared-sensitive television camera for nocturnal use which became part of the British anti-aircraft defense in 1929.[28]

After World War II, research on infrared technology was expanded and perfected in various programs primarily in the industrialized states, such as Germany, Great Britain, and the United States. For example, the US Air Force began to equip its air-to-air missiles with infrared homing systems. Research on such applications began as early as 1949 and culminated with the air-to-air missile "AIM-26 Falcon" (Figure 4.11.2), which was completed in 1959 but was still considerably hampered by technical faults.[8 p98]

Fig. 4.11.2. "Air-to-Air Missile AIM-26 Falcon" by the Hughes Aerospace Group, 1959, the first infrared guided missile of the United States. Press image from Hughes Aerospace Group. *Credit*: History of science collection of A. Haka.

Thermocameras were later also developed for the civilian sector, although quite hesitantly at first. The Swedish company *FLIR Systems* — then still under the company name *AGA-Bofors* — developed the first thermocamera for the military in 1958 and eventually also marketed it for civilian use in 1968. The purpose of this first camera was to detect faulty components in circuit lines.[9 p1]

Nondestructive Testing of Manufacturing Materials by Thermography

Nondestructive testing procedures are based on the stimulation of specific components.[14 p1–11] After a component is stimulated, observing its response behavior permits an analysis of its properties. Furthermore, unlike destructive tests, nondestructive inspection does not impair the operational properties of the tested specimen. Nondestructive testing can additionally be performed on the tested part under diverse operational states, whereas destructive inspection procedures can only be performed in a nonoperational state.

A number of nondestructive testing procedures exist — for example, irradiation, visual inspection and thermography. Thermography is a technique involving nondestructive assessment, processing, and imaging of the distribution of electromagnetic radiation emitted by a surface as registered by an infrared-sensitive detector system.[34 p10]

In thermography the stimulation of the component part occurs thermally without direct contact, by the absorption of radiation. The reflected thermal radiation is likewise registered by an infrared detector without contact. Inhomogeneities inside the component cause correlated differences in the manner of reflection and thus faults in the part can be detected.

Variant Thermographic Approaches

Different kinds of thermography can be distinguished on the basis of the stimulation and processing techniques used. Thermographic stimulation techniques come in both "active" and "passive" varieties.[17]

Materials testing preferentially works with "active" (dynamic) thermography.

Passive thermography merely exploits the radiation flow formed by the intrinsic warmth of the test object. This includes internally generated heat (for example, electrical components or live organisms) as well as the energy externally introduced by natural sources of heat (for example, heating by solar irradiation). Passive thermography is also the older application technology. Powerful computer technology was needed to make processing and analysis of the large amounts of data resulting from dynamic thermography applications realistically feasible.

In dynamic thermography, added energy sources (for example, light, ultrasound, hot air) produce heat flow inside the specimen. These additional heat sources can be made to act once, periodically, or continuously.

Depending on the stimulation technique, we distinguish:

- transient thermography (for example, impulse, cooling down)
- modulation thermography (for example, lock-in thermography, burst)

Depending on the processing technique, we distinguish:

- phase angle thermography (or lock-in thermography)
- temperature variation thermography.

Dynamic Thermography Exemplified by Impulse and Lock-In Thermography

Dynamic thermography is essentially based on specifically generating a variation in temperature and analyzing its distortion. A distortion in the field temperature indicates hidden faults. Forms of dynamic thermography include lock-in or impulse thermography.[3 p8] The latter supplies heat to the surface of the test object in the form of intense pulses of light produced by a flash bulb. An infrared camera registers the radiation emitted from the surface. Here a synchrony between the flash and the image recording, as well as the registration and analysis of the elapsed time is crucial. Afterwards the exposures can be evaluated in a computer system.

Lock-in thermography works similarly. A modulated heat source is used.[7] The temperature input introduced by this source into the component being tested then spreads as a "thermal wave" from the surface into the interior of the component. These waves are reflected at interfaces where the temperature conductivity is different; they return to the surface where they alter the temperature field by interference effects from the incoming thermal wave. An evaluation of this data permits the detection of internal defects in the component.

Similar to thermal stimulation, absorption of elastic waves from an ultrasound transmitter also leads to heating inside a test component. The hysteresis effect, which converts mechanical energy into heat, is not equally intense everywhere, however. In faulty areas inside a component, for example, as a consequence of concentrations of tension or crack-edge friction, the defects are selectively heated more than intact areas and are thereby detectible.

Optical Thermography as One Example of Lock-In Thermography

In an experiment to study the temperature conductibility of metals in 1863, the Swedish physicist Anders Jonas Ångström (1814–1874) determined the phase shift of a heat wave by letting hot steam and cold water act on a metallic material.[2] This procedure involved direct contact of the mentioned media on said material, though. It took until the beginning of the 1970s for the first photothermal radiometric measurements to be performed on a manufacturing material.[19]

In optical thermography, a powerful luminous source, such as a halogen lamp, is used to irradiate a component briefly at an intensity comparable to the temperatures produced during a hot summer day. An infrared camera takes a predetermined number of exposures during this input of heat. The connected computer then generates an amplitude and phase image.[14 p1] Optical thermography only reacts to changes in temperature that are caused by this periodic illumination, which then yield a temporally oscillating temperature field.

With the aid of Fourier analysis of each individual picture point throughout the resulting thermographic sequences of images, these data are concentrated to generate an amplitude image and a phase image. If the test specimen manifests defects, they influence the amplitude or phase of the detector signal. In evaluating the data, one can infer from the amplitude image the thermal structure, the optical structure of the surface, the structure of the emission coefficient in the infrared spectral region, and finally also the structure of the object illumination. But ultimately the thermal structure is decisive in defect detection during materials testing.

The phase image represents the local temperature course and the decisive thermal structure of the component. Optical thermography can effectively perform nondestructive, non-contact inspection of approximately one square meter per half-minute, depending on the makeup of the component part.

Other lock-in thermography approaches include, for example, ultrasonic lock-in thermography, which is commonly used to identify ruptures in metallic components, and induction lock-in thermography, which is also well suited for probing metallic construction materials, although it is rather unsuitable for analyzing materials such as CFRP (carbon fibre reinforced plastic).

References

1. Abbot CG. Samuel Pierpont Langley. Astrophysical Journal. 1906;23(4): 275ff.
2. Ångström AJ. New method of determining the thermal conductibility of bodies. Philosophical Magazine. 1863;25:130–42.
3. Bauer N, editor. Leitfaden zur Wärmefluss-Thermographie. Zerstörungsfreie Prüfung mit Bildverarbeitung. Erlangen: Fraunhofer-Allianz Vision; 2005.
4. Belling L. Means for detecting the presence at a distance of icebergs, steamships and other cold or hot objects. United States Patent 1158967. 1913.
5. Brüche E. Elektronenmikroskopische Abbildung mit lichtelektrischen Elektronen. Zeitschrift der Physik. 1933;86:448–50.
6. Brügel W. Einführung in die Ultrarotspektroskopie. Berlin: Springer; 1962.
7. Busse G. et al. Thermal wave imaging with phase sensitive modulated thermography. Journal of Applied Physics. 1992;71(3962):3962–65.

8. Chong A. Flying wings and radical things: Northrop's secret aerospace projects. Forest Lake, MN: Specialty Press; 2016.
9. FLIR System. FLIR system reaches production milestone; Sweden facility produces 25,000th commercial infrared camera. Wilsonville; 2004.
10. FLIR Systems. Users' manual. FLIR B series. FLIR T series. Publ. No. 1558797. Rev. a288. 2008.
11. Hentschel K. Unsichtbares Licht? Dunkle Wärme? Chemische Strahlen? Eine wissenschaftshistorische und -theoretische Analyse von Argumenten für das Klassifizieren von Strahlungssorten 1650–1925 mit Schwerpunkt auf den Jahren 1770–1850. Diepholz: Verlag für Geschichte der Naturwissenschaften und der Technik; 2007.
12. Hentschel K. Macedonio Melloni über strahlende Wärme. NTM International Journal of History & Ethics of Natural Sciences, Technology & Medicine. 2005;13:216–37.
13. Kaechele P. Zerstörungsfreie Prüfung von Strukturen mit der Lockin-Thermographie. In: LTH-Koordinierungsstelle, editor. Luftfahrttechnisches Handbuch. FL 53 800-17, 2006.
14. Kutzscher E. Über den Nachweis Ultraroter Strahlen auf große Entfernung und ihre Schwächung durch Nebel. PhD dissertation. Berlin: University of Berlin; 1933.
15. Langley SP. The Bolometer. Nature. 1898;57:620–22.
16. Laski G. Ultrarotforschung. Ergebnisse der exakten Naturwissenschaften. 1926;5:86–115.
17. Maldague XP. Theory and practice of infrared technology for nondestructive testing. New York: Wiley-Interscience; 2001.
18. Murray W, Millett AR, editors. Military innovation in the interwar period. Cambridge: Cambridge University Press; 1996.
19. Nordal P-E, Kanstad SO. Photothermal radiometry. Physica Scripta. 1979;20:659–62.
20. Rogalski A. Infrared detectors. Boca Raton: CRC Press; 2011.
21. Rubens H. Die Anwendung des Bolometers zur quantitativen Messung der Hertz'schen Strahlung. Verhandlungen der Physikalischen Gesellschaft zu Berlin, 1891;9:27–32.
22. Schaefer C, Matossi F. Das Ultrarote Spektrum. Berlin: Springer; 1930.
23. Scheele CW. Chemische Abhandlung von der Luft und dem Feuer. Oswalds Klassiker der exakten Wissenschaften 58. Upsala: Magn. Swederus; 1894.
24. Schneider D. Einführung in die praktische Infrarot-Thermografie. Aachen: Shaker; 2012.
25. Schuster N, Kolobrodov VB. Infrarotthermographie. Weinheim: Wiley-VCH; 2004.
26. Splinter R. Illustrated encyclopedia of applied and engineering physics, vol. 3. Boca Raton: CRC Press, 2017.
27. Thiele O. Blohm & Voss 143: Urahn der Sea Skimmer. Books on Demand. Hamburg: Blohm & Voss: 2012.
28. Tihanyi K. Automatic sighting and directing devices for torpedoes, guns and other apparatus. United Kingdom Patent 352,035. December 16, 1929.

29. Tolksdorf S. Gerda Laski. Nachruf. Physikalische Zeitschrift. 1929;30(13): 409–11.
30. Webel J. Von der Glas- zur Elektronenlinse: Die Anfänge der Elektronenmikroskopie. In Haka A, editor. Geschichte und Praxis von Forschungstechnologien. Stuttgart; 2018. Forthcoming.
31. Weiss B. Rüstungsforschung am Forschungsinstitut der Allgemeinen Elektricitäts-Gesellschaft bis 1945. In Maier H, editor. Rüstungsforschung im Nationalsozialismus. Organisation, Mobilisierung und Entgrenzung der Technikwissenschaften. Göttingen: Wallstein; 2002.
32. Wien W, Lummer OR. Methode zur Prüfung des Strahlungsgesetzes absolut schwarzer Körper. Annalen der Physik, 1897;292:451–56.
33. Zimmermann D. Britain's shield: radar and the defeat of the Luftwaffe. Stroud: Sutton; 2010.
34. Zimmermann T, Zimmermann M. Lehrbuch der Infrarotthermografie. Stuttgart: Lehrbuch der Infrarotthermografie; 2012.

CHAPTER 4.12

NEUTRON SOURCES

OLOF HALLONSTEN AND THOMAS KAISERFELD

Neutrons were first used for studies of materials in the 1940s, as a minor auxiliary activity at nuclear weapons labs. Since then, as nuclear reactor technology has improved, the use has spread to wider areas of the field of materials science, including applications with commercial potential. Its detachment from military research and development (R&D) activities was pushed by the broadening of the range of applications, but also by a drift of the global center of gravity of neutron-beam use from North America to Europe in the 1970s and 1980s, and more recently by the emergence of new and superior techniques to complement and eventually replace reactor-based sources, namely spallation sources.[4 p67]

The neutron was detected in 1932 by the British physicist James Chadwick, and it was shown only a year later that neutrons could be diffracted and hence be used as a probe to study materials.[2] As part of the World War II R&D efforts in the United States and several other countries, fission reactors of various types were constructed that accidentally produced neutron beams that could be studied and characterized, and used.[6 p27] The results of the world's first neutron-diffraction experiments were published right after the end of the war, and at the US National Labs in Argonne and Oak Ridge neutron beams from reactors were soon used to achieve important results that demonstrated the potential of the technique to the global physics and chemistry communities.[10 p31]

Several countries in Europe, North America, and Asia began ambitious nuclear research programs at this time and devoted vast resources to reactor construction. Therefore, there was no shortage of neutron beams for physicists to use. The breakthrough results of neutron diffraction for studies of magnetism and the mapping of the atomic structure of sixty elements and isotopes would several decades later garner Clifford Shull the 1994 Nobel Prize in Physics, together with Bertram Brockhouse, "for pioneering contributions to the development of neutron scattering techniques for studies of condensed matter."[4 p68] These results led to a surging interest in the technique and the exploitation of neutrons accidentally produced by reactors across the globe.

As the 1950s and 1960s passed, reactor development progressed swiftly and the average neutron *flux* (flow rate per unit area) increased by two orders of magnitude in a mere two decades. Both the Atoms for Peace policies launched by the presidential administration of Dwight D. Eisenhower and the *Sputnik* crisis fed into the development by making available nuclear material and technology for peaceful uses and by provoking radical increases in national R&D expenditures in many countries. The use of neutron beams for materials research was still an auxiliary activity, but the supply of neutron beams was generous because of the many reactors available. By the early 1960s, neutron use had reached a position as "a well established branch of international science," with users mostly in solid state physics and solid state chemistry, but with a clear course set towards "crossing of the threshold into biology."[2 p5]

The United States kept a global lead by a series of investments in new reactors in the National Labs system in the wake of the *Sputnik* crisis.[11 p121–122] But in Europe, a user community had started to form around prominent research groups at national R&D laboratories, most of all the labs of Heinz Maier-Leibnitz in Munich and Louis Néel in Grenoble, and a political process of launching a new major European, collaborative, applications-oriented research facility had also begun (Figure 4.12.1).

The French–German political axis that had formed in the early 1960s became the basis of many bi- and multilateral agreements

Fig. 4.12.1. Research reactor north of Munich in Germany inaugurated in 1957 and called the Atomic Egg for obvious reasons. *Credit*: ©2018 INTERFOTO/Alamy Stock Photo.

(including, decades later, the EU Common Market and the Euro).[12 p138–141; 7 p145] In 1967, the French and West German governments signed an agreement to build the reactor-based neutron facility Institut Laue Langevin (ILL) in Grenoble. Upon its completion, in 1974, the United Kingdom joined as an equal third partner in the collaboration.[14 p254; 1 p146] The ILL reactor had a global competitive advantage already at the design stage, with its projected record-high neutron flux, but its tremendous subsequent success was also secured by the organization of the lab as essentially a service facility for external users. This made the ILL not only a resource for the scientific communities of its partner countries and the rest of Europe, but also the driver of scientific and technological development in neutron use in Europe and a key asset in maintaining and cultivating a large neutron user community across Western Europe.[9 p137–139; 13 p140] Nothing similar existed in the United States, where neutron scattering facilities were auxiliary to reactors and served mostly scientists employed internally at the National Labs. Complaints were heard among directors of US National Laboratories in the late 1970s that the Europeans were taking over the global lead in the use of neutrons for research on materials.[3 p45] In 1993, a report by a special panel on neutrons commissioned by the US Department of Energy (DOE) concluded that "the United States has fallen alarmingly behind the European scientific community in the availability of up-to-date neutron sources and instrumentation."[15]

The most important technical development for the use of neutrons as a probe to study materials was the vast flux increase, but dramatic improvements in instrumentation, sample handling, and data taking/analysis also contributed during the postwar period. Still, though, reactors were limited in performance and could only produce continuous neutron beams, which (among other things) made time-resolved studies impossible. Another technique that gained ground in the future planning of neutron sources in Europe and the United States in the 1970s, which could provide pulsed neutron beams and would also lead to further flux increases, was *spallation* (Figure 4.12.2).[7 p29] Exploratory work on spallation techniques had begun in North America in the 1950s for the purpose of producing fissile material without reactors. The basic principle is that a neutron-rich element is bombarded with protons of high energy, which knock neutrons out in large quantities and with great intensity (flux). Spallation sources produce much less waste energy (heat) than reactors, and require fewer safety and security controls and regulations, but they are also slightly less reliable.[8] The first spallation neutron sources began operation at Argonne National Lab in 1981 and at the Rutherford Appleton Lab in the United Kingdom in 1985.[14; 16]

During the 1980s, many reactors in Europe and North America began showing severe signs of aging; soon enough a looming "neutron gap" between supply and demand of neutrons was identified.[13 p31] A European Commission panel on future European neutron sources delivered its report in 1991, recommending that several European

Fig. 4.12.2. The Materials Test Accelerator constructed at the Lawrence Livermore site 1950–1954 to produce fissile material through spallation. Note the rails going into the vessel, which was thought to be the largest evacuated space in the early 1950s. Operation was terminated before spallation was achieved as costs escalated out of control. *Credit*: ©2018 The Regents of the University of California, through the Lawrence Berkeley National Laboratory.

countries jointly develop a new pan-European spallation source. In 1993, the European Spallation Source (ESS) Council was formed in collaboration between the Rutherford Appleton Lab in the United Kingdom and the Forschungszentrum Jülich (Research Center Jülich) in Germany to carry out design studies and make plans for a new future source.[7 p29] The ESS was endorsed by the OECD Megascience Forum in 1999, together with recommendations that counterparts be built in North America and East Asia, which led the United States and Japan to press ahead with the funding and construction of new spallation sources. While these two began operations in 2007 and 2009, respectively, the European efforts were delayed by political inability to reach a solution for the funding and localization of the ESS facility. Only in 2014, more than two decades after planning began, was ground broken for the facility in Lund, Sweden. The ESS is set to open for scientific use in 2023.[5 p415–417]

The delay of the ESS has so far not meant that the center of gravity of neutrons has shifted back to the United States. Authors have claimed that the US Department of Energy, in its planning and construction of new large user facilities in the US in the 1970s, prioritized synchrotron radiation (and particle physics) over neutrons, if not on purpose then by default.[13 p141–142; 3 p51] The effects of such long-term neglect might be one explanation as to why the new Spallation Neutron Source (SNS) at Oak Ridge National Lab has not unequivocally returned the global lead in neutrons back to the United States, even though the ILL is aging and the ESS was delayed several years. Also, the localization of the facility to Oak Ridge, which lacks a strong tradition in neutron research but which was apparently chosen as host out of political reasons (of renewing the mission of the lab with a major facility project), likely played a role for making the SNS less of the success it could have been.[4 p75; 13 p147] Although some years remain until the ESS is taken into scientific operation, it seems Europe is on track to continue and reinforce its global lead, building on the successes of ILL and the strong community that has been cultivated in symbiosis with the Grenoble facility.

Reactor facilities and spallation sources are expensive, complicated technical systems and neutrons are hard to control and manipulate. The history of the use of neutron beams for studies of materials is therefore a history of a comparably small number of megaprojects, also in comparison with, for example, the use of X-rays from accelerators and storage rings (synchrotron radiation), where minor labs have also brandished some success and user communities have been able to grow also in the shadows of major R&D labs. Similar to synchrotron radiation, however, the history of the use of neutron beams for studies of materials has been characterized by a broadening of areas of use, a certain routinization of experimental work and lowering of thresholds for users with no experience of large-scale equipment, which has been important in order to attract users beyond smaller groups of enthusiasts.[4 p76]

Materials research with neutrons has been detached from weapons R&D in a gradual broadening of its use and a similarly gradual improvement on the technological side. Interest in the use of neutrons for experimental work continued to diversify from solid state physics over structural and dynamical characteristics of materials to biological applications in which macromolecules and molecular assemblies were analyzed. The process of expanding applications evolved further in the 1990s when even broader sets of uses proved possible in medicine and nanotechnology. The widening ranges of civil applications of the neutron had undoubtedly contributed to its beginning neutralization toward the end of the Cold War.

References

1. Atkinson H. Commentary on the history of ILL and ESRF. In Krige J, Guzzetti L, editors. History of European scientific and technological cooperation. Luxembourg: Office for Official Publications of the European Communities; 1997. Chapter 4.
2. Bacon GE. Introduction: the pattern of 50 years. In Bacon GE, editor. Fifty years of neutron diffraction: the advent of neutron scattering. Bristol: Adam Hilger, 1986. Chapter 2.
3. Crease RP. Anxious history: the high flux beam reactor and Brookhaven National Laboratory. Historical Studies in the Physical and Biological Sciences. 2001;32(1):41–56.
4. Hallonsten O. Big science transformed: science, politics and organization in Europe and the United States. Basingstoke: Palgrave Macmillan; 2016.
5. Hallonsten O. Unpreparedness and risk in big science policy: Sweden and the European Spallation Source. Science and Public Policy. 2015;42(3):415–26.
6. Hewlett RG, Anderson OE. A history of the United States Atomic Energy Commission, vol. 1, the new world, 1939–1946. Pittsburgh: Pennsylvania State University Press; 1962.
7. Kaiserfeld T. ESS from neutron gap to global strategy: plans for an international research facility after the Cold War. In Kaiserfeld T, O'Dell T, editors. Legitimizing ESS: big science as a collaboration across boundaries. Lund: Nordic Academic Press; 2013. Chapter 1.
8. Lander GH. The future with accelerator-based sources. In Bacon GE, editor. Fifty years of neutron diffraction: the advent of neutron scattering. Bristol: Adam Hilger; 1986. Chapter 9.1.
9. Maier-Leibnitz H. The birth of the Institut Max von Laue-Paul Langevin in Grenoble. In Bacon GE, editor. Fifty years of neutron diffraction: the advent of neutron scattering. Bristol: Adam Hilger; 1986. Chapter 5.3.1.

10. Mueller MH, Ringo GR. Early work at the Argonne. In Bacon GE, editor. Fifty years of neutron diffraction: the advent of neutron scattering. Bristol: Adam Hilger; 1986. Chapter 3.3.
11. Passell L. High flux at Brookhaven. In Bacon GE, editor. Fifty years of neutron diffraction: the advent of neutron scattering. Bristol: Adam Hilger; 1986. Chapter 5.1.
12. Pestre D. Prehistory of the Franco-German Laue-Langevin Institute. In Krige J, Guzzetti L, editors. History of European scientific and technological cooperation. Luxembourg: Office for Official Publications of the European Communities; 1997. Chapter 4.
13. Rush JJ. US neutron facility development in the last half-century: A cautionary tale. Physics in Perspective. 2015;17(2):135–55.
14. Stirling GC. ISIS–The UK pulsed spallation neutron source. In Bacon GE, editor. Fifty years of neutron diffraction: the advent of neutron scattering. Bristol: Adam Hilger; 1986. Chapter 9.2.
15. US Department of Energy. Neutron sources for America's future: report of the Basic Energy Sciences Advisory Committee Panel on Neutron Sources. Washington, DC: US Department of Energy Office of Energy Research; 1993.
16. Westfall C. Surviving to tell the tale: Argonne's intense pulsed neutron source from an ecosystem perspective. Historical Studies in the Natural Sciences. 2010;40(3):350–98.

CHAPTER 4.13

NUCLEAR MAGNETIC RESONANCE SPECTROSCOPY

JOSEPH D. MARTIN

The introduction to this section discusses how spraying things at materials can be a method for learning about their properties. Nuclear magnetic resonance (NMR) spectroscopy works in a similar, but subtly different way — it induces the materials themselves to do the spraying. NMR is a phenomenon that occurs when the nuclei of atoms in a strong magnetic field resonate with small oscillations in the near field (the portion of the magnetic field close to them) generated with radiofrequency (RF) pulses (Figure 4.13.1). The resonance causes nuclei to give off electromagnetic radiation, which carries information about the properties of those nuclei — in particular their magnetic moment, which can identify elements and give hints about how atoms are arrayed in a crystal or molecule. This information can be invaluable for determining molecular and crystal structure, monitoring chemical reactions or biological processes as they unfold, analyzing the purity of samples, and for applications like imaging. NMR has therefore become a widespread technique for investigating physical, chemical, and biological materials.

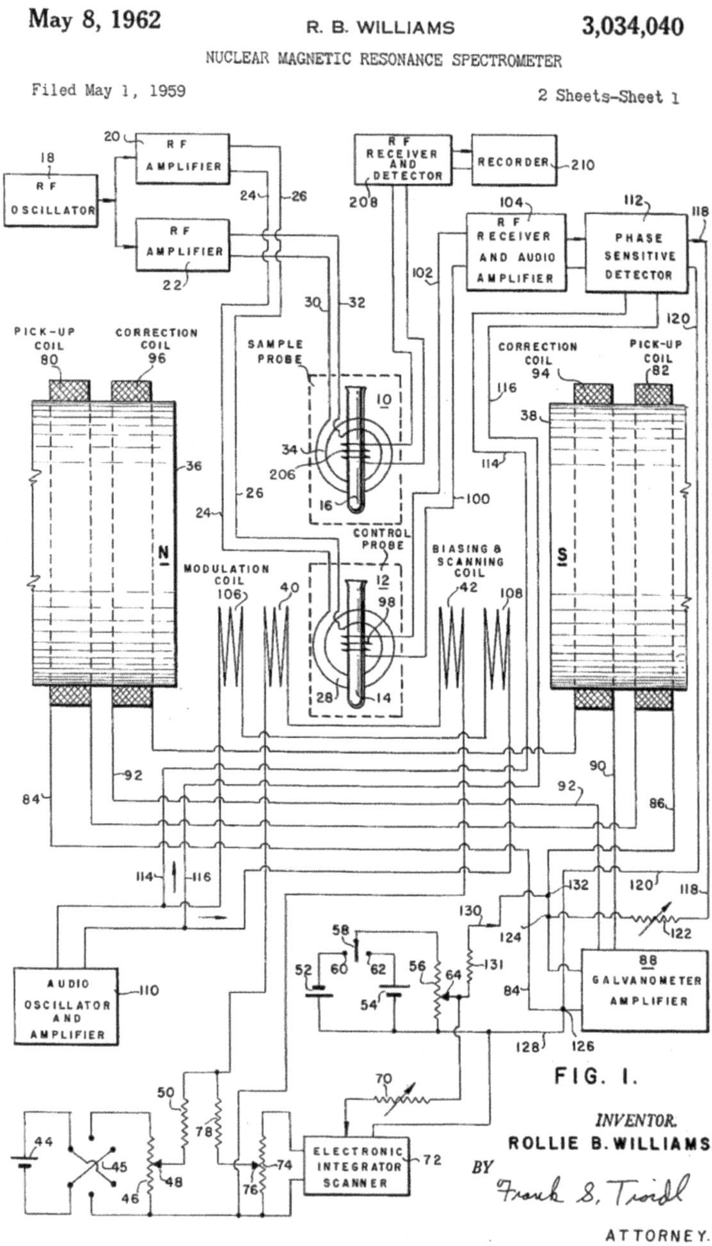

Fig. 4.13.1. A schematic design for an early NMR spectrometer. *Source*: Rollie B. Williams, "Nuclear magnetic resonance spectrometer," patent no. US3034040A.

Swords into Plowshares (into Swords, into Plowshares)

Today, we think of NMR spectroscopy as a civilian technology, but it owes much of its early development to wartime, following what Paul Forman calls a "swords into plowshares" trajectory.[4] The phrase refers to a line from the biblical book of Isaiah: "they shall beat their swords into plowshares, and their spears into pruning hooks: nation shall not lift up sword against nation, neither shall they learn war any more." World War II inspired vast investment in science and technology for military applications, in particular in the United States where NMR techniques were first developed, and from where the early instruments and expertise disseminated. After the war, opportunities emerged to find new applications for much of this expertise. NMR spectroscopy was one of the techniques that owes a considerable debt to the fruitful program of radar research conducted during the war.

But it is also a reminder that the swords-to-plowshares transition is not a one-way street; plowshares can equally well be beaten into swords. Isidor Isaac Rabi's research on magnetic resonance in molecular beams at Columbia University, which laid the groundwork for NMR techniques and for which he won the 1944 Nobel Prize in Physics, dates to 1937.[10] Rabi's work in the 1930s was fairly abstract. He was interested in measuring the quantum spin of atomic nuclei. Because the spin of an atomic nucleus is related to its magnetic moment, this work was of relevance to fundamental questions about both nuclear physics and magnetism.

In addressing these questions, Rabi and his team also developed what proved to be a fundamental technique. They began with what was known as a Stern–Gerlach experiment, in which a beam of atoms (usually silver atoms) is passed through an inhomogeneous magnetic field. The splitting of the beam demonstrates that the quantum spin of the atoms can take only one of two orientations. The insight that turned this rather crude apparatus into a delicate tool for measuring nuclear magnetic moment was that, by introducing small, carefully calibrated fluctuations to the magnetic field with RF pulses, it was possible to locate a nucleus's resonance frequency, where it would absorb RF energy.[14]

It was this fundamental mechanism — the sensitivity of nuclear magnetic moment to small magnetic-field fluctuations — that would lead directly to NMR techniques. But such experiments were delicate and called for the cultivation of considerable instrumental expertise. Interpreting the results of such an experiment required precise knowledge of both the magnetic field strength and the magnitude of its fluctuations, which in turn demanded exacting calibration. Wartime radar work, which inspired close attention to the way that radio waves interact with both matter and magnetic fields, promoted just that sort of expertise. So, in fact, NMR was not a direct result of military radar research, but rather is "generally indebted to the advance of radio-frequency instrumentation and methods over the course of the war — and to theoretical as well as experimental physicists' familiarity with them."[4 p430] The Radiation Laboratory, or Rad Lab, at MIT incubated considerable advances in the generation, detection, and interpretation of various types of electromagnetic radiation, and so although radar research was not directly concerned with the magnetic moments of nuclei, the instrumental expertise it encouraged created the conditions friendly to the emergence of NMR techniques.

A similar process unfolded on the opposite coast. Before the war, Stanford University was at the forefront of developing the type of equipment that would prove essential for radar technology. Felix Bloch, who would share the 1952 Nobel Prize in Physics with the Rad Lab's Edward Mills Purcell for their independent development of NMR spectroscopy techniques, recalled "acquaintance with radio techniques during the war suggested to me still another and much simpler way," to detect changes in the orientation of nuclear magnetic moments.[3 p427] But the physicists who developed it, although benefitting from the equipment, concentration of expertise, and instrumentation the war provided, pursued NMR as a matter of fundamental interest — in particular the extension of quantum mechanics to provide a description of the atomic nucleus.[13] By 1960, Bloch would report that "nuclear magnetic resonance has become one of the standard tools of investigation and has furnished one of the most striking examples of the close link between basic research and peaceful applications during the post-war period."[2]

From Physics to Chemistry and Biology

NMR began as a firmly physical undertaking. Both the instrumentation of radar research and the study of the nucleus were territory claimed by physicists. The birth of NMR also coincided with the postwar emergence of solid state physics as a new and lively subfield.[7] The earliest NMR investigations undertaken by Purcell's group at Harvard University were simple — identifying the presence of hydrogen nuclei in a sample of paraffin by showing an absorption spike at the predicted resonance frequency for hydrogen's magnetic moment. But the technique soon demonstrated its superiority over existing tools for particular crystal-structure determinations — the characteristic resonance spikes for certain nuclei would change depending on their structural environment, permitting inferences about that environment. Such structural determinations became more and more germane in the wake of the invention of the transistor at Bell Laboratories in 1947, after which semiconductor physics took off as an area of research. Shortly thereafter, NMR proved its mettle as a tool for assessing superconducting materials as well.[16] From the first, therefore, NMR was a physical tool for physical investigations.[9]

But such an interpretation is necessarily incomplete. As many contributions to this volume demonstrate, instruments have a way of behaving obstreperously with respect to disciplinary boundaries. NMR also had some of its earliest and most notable successes as a chemical technique. Cambridge, Massachusetts, where Purcell brought Rabi's ideas to fruition, had been a center of molecular theory and an incubator of chemical physics since before the war. NMR was therefore born into an environment with a well-developed community and culture organized to investigate molecular structure. It functioned not as a replacement for, but as a complement to existing techniques like X-ray diffraction and IR spectroscopy, and entered an instrumental community built around those techniques.[8; 11]

If NMR offered a new way to apprehend crystal structure, it was a similarly fruitful approach to chemical structure, an area pioneered by the physical chemist Herbert S. Gutowsky.[12] NMR proved particularly useful in this arena once chemists realized that, not only could NMR provide a wealth of data about the overall structure of a molecule, but certain chemical substructures also exhibited characteristic NMR signatures, meaning that NMR data was an invaluable supplement to X-ray diffraction or IR spectroscopic data.[16] Using these techniques in conjunction vastly expedited the sometimes finicky task of puzzling out a chemical's three-dimensional structure.

The wealth of molecular NMR data that accumulated through the 1950s and 1960s paved the way for the application of the technique to more complex molecules — in particular to organic molecules. NMR's potential to illuminate protein structure, function, and dynamics was long known. The rise of molecular biology through the 1970s focused attention on proteins, and how both their structure and the ways they assume that structure is related to their function, and NMR's in-principle ability to trace such processes in real time made it well adapted for such investigations. But the technique did not develop sufficiently to realize that potential until the 1980s, when new techniques like pulse Fourier transform technology made it possible to interpret the often-messy results of such experiments.[17] Here as well, NMR was used in conjunction with other techniques, in particular X-ray crystallography, to create a synthetic picture that no one instrument could provide on its own.[12]

Commercializing NMR

Like many other instruments and techniques discussed in this volume, NMR hatched with the help of government investment in science, fledged in a nest of university research, and then found a new home within commercial enterprises. In contrast to instruments optimized for making materials, however, NMR had a somewhat slower route to commercialization. In its early days, it was an expensive, boutique technique, and although the information it provided was scientifically valuable, it was often not directly germane to commercial applications. One reason the NMR community became such a strong, coherent group of researchers, therefore, is because NMR research for some time required a strong command of instrument design and operation of the type that was rapidly becoming less relevant for the operation of black-boxed instruments.

The first company to develop a commercial NMR spectrometer was Varian Associates, which had been founded near Stanford in 1948, in what would shortly become Silicon Valley, to commercialize the klystron. As one of the leaders in tube manufacture, Varian was well positioned and made a few early instruments that made use of the NMR principle. The company actively promoted the new technique with the aim of expanding its market. It organized a series of annual workshops, the first held in October 1957, with the goal of galvanizing the community of NMR researchers and spreading familiarity and facility with the technique.[5; 6 p188] When its practical off-the-shelf spectrometer, the A-60, became available in 1962, demand was high and it sold over 1,000 units internationally, most of them to chemical laboratories, before being superseded by newer models in the late 1960s and early 1970s.[1]

The black-boxed machine changed the nature of NMR analysis. Tasks that once took hours or even days could now be completed in minutes. Rather than spending time tinkering with shop-built machinery and fussing over samples, chemists could focus more squarely on the data that machinery generated. Instrument operation, once the province of research chemists, was transferred to technicians. NMR was therefore representative of a broader change in chemical and materials science practice that occurred in the 1960s and 1970s, when the guts of common tools retreated behind the smooth sheet-metal and gleaming dials of prefabricated instruments.[15]

Conclusions

NMR spectroscopy is indicative of a number of themes that are relevant for understanding late-twentieth-century materials science more generally. NMR techniques came within reach in the context of the military potential radar offered during World War II and it matured against a Cold War context in which knowledge of materials was recognized as a strategic bottleneck. It crossed disciplines freely, finding a ready welcome wherever the data it provided could be useful, even if its roots rested firmly in the world of physics. It responded to the subtle give-and-take between instrumental communities and the commercial firms that both served their needs and sought to package the fruits of their labors for new audiences. And the instrument itself became increasingly standardized, its use increasingly routinized — in short, it became increasingly black-boxed as the twentieth century wore on.

We can also see through the story of NMR how interdependent tools for materials research became. The goals of NMR, and the most favorable trajectories for its development, were defined not just by the nature of the phenomena under investigation, but also by the advantages and limitations of existing techniques. Its value was not measured by the information it could extract from samples alone; it was measured too by how that information compared to that generated by X-ray spectroscopy and other techniques. This simple example illustrates the rich, complex instrumental ecology that emerged in the second half of the twentieth century, in which any new instrument, to succeed, had to find its niche.

References

1. Becker ED. A brief history of nuclear magnetic resonance. Analytical Chemistry. 1993;65(6):295–302.
2. Bloch F. Letter to Sam Brand and Associates. October 3, 1960. Felix Bloch Papers, Stanford University Archives. Box 2, Folder 1.
3. Block F. The principle of nuclear induction. Science. 1953;118(3068):425–30.
4. Forman P. "Swords into ploughshares:" breaking new ground with radar hardware and technique in physical research after World War II. Reviews of Modern Physics. 1995;67(2):397–455.
5. Lécuyer C. Making Silicon Valley: innovation and the growth of high tech, 1930–1970. Cambridge, MA: MIT Press; 2006.
6. Lenoir T, Lécuyer C. Instrument makers and discipline builders: the case of nuclear magnetic resonance. Perspectives on Science. 1993;3(3):276–345.

7. Martin JD. Solid state insurrection: how the science of substance made American physics matter. Pittsburgh: University of Pittsburgh Press; 2018.
8. Mody CCM. Instrumental community: probe microscopy and the path to nanotechnology. Cambridge, MA: MIT Press; 2011.
9. Pavlish U. Robert Vivian Pound and the discovery of nuclear magnetic resonance in condensed matter. Physics in Perspective. 2010;12(2):180–89.
10. Rabi II, Zacharias JR, Millman S, Kusch P. A new method of measuring magnetic moments. Physical Review. 1938;53(4):318.
11. Reinhardt C. Chemistry in a physical mode: molecular spectroscopy and the emergence of NMR. Annals of Science. 2004;61(1):1–32.
12. Reinhardt C. "This other method:" the dynamics of NMR in biochemistry and molecular biology. Historical Studies in the Natural Sciences. 2017;47(3):389–422.
13. Rigden JS. Quantum states and precession: the two discoveries of NMR. Reviews of Modern Physics. 1986;58(2):433–48.
14. Rigden JS. Rabi, scientists and citizen. Cambridge, MA: Harvard University Press; 1987.
15. Roberts JA. Negotiated identities of chemical instrumentation: the case of nuclear magnetic resonance spectroscopy, 1956–1969. Annals of the New York Academy of Sciences. 2003;988(1):257–68.
16. Slichter CP. The golden anniversary of nuclear magnetic resonance — NMR: fifty years of surprises. Proceedings of the American Philosophical Society. 1998;142(4):533–56.
17. Torchia DA. Solid state NMR studies of protein internal dynamics. Annual Review of Biophysics and Bioengineering. 1984;13:125–44.

CHAPTER 4.14

SYNCHROTRON LIGHT SOURCES

ROBERT P. CREASE

A synchrotron light source is a facility that produces multiple tunable and intense beams of light, from infrared to hard X-rays, able to be put to a range of uses. Such sources have become a mainstay of modern materials-science research, across a surprisingly diverse and flexible array of topics and applications. For example, in 2001, organizers of one symposium on the application of synchrotron light to materials science were startled to discover they had enough papers to devote an entire session to cement. Similarly, a synchrotron in Trieste was used to study a Stradivarius violin, revealing its structure as well as details about its manufacture, in a non-invasive way that did not affect the sound.[5]

This article will briefly cover the history of synchrotron light sources, the various techniques and applications used at them, and their impact on the character of materials science research.

History

Synchrotron radiation was initially regarded as a nuisance, but rapidly grew into an indispensable scientific tool. When synchrotrons — accelerators which use carefully configured magnets to bend charged particles in a closed circular loop — were first built shortly after World War II, scientists were aware that the particles would radiate energy tangential to their orbit, but the effect was thought to be negligible to the synchrotron's operation. As electron synchrotrons grew in size and energy, however, the radiation siphoned off an ever-larger amount of energy, threatening to pose diminishing returns for such accelerators. In 1947, scientists at the General Electric Research Laboratory in Schenectady, New York, installed a mirror in a synchrotron they were building and were able to visibly observe a bright speck of the light. The phenomenon was baptized "synchrotron radiation," not because of any intrinsic connection with synchrotrons but because a synchrotron was where the phenomenon was first noticed.

Within a few years, scientists had figured out how to construct beam lines to carry off synchrotron radiation and measure its properties, and then to use the radiation for experimentation. By the early 1970s, enough uses for such radiation had emerged that some abandoned storage ring accelerators were transformed into devices to produce synchrotron radiation (Tantalus, for instance, at the University of Wisconsin–Madison, operational in 1968), or attachments were made to electron accelerators to create such sources (SPEAR, for instance, at the Stanford Linear Accelerator Facility, operational in 1972). These parasitic facilities, which relied on bending magnets to create the synchrotron radiation, are now called first-generation light sources.[6]

By the mid-1970s, uses for synchrotron radiation across fields from biology to physics had boomed, and a second generation of facilities were designed from scratch specifically to produce synchrotron radiation for research. The first of these, commissioned in 1981, was the National Synchrotron Light Source (NSLS) at Brookhaven National Laboratory (BNL). Its two designers, Kenneth Green and Renata Chasman, realized that, whereas high energy physicists are interested in maximizing the energy and luminosity of an accelerator's beam, users of synchrotron radiation are mainly interested in a property called brightness, the components of which are intensity, spread in angle, and volume. They therefore designed a lattice, or magnet configuration, to contain the beam in its circular path, which would optimize brightness. The Chasman-Green lattice continues to be fundamental to synchrotron design. The NSLS was unique in that it had

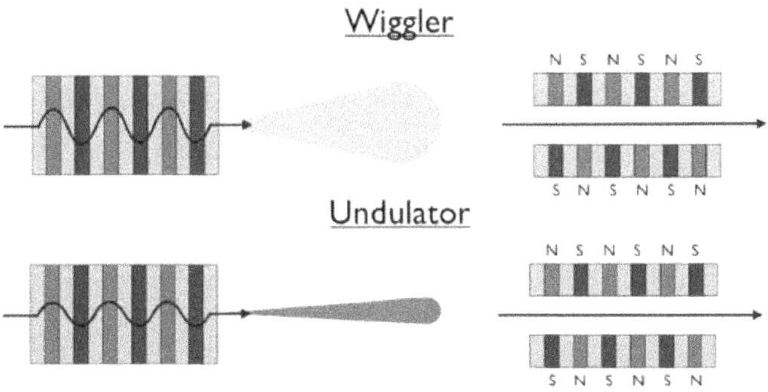

Fig. 4.14.1. Insertion devices. *Source*: Cereser A. Nanofocused X-ray analysis of semiconductor nanowires [MSc thesis]. Copenhagen: University of Copenhagen; 2011.

two rings to produce X-ray and ultraviolet light separately. It was also deliberately called a "light" rather than a "radiation" source because of the growing negative connotations of the latter term.[1;2]

The first two generations of synchrotron light sources relied mainly on bending magnets to produce light; subsequently, special devices were developed that were designed to be inserted in straight sections in the lattice to produce even more brilliant synchrotron light (Figure 4.14.1). These were of two main kinds: wigglers and undulators. In wigglers, an array of magnets is made to jiggle the beam back and forth to get it to shed more X-rays, with each change in direction producing its own cone of light. These cones are independent of one another, and pile up to create greater intensity than bending magnet radiation. In undulators, the electrons make smoother bends, but these bends are spaced so that the cones interfere with each other to produce a narrow, more intense beam. These insertion devices allowed such an improvement in brightness that a third generation of synchrotron sources was designed to take advantage of them. The first third-generation source was the European Synchrotron Radiation Facility (ESRF, 1994) in Grenoble, followed by the Advanced Photon Source at Argonne National Laboratory (1996), SPring-8 in Japan (1997) and Max IV in Sweden (2016).[8]

The brightness of third-generation synchrotron sources is limited by the fact that their electrons thermalize (bounce around in the beam). This limitation is surmounted in free-electron lasers (FELs) and energy-recovery linacs (ERLs). FELs deliver a coherent, straight-line, laser-like beam of electrons to wigglers and undulators, while ERLs send electrons around in a single orbit but then recover their energy after their return to a linac injector. FELs and ERLs may make possible a fourth generation of synchrotron light sources, whose brightness may be orders of magnitude greater than the third generation.[7]

Techniques and Applications

The intense and tunable light produced by synchrotron sources, from the infrared to ultraviolet and X-rays, is amenable to a variety of research techniques (Figure 4.14.2). These include: *spectroscopic* techniques, which involve studying the energies of particles absorbed or emitted by atoms when exposed to X-rays; *scattering* or *diffraction*

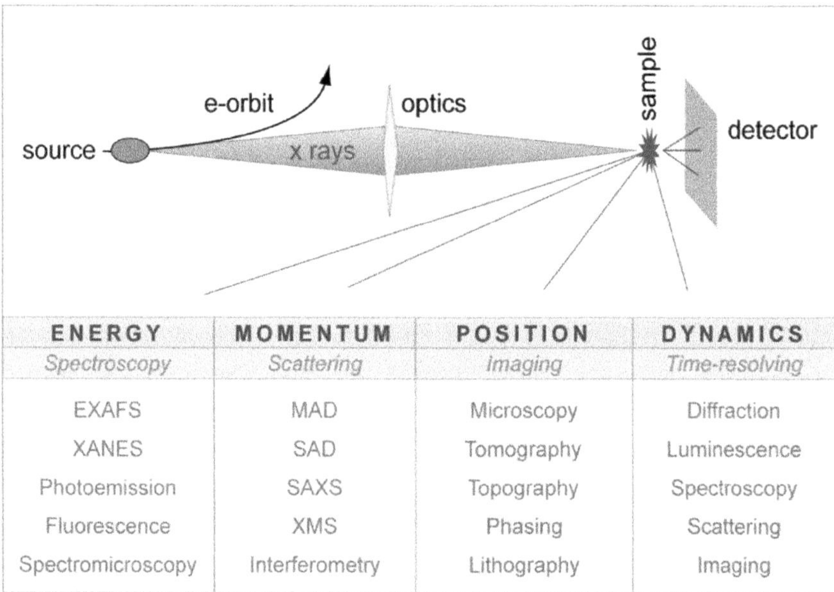

Fig. 4.14.2. X-ray absorption spectroscopy techniques. *Credit*: Courtesy of the Advanced Light Source, Lawrence Berkeley National Laboratory.

techniques, which involve studying the patterns produced when X-rays are deflected by arrays of atoms; and *imaging* techniques, which use X-rays to picture or etch materials.

Regarding *spectroscopic* techniques, X-ray absorption spectroscopy (XAS) is a technique widely used for materials research at synchrotron sources. It involves using a monochrometer to tune the X-ray beam to a particular "edge" at which they will excite particular core electrons, detection of which allows characterization of both the chemical and electronic features of matter. A family of XAS techniques have been developed, which include extended X-ray absorption fine structure (EXAFS), X-ray absorption near edge structure (XANES), and diffraction anomalous fine structure (DAFS). EXAFS probes the environment of an atom, XANES probes the absorbing atom, and DAFS probes the environment like EXAFS, but uses a different signal. Whereas regular X-ray diffraction provides long-range information about an entire structure, XAS techniques study the local structure near an atom.

Scattering or *diffraction* techniques common at synchrotron light sources include small-angle X-ray scattering (SAXS), which is used to characterize the size and shape of macromolecules, and powder diffraction, which is used to characterize materials, resolve unit cell dimensions, and determine sample purity.

Imaging techniques found at light sources include X-ray microscopy, which complements optical and electron microscopy, in which X-rays are focused to a point and scanned to image an area. This method is used to examine fine structures of materials. The scanning transmission X-ray microscope (STXM) developed at BNL and used at the NSLS in the 1990s is an example of an instrument designed for this purpose. X-ray tomography is used to image the three-dimensional microstructure of materials. Synchrotron sources have also been used for X-ray lithography to etch the surfaces of materials in micromanufacturing, particularly in the electronics industry.

These, however, are only the most common general categories of techniques used in materials science research at synchrotron light sources; other techniques combine aspects of these three. One example is X-ray topography, another family of techniques that combines diffraction and imaging, used to probe the character and distribution of crystallographic defects and phase transitions.[4] All of these techniques are used in the study of surfaces, interfaces, electronic materials, phases, metal oxides, films, glasses, polymers, nanoparticles, metal composites, multilayer metal systems, contaminates, and other areas of materials research.

Impact on Research

In the 1990s, synchrotron light sources began to replace high energy physics accelerators as marquee projects at basic-research laboratories in the United States, and began to become a major concern for European laboratories as well. They are such a mainstay of modern materials science research that over sixty have been built worldwide.

Synchrotron light sources have introduced important changes in the character and culture of the research ecosystem of materials science, making it not just bigger in scale but more interdisciplinary, international, and focused on applied research. The resulting phase transition has been baptized the "New Big Science."[3]

In this phase change, materials science involves more fields and subfields, as well as a wider variety of instruments, more connections between seemingly disparate research programs, a broader spectrum of users, and a faster turnover of programs. Research cultures at such facilities now resemble ecosystems, comprised of complex and evolving interactions between individuals, institutions, and global commercial markets. It involves new networks of experiments and laboratories, as well as of participating universities, industries, and governments. The change is accompanied by new problems involving use of the facilities, funding of experiments, and intellectual property issues. It has transformed the nature of invention and innovation spaces, and affected how researchers must be trained to operate in them.

An emblem of their regional indispensability is SESAME, an accelerator recently completed in Jordan. SESAME is a joint project of nine Mideast partners including Cyprus, Egypt, Iran, Israel, the Palestinian Authority, and Turkey, who collaborated on the $110 million machine because of its bread-and-butter applications to such purposes as semiconductors, medicine, the crystallography of macromolecules such as proteins, environmental research, art history, and archaeology. The diversity and widespread character of such applications reveal how useful synchrotron radiation is throughout experimental science. It is now an important tool even to researchers who do not think of themselves as synchrotron-radiation scientists, which is a tribute to those who played a role in its development.

References

1. Crease RP. The National Synchrotron Light Source, part I: bright idea. Physics in Perspective. 2008;10(4):438–67.
2. Crease RP. The National Synchrotron Light Source, part II: the bakeout. Physics in Perspective. 2009;11(1):15–45.
3. Crease RP, Westfall C. The new big science. Physics Today. 2016;69(5):30–36.
4. Dudley M. X-ray topography. Materials Research Society Symposium Proceedings. 2011;307:213–24.
5. Echard J-P et al. The nature of the extraordinary finish of Stradivari's instruments. Angewandte Chemie. 2009;49(1):197–201.
6. Hallonsten O. The parasites: synchrotron radiation at SLAC, 1972–1992. Historical Studies in the Natural Sciences. 2014;45(2):217–72.
7. Heinze T, Hallonsten O, Heinecke S. Turning the ship: the transformation of DESY, 1993–2009. Physics in Perspective. 2017;19(4):424–51.
8. Westfall C. Institutional persistence and the material transformation of the US national labs: the curious story of the advent of the Advanced Photon Source. Science and Public Policy. 2012;39(4):439–49.

INDEX

accidents: happy, 172, 243, 294, 553–54; unhappy, 21, 130–31, 183, 321, 346
acetylene, 87, 295–97
acid: acetic, 88; amino, 302–3, 309, 321, 430; citric, 301, 303; ethylenediaminetetraacetic, 352; fatty, 323; lactic, 301; lithography and, 327–30; nitric, 102, 290; nucleic, 303, 397, 462; organic, 302, 307; phosphoric, 102–3; refrigeration and, 276; safety of, 131; sulfuric, 98, 274
Advanced Research Projects Agency (ARPA), 115, 122–24, 127, 486–87, 489
aerospace, 122, 148, 232, 235, 248–49, 255, 269, 335, 346–47, 383, 395, 541, 546f
Africa, 44, 45f, 47, 235, 257, 384
agriculture, 15, 97–100, 103, 106, 378, 391–92, 399
alchemy, 7, 19, 45–49, 48f, 152f, 257–59, 258f
alcohol, 33, 36, 65, 69–70, 85, 87, 91, 273, 292, 302
alloys, 6, 15, 44, 232, 249, 252, 259, 260, 292, 427, 535
aluminum, 73, 145, 214, 232, 288, 499
American Chemical Society, 75, 173
American Society for Testing Materials, 174, 184, 190, 222
American Vacuum Society, 75, 174
ammonia, 101–5, 282f, 287–88, 290–94, 296
amorphous solids, 4, 113–14, 266, 420
ancient world, 7–9, 15–16, 21, 43, 46, 53–54, 155, 257, 273
anthropology, 17, 44, 257, 421
antibiotics, 300, 302–5, 307
antiquity, *see* ancient world
applied research, 5, 250, 576
archaeology, 44–45, 246, 257, 388, 577
architecture, 49, 117, 121–29, 127f, 266, 539
argon, 87, 346
Argonne National Laboratory, 378, 389, 478–79, 518, 522, 553, 556, 573
ARPA, *see* Advanced Research Projects Agency
artisans, 17, 34, 141
assaying, 54–55, 182, 301, 307–9, 466
Atomic Energy Commission, 123, 155, 378
Atoms for Peace, 378, 554
Australia, 279, 384
Austria, 234, 319, 397, 407, 437, 543
automation, 99, 103, 175, 209, 224, 309–10, 319, 322–23, 369, 451, 469, 521f, 543
aviation, *see* aerospace

bacteria, 251, 282, 304, 208, 311, 435, 444
balances, 6, 37, 53–61

barometers, 22, 30
basic research, 305, 360, 378, 453, 565
batteries, 136f, 349–55
Beckman, Arnold O., 457–59, 458f, 465–70
beer, *see* brewing
Belgium, 144, 293f, 506
Bell Laboratories: as disseminator of techniques, 247; chemists at, 265, 267–68; instrument development at, 334, 336; interdisciplinary research at, 268; involvement in surface science, 147, 497; microelectronics development at, 342, 359–60, 362–64, 565; patents, 267–68, 267f; Raman spectroscopy at, 438
Berzelius, Jacob, 22, 25
big science, 247–48, 310, 336
biochemistry, 66, 71, 308–9, 315, 321, 427
biomaterials, 4, 64, 147–48, 249–51, 254–44, 300, 310, 425–33, 501
biomedicine, 6, 66, 250, 254–55, 321, 232
black box, 3–4, 21, 33, 207, 212, 216, 259, 264, 444, 657–68
Black, Joseph, 36–37, 49
Bloembergen, Nicolaas, 484–86, 488
blood, 34, 65, 391, 396–97, 399
boiling point, 28, 34–37, 83, 86, 281
bolometers, 540–41, 544
Boyle, Robert, 34, 49
Bragg, William Henry, 414, 416, 425–26
Bragg, William Lawrence, 414, 416, 425–27, 509
brass, 39, 55f, 214, 283
brewing, 41, 90, 274, 276, 279, 282
Brock, David C., 10, 267
bronze, 44, 214
Brookhaven National Laboratory, 336–37, 337f, 378, 478, 522, 559, 572
buckminsterfullerene, 243–47, 256
buildings, 1–2, 6–7, 117, 121–28, 154–54, 241, 266, 273, 539
Bunsen burners, 43, 50, 260, 417
Bunsen, Robert, 50, 416–17, 417f, 461
Bureau International des Poids et Mesures, *see* International Bureau of Weights and Measures
bureaucracy, 183, 185–86, 189

calibration: constraints on buildings, 122, 127; microscopy and, 496; mechanical testing and, 223; mundanity of, 181; nuclear magnetic resonance and, 563–64; of FTIR, 451, 453, 464f; of scales, 58–59; of thermometers, 33–36, 39, 40
California Institute of Technology, 198, 227, 459, 498, 499f
caloric, 37, 49
calorimeters, 37, 49, 283, 309
Caltech, *see* California Institute of Technology
Cambridge Instrument Company, 334, 510
Canada, 175, 387, 506, 509
cancer, 66, 117, 131, 182, 302
carbon: calcium carbide, 260; chemical analysis and, 22–24; chip manufacture and, 373; compounds of, 243–47; conductivity of, 113, 261; dioxide, 27–28, 87, 280–81; fibre, 549; fermentation and, 308; graphite, 246, 352–53; hydro, 87, 96, 533; microscopy and, 522; monoxide, 27–28, 292; nanotubes, 245–47, 470; reactions including, 295; steel production and, 50, 289
catalogues, 2, 6, 112, 159–67, 175–77, 453
cement, 75, 79, 183, 260, 571
centrifuges, *see also* ultracentrifuges 6, 93, 292, 391–99, 393f, 395f, 389f, 478

ceramics, 44, 74, 76f, 257, 259–61, 265, 288, 351, 425, 501, 535
cesium, 461, 465
charcoal, 44, 143, 258, 283
chemical engineering, 77–78, 88, 91, 87, 103, 242, 251, 292, 296
chemistry: academic, 123, 125; aesthetics of, 113; basic, 292, 308, 453; batteries and, 349–54; colloid, 7; combinatorial, 309–10; communities, 316, 553; condensed matter, 514; conferences, 114, 254, 479; history of, 29–30; industrial, 76–77, 151, 263–65, 291–92, 457; inorganic, 2; interdisciplinary collaborations and, 124, 132, 249, 565; laboratories and facilities, 22–29, 115, 234, 388, 543; methods in, 18, 22–23; modern 47, 49, 324, 406; Nobel Prize in, 113, 142, 245, 294, 311, 396, 541; of wool, 427; organic, 25–26, 204, 261, 299, 300, 302, 304, 315, 318–20, 394, 396, 430, 462, 532–33; pedagogy, 6, 35, 438; physical, 315, 394; polymer, 7; practitioners of, 17; quantum, 198–200, 203; solid state, 554; synthetic, 249; women in, 77
China, 16, 93, 296, 354, 384
chromatography, 247, 254, 307, 309, 315–25
circulation, *see* communication
clean rooms, *see also* cleaning, 134, 138, 154–58, 381
cleaning, *see also* clean rooms, 133, 151, 154, 156–57, 392
Cold War: American hegemony and, 269; coldness of, 255; funding environment in, 343, 347; high-temperature research in, 265; materials bottleneck and, 122, 248, 568; microelectronics and, 335; missile defense and, 485; neutron sources during, 559; nuclear engineering in, 378, 383; strategic importance of materials research during, 122, 268, 335, 421–22, 483; surface science in, 147
collaboration: academia-industry, 125, 143, 244, 334, 509, 522; among institutions, 182; among libraries, 175; expansion of, 117; friction in, 112; fruitfulness of, 27–28, 279, 319, 414, 416, 429, 438, 466, 488, 514; government, 189, 544; interdisciplinary, 121–27, 197, 347; international, 4, 301, 305, 554–55, 557, 577
Columbia University, 250, 359, 563
commerce, *see also* markets: academia and, 181; applications and, 102, 283, 310, 359, 392, 450; devices and, 268, 338, 347, 352–53, 357, 370; everyday conduct of, 17, 53; incentives, 26, 327, 329–30, 332–33, 343–44, 373, 397; interests, 4, 139, 553; links to the state, 46; materials production and, 69, 88, 113, 165, 229, 246, 265, 288, 295; research culture and, 576; strategies, 320; tool production and, 92, 144, 199, 212, 215, 243–44, 278–79, 334, 369, 451–53, 457–67, 518, 566–68; tool use and, 56, 94, 247, 266, 322, 394, 447, 500, 506, 510, 522; utilities and, 227
communication: channels, 201; circulation of knowledge, 17; companies, 359, 369; digital, 330; flow, 116; modes and methods of, 73, 111; networks, 406; personal, 178–79, 323; secure, 541; speed of, 177; technology, 335; tele-, 122, 357, 369
computers: chips, 329–332, 334, 338, 370; code, 179, 201; computational materials science, 133; control, 319, 334, 336, 338, 369, 372, 454,

528; data storage and display, 211, 216, 319, 451; generic, 169, 176, 197, 212, 248, 255, 265, 358; growth in computing power, 196, 199–204, 216, 255, 327, 428, 430, 443; modeling and simulation, 92, 170f, 181, 195–204, 245, 302, 311, 388, 429, 430; scientific communication, 179, 201, 330; searching, 174; solving equations and processing data, 92, 196–204, 319, 428, 438, 447, 451, 454, 530, 547, 548
condensed matter physics, 7–8, 198, 202, 406, 514
conferences, 114, 116–17, 173–74, 178, 190, 244, 252, 254, 359, 363, 447, 467, 478–79
Connes, Janine, 447, 449–51
consumption, 17, 74, 105, 137, 159–165, 171, 177, 255, 268, 269, 274, 279, 334, 341, 347, 352, 353
contamination, *see also* dirt: airborne, 145; control, 151–56; de-/anti-, 381f, 382f; identification of, 534; laboratory procedure and, 78; of beams, 333; of food and drink, 435; of gases, 28, 87; of heating apparatus, 50, 261; of spacecraft, 322; of vacuum, 358; of water 273
continuity, 4, 123, 255, 317
cooling, *see also* refrigeration 6, 273–74, 276, 278, 281, 378, 395, 547
copper, 15, 40, 44, 279, 295, 328, 330, 361, 396
Cornell University, 2, 123, 363
corrosion, 46, 112, 231, 227–29, 231–36, 346, 535
craft knowledge, 2, 17, 19, 98, 258, 327
CRC Handbook of Chemistry and Physics, 74, 174
Crookes tube, 407–11, 410f, 414, 418
Crookes, William, 407–9, 408f
cryogenics, 86, 282–83
crystallography, 65, 425–31, 566, 576–77
crystals: analysis of, 63, 266, 350, 413–16, 425–30, 438–39, 444, 484, 522, 561, 565–66, 576; growing, 47, 51, 63, 253, 260–62, 265, 268, 274, 317, 338, 358–61; in instruments, 351, 485; organic, 265, 427–30; polycrystals, 15, 267f; preparing, 64, 341–42, 345, 362, 392; purifying, 26, 51, 151; regularity of, 4, 113, 264, 420; single, 51, 249, 262, 265, 267, 535; theory of, 198–99, 202, 420, 426–27, 475–78
Curie, Marie, 131, 407

DARPA, *see* Advanced Research Projects Agency
Department of Defense, *see* United States Department of Defense
Department of Energy, *see* United States Department of Energy
dewars, 282–83, 284f
diagrams, 50, 171, 187–88, 208, 224, 242, 251
diffraction: electron, 409; general phenomenon of, 2, 403; gratings, 439, 443, 459–60, 466–67; limit in microscopy, 338, 515–16, 519; neutron, 553–54; X-ray, 4, 125, 265–66, 350–52, 362, 406, 413–416, 415f, 420, 425–431, 565–66, 574–76
diodes, 113–14, 342, 362, 370, 470
dirt, *see also* contamination 153, 155
disciplines: bridges between, 122, 246, 324–25, 388, 501, 568; cohesion of, 74–75, 177; constituent fields of materials research, 7, 132, 249; disciplined practice, 134, 181; formation of, 113, 151, 174; jurisdictions of, 73, 77, 79, 114, 151, 181, 249, 251, 299, 565; laboratories and, 27, 121–22, 247;

sites of tool development, 8, 17, 64, 319, 358; texts associated with, 73–75, 79, 487; training in, 122, 124, 127, 250, 263, 501
discontinuity, 4, 317
distillation, 6, 15, 19, 26, 46–47, 49, 83–96, 247, 251
DNA: as material, 250, 256f; biotechnology, 300, 309–11, 430; molecule, 4, 65, 70, 302, 399, 435, 468; structure of, 321, 396, 397, 428–30, 432f
doping, 41, 116, 156, 330, 342, 345, 349, 362–63, 486
Dresselhaus, Mildred, 245, 415f
DuPont, 114, 145, 160–61, 163, 165
dyes, 64–66, 68–69, 242, 294, 392, 421, 459

economics, 9, 11, 16, 13, 139, 183, 259, 263, 310, 318, 354, 388, 406
education, 29, 74, 78–79, 112, 123, 134, 388, 436
Egypt, 46, 53–55, 54f, 55f, 576
Einstein, Albert, 394, 396, 419, 479
Eisenhower, Dwight D., 122, 231, 378, 554
electricity, 51, 99, 113, 121, 227, 261, 264, 292, 378, 404, 407–9
electrochemistry, 228, 349–54, 465
electromagnetic radiation, 252, 419, 457, 485, 513, 515, 539, 540, 546, 561, 564
electron microscopes, *see* microscopes, electron
electrons: in beams, 210, 252, 253f, 261, 332–34, 336, 403, 495–97, 503, 504f, 505, 514–16, 522, 573–74; in crystals, 360–62, 426, 428; in molecules, 114, 488; physics of, 142, 408–10, 418–20, 479, 484, 491, 510, 513, 515–16, 521, 575; theory of, 113, 198, 198–202, 403–404, 419, 485
engines: generic, 38, 263; internal combustion, 261, 361; jet, 20f, 248, 249, 341; steam, 38, 49, 98, 258; turbine, 395
England, *see also* Great Britain, United Kingdom, 34, 106, 153, 229, 292, 294–95, 330, 359, 392, 421
ENIAC, 197, 330
ethanol, 83, 305, 316
Ettre, Leslie, 320, 323
Europe: economy of, 310; facilities in, 384, 553–58, 573, 576; funding in, 519; industry in, 292, 301; international collaboration and, 190; instrument making in, 23, 25; instrument sales in, 27; historians' focus on, 10; practices in, 105–6, 258, 265, 274, 334, 406–7, 540; standards in, 55; traditions of, 257; travel to, 436
explosives, 26, 87, 129, 137, 143, 165, 195–96, 281, 289, 301, 440, 469
Exxon, 86, 243–44, 246, 293, 351

fabrication: chemical 277; of instruments, 34, 279, 342, 510; of metals, 6; of nanostructures, 254, 491; of semiconducting materials and device, 341, 343–45, 347, 357, 373; of textiles, 156; pre-, 2, 241, 567; safety of, 129, 133; sample testing and, 364
factories, 132–33, 137, 184, 268, 301, 337, 344, 345, 509
Fahrenheit, Daniel, 34–36, 49
Fairchild Semiconductor, 268, 344
Faraday, Michael, 22, 30, 50
farming, *see* agriculture
fermentation, 35, 90, 299–305, 307–8
fertilizer, 97–107, 287, 290, 296
filtration, 151, 153–55, 158, 316–17, 370
First World War, *see* World War II
flasks, 2, 21, 27, 273–84, 284f

food, 18, 43, 155, 273, 277–78, 282–83, 296, 299–301, 435, 463, 468, 510
Fourier transform, 428, 439–40, 444–48, 451, 454, 529, 549, 566
France: circulation of techniques to, 34; distinctiveness of, 268; fertilizer in, 105; industry, 99, 101, 363; interwar, 263; institutions in, 38, 212; theoretical developments in, 287, 447; tools, methods, and processes developed in, 209, 291, 311
Franklin, Rosalind, 428–29, 428f
fullerenes, 243, 245–46, 533
furnaces, 27, 44–51, 45f, 258–69, 258f, 260f, 262f, 342, 368–74, 369f, 371f, 372f, 374f
fusion: cold, 177; con-, 515, 516f; dif-, 143, 247, 322, 342–43, 435, 451–53, 535; flame, 261–62; nuclear, 47, 147; trans-, 397

gallium arsenide, 360–61, 520
galvanometers, 40, 209, 212–14, 214f, 465
gasoline, 293–94, 296, 301
gels, 1, 264–65
General Electric, 2, 90, 378, 386, 463, 572
genetics, 107, 308–9, 397–99
geology, 63, 383, 435, 439, 532
germanium, 200, 267f, 509, 522
Germany: distinctiveness of, 268–69, 541; economy of, 301, 310; facilities in, 337, 545–55, 557; industry of, 103, 105, 231–32, 269, 288, 496, 543; infrastructure in, 227–28; instrument making in, 23, 25–27, 50, 283, 478, 505–6; military, 231–32
Giessen, Germany, 23, 25, 29
glass: as material, 21, 29, 229, 246, 265–66, 535, 576; glassware, 2, 6, 7, 15, 19, 21–30, 134, 235, 274, 395–96; heat needed to make, 44, 46, 257, 259–61; sealed tubes, 26, 141–42, 144–45, 407; used for electrical properties, 459, 465; used for optical properties, 22, 63, 69, 70, 329, 380, 416, 486, 495, 514–15; used for thermal properties, 34, 38, 283
glassblowing, 3–4, 22–23, 25–30, 117, 142, 283, 407
global, *see* international
gold, 19, 189, 497
Goodenough, John B., 265, 350–52
government: agencies of, 76, 116, 154, 174, 544; as employer, 436; involvement in professional societies, 143; laboratories, 169, 189, 195, 335; links with industry and academia, 46, 222, 576; sponsorship of research, 1, 122, 292, 302, 353–54, 357, 483, 566; standardization and, 183–84, 186
granulation, 102–7
graphite, *see* carbon
Great Britain, *see also* England, United Kingdom, 16, 57, 105, 227–29, 233, 279, 293, 301, 407, 447, 506, 509, 541, 544–45

Haber, Fritz, 228, 288, 289f, 291–92, 294
handbooks, 15, 17, 19, 63–64, 73–80, 91, 174–75
Hartree–Fock method, 199–200, 202
Harvard University, 154, 323, 479, 484, 488, 565
Harwell Laboratory, 90, 478–79
heating: apparatus, 370, 417; as byproduct, 548; chemical procedures and, 76; chromatography and, 317; complex, 247, 251–51, 257–69, 267f; distillation and, 90; granulation and, 102; natural, 547; semiconductor manufacture

and, 342; simple, 43–52, 48f; spectroscopy and, 478; quality control and, 132
hemoglobin, 396, 459–60
high school, 78, 436
high-tech, 7–8, 112, 264, 315
histochemistry, 68–69, 71
Hodgkin, Dorothy, 427, 429
hormones, 302, 427, 457, 466
Horvth, Csaba, 320, 323–24
Hughes Aircraft Company, 343–44, 347, 483, 486, 545f
hydrocarbons, 87, 533
hydrogen, 22, 39–40, 287, 289, 292–94, 305, 395, 418–19, 465, 467, 490, 565

IBM, 114, 147, 199, 200, 334, 336, 338, 359–362, 411, 447, 450, 497–499, 501, 521
implants, medical, 6, 250–51
India, 11, 296, 384, 436–37
Industrial Revolution, 183, 258, 406
industry: aviation, 232; chemical, 229, 232; communications, 359; early modern, 259; employment in, 143; electronics, 268, 459; fertilizer, 97–98, 101, 106; French, 263; German, 26, 231, 269; heavy, 269; instrumentation, 460; iron, 258; laboratories, 154, 209, 212; links with academia and government, 46, 124–25, 184, 228, 259, 269, 324, 354, 422, 502; materials production and, 51, 77, 260, 292; metallurgical, 462; microelectronics, 147, 248, 252, 259, 267, 269, 345, 370; New Big Science and, 337; nuclear, 269, 378; petrochemical, 246; pharmaceutical, 309; researchers, 220; semiconductor, 268, 334, 341, 344, 346–47, 444, 509; space, 269; shipbuilding, 229; steel, 50, 265; sugar, 392; tool use in, 85–86, 283, 320, 322, 373, 430, 444, 453–54, 462, 510, 575; trade catalogs in, 159–61, 167
information display and recording, 4, 6, 112, 207–17, 208f, 211f, 213f, 214f, 215f, 216f, 217f
information storage, 211, 469
infrastructure: as tool, 6–9, 111–13; 116–18; bureaucratic, 189; mass spectroscopy and, 534; railroad, 220; social, 184, 188; utility, 129
innovation: computation and, 197; knowledge production and, 349; of materials, 25, 181–82, 189, 191; of techniques, methods, and processes, 44–46, 49–52, 88, 94, 106–7, 342, 507; of tools and apparatus, 18, 27, 29, 47, 57, 143, 148, 235, 338, 454, 478, 498–501, 529; prestige associated with, 9, 16; safety and, 136f; sites of, 343, 354, 576; society and, 15; systems of, 248; teaching and, 23
instruments, *see also* measurement: accuracy and reliability of, 38–40, 59, 189, 264, 438, 450–51. 453, 488; bespoke, 243; black-boxing of, 4, 21, 33, 207, 212, 444, 567; buildings as, 2, 121–28; calibration of, 181; collection, demonstration, and display of, 35, 57; communities around, 263, 266, 364, 447, 565, 567–68; design and development of, 4–5, 11, 25, 34–35, 37, 144, 283, 315, 319–22, 324, 351–54, 440, 446, 463, 506–12, 519–22, 575; display and recording, 209–18, 213–16f; distinction between devices and, 349–50; distribution of, 160; engineering and, 360; expertise and, 343, 564; funding for, 266, 519, 563, 566; historiography of, 7–8, 11, 207, 269; industry and, 210, 320–21, 457, 459–60, 462,

469–70, 501; makers, 56–57, 213, 541; markets for, 310, 325; materials testing and, 232, 235, 397; nuclear, 341, 383, 556; optical, 22, 443–44, 447–49, 455, 466–69; precision, 22; publication and, 461, 465, 469; safety of, 132; sharing of, 117; specialization and, 97; stability of, 50, 505; standardization and, 181, 190, 568; use and usefulness of, 301, 349, 395; variations in, 528–30, 532, 534–35, 576; visibility of, 118
integrated circuits, 147, 200–202, 247–50, 268, 330, 332, 341–45, 370
Intel, 255, 344
intellectual property, *see also* patents, 4, 369, 576
intercalation, 264, 352–53
interdisciplinarity, 115, 117, 122–28, 174, 177, 268, 305, 349, 422, 501–2, 576
Interdisciplinary Laboratory program (ARPA), 115, 122–27
interferometers, 440, 443–47, 443f, 451, 454
international, 4, 59, 181–91, 231, 264, 436, 540, 554, 567, 576
International Atomic Energy Agency (IAEA), 384, 385f, 388–89
International Bureau of Weights and Measures (BPIM), 59, 188, 190
International Prototype Kilogram (IPK), 59–60, 60f
invention: contexts of, 27, 94, 248; influence of, 44, 85, 91, 93; historiographical focus on, 16, 19; funding for, 112; in United Kingdom, 11; in United States, 11; of apparatus, 23, 47, 50–51, 87–89, 105, 142, 169, 209–10, 213, 219, 234, 495–500, 545; of computer languages, 201; of devices, 330, 351–52, 358–59, 451, 465, 485, 565; of science, 148; of techniques and processes, 254, 268, 288, 328, 392; sites of, 16, 143, 576; success of, 244
inventors, 88, 178, 244, 278, 280, 291, 323, 342, 350, 501, 545
ion implantation, 248, 254, 341–47, 534
Iran, 384, 576
iron: as catalyst, 288, 291; cast, 184, 258, 263; corrosion, 227, 230–32; in hemoglobin, 396; melting of, 260; Mössbauer spectroscopy and, 478–79; pig, 263; production of, 212; smelting of, 44–45, 45f
isotopes, 247, 351, 383, 391, 398, 399, 533, 535, 554
Israel, 382, 576
Italy, 16, 291, 316, 324

Japan: fermentation research in, 299, 301, 303–5, 307, 310; fullerene research in, 245; industry, 292, 354; infrastructure in, 227; instrument development in, 519, 522, 557, 573; reactors in, 384; synchrotron facilities in, 337, 573
Johnson, Ann, v, viii, 10, 204
journals, 25, 74, 116–17, 144, 169–78, 231, 249, 392, 407, 436, 461, 469, 487, 500–501

Kaliapparat, 23–29, 24f, 26f
Karlsruhe Technische Hochschule, 142, 228, 288
kilns, 39, 44, 46, 49
Kirchhoff, Gustav, 416–17, 417f, 461, 540
Knoll, Max, 505–6, 513–14
Kyoto University, 115, 354

labor history, 17, 129–39, 219–25
laboratories: academic, 121–28, 127f, 345, 378, 387, 509–10, 521, 544; alchemical, 152; animals in, 318; apparatus in, 58; biological,

299–308, 310, 510; buildings, 7, 117, 121–28, 127f; centrifuges in, 394–97; chemical, 21, 23, 25, 26–28, 56, 78, 234, 315, 320, 567; chromatography in, 315–16, 318–20, 323–34; clandestine, 440; clinical, 309; computing in, 195–96, 198, 200; contamination control in, 151–56, 152f; cookbook, 79f; experiments in, 57, 246; glassblowers in, 288, 142; government, 102, 169, 316, 359, 369, 378, 388, 478, 518–19, 522, 543, 554, 556, 576; historiography of, 141, 247; hot, 381–82; industrial, 169, 330, 342–43, 351, 363–64, 453, 459, 469, 483, 517, 572; information display and recording in, 212, 215; lasers in 485, 488–89; living, 299; materials, 1, 39, 56, 59, 113, 115, 122, 127, 127f, 145, 181, 248, 266; materials production, development, and synthesis in, 69, 77, 263, 265, 268, 294, 363–64, 373; physics, 451, 454; private, 506, 514; procedures and practices in, 6, 74, 288, 404; safety in, 129–31, 133–38, 135–36f, 152–54; spectroscopic tools in, 451, 453–54, 459, 461, 464, 467–69, 478–79; standards in, 181–83, 185–88, 190; teaching and training in, 23, 50; technicians in, 77–78; technologies used in, 9; testing, 75, 76f, 219, 233; tool development in, 260, 564; trade catalogs in, 160, 163, 165; users of, 337; X-ray crystallography in, 426–27, 429–30

Laboratory for Research on the Structure of Matter, 115, 117, 124–27, 127f

lasers: consumer applications of, 364; for heating, 43; for characterizing materials, 435, 438–440, 451, 453, 470, 483–493, 500, 531, 535, 574; for materials manufacture, 243, 254, 335

Lavoisier, Antoine, 22, 37, 49, 55–56

Lawrence Berkeley National Laboratory, 522, 574f

Lawrence Livermore National Laboratory, 198, 557f

lead, 19, 44, 154, 302–3

lenses: contact, 501; electromagnetic, 495, 503, 504, 513–520; glass, 22, 63, 64, 417, 495, 514, 515

libraries, 116, 160–61, 166–67, 169–79, 309–11

Liebig, Justus, 23–27, 29–31

Lincoln Laboratory (US), 256, 335

lithography, 6, 327–40, 328f, 333f, 341, 343, 345, 495, 575

London, England, 22, 34, 40, 227, 261, 283, 427

Los Alamos National Laboratory, 378, 478

lost-wax casting, 15–16

macromolecules, 173, 302, 321, 396, 427–29, 534, 559, 575, 577

magnesium, 66–67, 232

magnetic resonance imaging, 11–12, 561

maintenance, *see also* repair 9, 60, 111, 133, 324

Manhattan Project, 154, 196–98, 247–48

markets: generic, 11, 25, 181, 555, 576; for instruments, 56, 243, 247, 310, 320, 322, 325, 466, 469, 500, 509–10, 546, 567; for products and technologies (not instruments or tools), 268, 282, 332; for materials, 116, 145, 165–66, 353; for tools and techniques, 94, 105, 147, 344, 368, 370, 388, 392; strategies and marketing tools, 159–60, 162f, 163, 184, 186, 388

mass spectrometry, *see* mass spectroscopy
mass spectroscopy, 5, 307, 309, 363, 454, 527–36
Massachusetts Institute of Technology 199, 234, 335, 342, 363, 378, 386–7, 415f, 463, 467, 487, 564
Materials Research Society 174, 241, 249, 299
materials research: antecedents of, 405, 408, 416, 420, 422; biology and, 148, 299, 310–11, 495; combinatorial, 309–11; communities in, 316, 341; definition of, 2–3; electron physics and, 142; funding for, 335; heating technology for, 45–51, 259, 265–66; historiography of, 1, 5–11; infrared spectroscopy and, 454; interdependence of tools in, 568; internationalism of, 4; invisible and infrastructural tools for, 111–18; libraries and, 169–71, 174; lithography and, 327, 335–36, 337f; mass spectroscopy and, 532–34; materials production and, 241–43, 246–49, 251, 254–55; materials testing and, 232; measurement in, 457; nuclear reactors in, 377, 388, 554, 559; power sources in, 350–52; probe microscopy and, 501–2; requirements for, 111–12; standards and, 182–83, 189–91; synchrotron light sources and, 575–76; trade catalogs and, 159–60, 162, 165, 167
materials science: applications of, 249; biology and, 254–55, 425; computer simulation in, 195, 197–98, 203–4; craft knowledge and, 19; definition of, 2–3, 241, 248; discipline of, 79, 113, 115, 151, 266, 532, 539, 553; funding for, 335; historiography of, 7, 10, 191, 405; imaging in, 514, 519; laboratories for, 121–28; 181; infrastructure for, 121–28; interdisciplinarity of, 252; materials used in, 114, 156, 530, 571; power sources in, 349–50, 353–54; safety in, 129–39; techniques and practices in, 41, 338, 527, 567, 571, 575–76
materials: agency of, 5, 18, 29; amorphous, 4, 113–14, 266, 420; as medium, 29, 43, 437; behavior of, 58, 137, 148, 195–97, 201–4, 219–21, 225, 252, 255, 391, 404, 419–20; biological, 4, 64, 66, 71. 147–48, 249–51, 254–44, 300, 310, 318, 425–33, 444, 501, 532–33, 561; bottleneck, 122, 248, 334, 422, 568; characterization of, 111, 182, 184, 254, 266, 372, 401–577; chemically combined, 101–4; construction, 83, 155, 196, 250; cost of, 75, 102, 106, 165, 189, 221, 332, 362; distinction between devices and, 349–50; durability of, 75, 347, 351; electronic, 147–48, 249, 309, 576; experimental, 141, 144, 179; familiarity of, 1; fascination with, 6, 116; fissile, 208f, 532, 554, 556, 557f; handling, 106, 129, 132, 235, 556; hazardous, 130f, 132, 135f; high-value, 51; historiography of, 11, 159; imaging of, 142, 189, 409, 421, 435, 480, 509–10, 514, 561, 575–76; inert, 21, 90, 101, 145, 295, 319; insulating, 253, 295, 330, 510; knowledge of, 9, 112, 148, 255, 568; laboratories, 1, 39, 56, 59, 113, 115, 122, 127, 127f, 145, 181, 248, 266; manipulation of, 1–2, 7, 29, 41, 63–64, 66, 71, 74, 98, 117, 144–45, 242, 251–52, 259, 319, 425; manufacture of, 5, 112, 186, 253–54, 239–400, 546;

metallic, 228–39, 231–32; modern, 204, 249, 251; new and novel, 17, 21, 37, 52, 114, 132, 161, 165–66, 181, 183, 203, 243, 245–46, 250, 258, 266, 295, 300, 309, 362, 386, 373, 422; omnipresence of, 159–60; organic, 1, 4, 396, 421, 427; post-production, 220–21, 225; prefabricated, 2, 241; preparation of, 98, 181, 252, 254, 454, 534; printed, 73–75, 91, 163, 175–76; production of, 52, 99, 137, 163, 182–83, 187, 189, 250, 259–61, 263; properties, 2, 6–7, 33, 41, 57, 60–61, 123, 183, 195, 197, 204–5, 225, 255, 374, 383, 405, 419, 420, 485; purification of, 46, 91, 257, 264–65, 268, 301, 427; radioactive, 133, 318, 380–82, 382f; raw, 16, 98–99, 106, 221, 241, 257; reference, 182–85, 187–88, 191, 534; science of, 1, 4, 422; semiconducting, 156, 195, 200, 359; simulations of, 196, 198, 200–204; solid, 98, 260, 324, 367; standard, 160, 222, 349; structure of, 4, 151, 200, 266, 350, 416, 435, 440, 561, 566; synthesis of, 2, 8, 44, 114, 132, 137, 246, 250–51, 254, 265, 300–301, 308–311; superconducting, 114, 497, 565; synthetic, 51, 64, 101–2, 160–61, 166, 204, 242, 264, 268, 295, 534, 543; testing, 112, 132–33, 184, 186–88, 219–25; 227–36, 384, 395, 487, 539, 546–47, 549; types of, 4, 44, 160, 411; use and usefulness of, 2, 5, 54, 64, 75, 151; volatile, 144, 351

McMullin, Dennis, 333f, 508, 510

measurement, *see also* instruments: chemical, 465; display and recording of, 207, 210–12, 235; data generated by, 427–28, 459; duration of, 466–68; experiment and, 457; improvement of, 460; of gamma rays, 477; of infrared radiation, 543; of material properties, 480, 491; photometric, 461, 463, 548; precision, accuracy, and resolution of, 181, 185–88, 446, 448, 450, 453, 466; safety of, 132, 138; temperature, 28, 34, 37, 40–41, 263, 283; weight, 56, 60;

melting point, 28, 102, 174, 141, 486

mercury, 34–36, 38, 234, 437–38

metallurgy: constituent of materials research, 113, 115, 123, 125; forerunner of materials science, 2, 44–45, 241, 248–49, industry and, 50, 147, 151, 259–60, 265, 268, 288, 462, 504, 510; institutionalization of, 212, 263; theory of, 242, 252, 427, 511; use of common laboratory equipment in, 6, 41, 266; use of specialized laboratory equipment in, 125, 212, 215–16, 267

metals: as components, 39–40, 88, 90, 155, 261, 283, 327, 352, 395, 518, 459, 567; as materials, 1, 148, 228–29, 231–33, 343, 352, 353, 548, 576; as medium, 407; assaying of, 55; base, 46; characterization of, 40, 56, 213, 243, 453, 462, 497, 501, 506, 535, 548; chemistry of, 281; manipulation of, 64; opacity of, 439; one-dimensional, 133; precious, 407; production of, 40, 44, 46, 50, 147, 227, 259, 262, 341, 345–46, 357; properties of, 252, 444; quality control and purification of, 214–15, 228–29, 231–33, 235, 257, 264; structure of, 64, 509; toxicity of, 300

methanol, 84, 88, 292

Metre Convention, 59, 190

metrology, 17, 59, 187, 188,

microscopes: antiquity of, 2, 6; corrosion analysis and, 232;

electron, 245, 125–26, 216, 217f, 332–34, 336, 353, 411f, 421, 503–511, 513–23; field emission, 142; field ion, 496; optical, 2, 6, 17–19, 63–71, 307, 409; optical fibers and, 439; philosophical attention to, 111; probe, 353, 495–502; structural analysis using, 266; X-ray, 575
microstructure, 6, 404, 535, 575
military: as users, 255; classification and, 4; engineering, 57; German, 231, 236; industry and, 268–69, 342, 347, 546; requirements of, 154, 186, 248, 332, 335, 350, 539, 541, 544, 553, 563, 568; research sponsorship by, 1, 147, 199, 264, 378, 483, 486
mineralogy, 241, 427, 532
minerals 63–64, 98, 259, 407
mining 46, 392, 407
mixing 15, 19, 36, 97–108, 132
molecular beam epitaxy, 254, 264, 357–64
molecules: carbon, 244–45; characterization of, 4, 200, 394, 404, 435; crystal growth and, 51; dynamics of, 198–99; forces on, 316; infrared spectroscopy and, 443–44, 453; laser spectroscopy and, 486, 488–90, 490f; life processes and, 250, 255, 425, 431; mass spectroscopy and, 527, 530, 532–34, 536; materials production and, 241, 252, 357–64; nuclear magnetic resonance spectroscopy and, 561, 563, 565–66; properties of, 66–67; quantum behavior of, 419; reality of, 396; Raman spectroscopy and, 435, 437–40; structure of, 28, 304, 321, 383, 430; synthesis of, 114; theory of, 198–203; ultraviolet spectroscopy and, 457, 459, 461–62, 466; vacuum and, 143, 145, 409
monochromator, 467, 469, 522
moonrocks, 323
Moore's Law, 254, 256, 270, 327, 345, 349
Mössbauer spectroscopy, 475–81
Mössbauer, Rudolf, 475–78, 476f, 480
MRI, *see* Magnetic resonance imaging

nanoscience, 148, 264
nanotechnology: as a field, 300, 310; funding for, 250, 357; medical research and, 182; organizations dedicated to, 126–128, 127f, 181, 246; reference materials for, 189–90; tools for, 148, 310, 501, 559;
NASA, *see* National Aeronautics and Space Administration
National Aeronautics and Space Administration (US), 15, 123, 245, 255, 323, 378, 379f, 380f, 381f
National Bureau of Standards (US), *see also* National Institute of Standards and Technology, 40, 76f, 174, 184–86, 227, 235, 463, 465, 489, 496, 498
National Institute of Standards and Technology (US), 40, 174, 184–86, 386
National Research Council (US), 134–35, 421
National Science Foundation (US), 123, 126, 519
National Synchrotron Light Source (US), 336, 337f, 572
Nature, 173, 407, 500, 521
neutrons, 337–78, 380–81, 383–84, 403–4, 419–20, 553–60
New Big Science, 336–37, 576
Newton, Isaac, 34, 46–47, 57, 416, 457
NIST, *see* National Institute of Standards and Technology
nitrogen, 50, 86–87, 101, 108, 28–89, 296, 308, 346

NMR, *see* Nuclear magnetic resonance
Nobel Prize: in Chemistry, 113, 142, 243, 245, 294, 311, 396, 541; in general, 9, 470; in Physics, 8, 342, 362, 364, 413, 414, 419, 435–36, 475–76, 485, 498, 501, 509, 554, 563, 564; in Physiology or Medicine, 301, 429
North America, 10, 27, 384, 387, 553–54, 556–57
nuclear energy, 132, 246–47, 377–78, 383, 384, 399
nuclear magnetic resonance, 4, 304, 307, 310, 352, 359, 454, 485, 561–69
nuclear power, *see* nuclear energy
nuclear reactors, 219, 247, 264, 377–88, 379, 380f, 381f, 385f, 387f, 397, 399, 480, 553–54
nylon, 156, 160–66

Oak Ridge National Laboratory, 134, 378, 479, 522, 553, 558
Oatley, Charles, 333–34, 333f, 508, 510
Office of Naval Research, *see* United States Office of Naval Research
oil, *see* petroleum
optics: discipline of, 436; electron, 503, 513–16, 519, 523, 525, 544; fiber, 470; non-linear, 483–88, 491, 503; quartz, 465–67, 470
Oregon State University, 357, 386
ores, *see* materials, raw
organic materials: chemical analysis of, 25; conducting, 113–14; growing interest in, 249–51, 254–55, 501; institutions dedicated to 147–48; raw, 98; research into, 1, 4, 383, 396, 425–31, 532–33; semiconducting, 203–4; synthesis of, 299–311
oscilloscopes, 210–11, 211f, 214, 216, 496
ovens, 43, 69, 293, 342, 358
Oxford University, 265, 421, 429, 544
oxidation, *see* corrosion
oxygen: chemical analysis and, 22; explosion of, 288; in blood, 396; in electron microscopy, 522; in materials manufacture, 253, 295; Lavoisier's theory of, 56; pressurized, 234; separation of, 83, 87, 142; steel production and, 50, 86, 346

Pakistan, 15, 384
paper (as material), 135, 154, 165, 170, 195, 209, 216, 316, 328, 414, 453, 468
paraffin, 64, 297, 565
Paris, France, 22–23, 25, 59, 260, 263
participant history, 10
particle accelerators, 8, 147, 336–37, 337f, 342, 480, 530, 532, 557f, 571–72, 576
patents, *see also* intellectual property: enforcement of, 118; for ammonia synthesis, 288; for centrifuges, 392–93, 393f; for distillation systems, 84–85, 85f, 87–91, 89f, 93f; for heating and cooling technologies, 267f, 277, 283; for semiconductor devices, 342; for sensing and metering equipment, 465, 541; industry and, 294; influence on research priorities, 114, 310, 320, 330, 470
pedagogy, *see* education
penicillin, 302, 304
Pennsylvania State University, 386, 465, 496
petroleum, 83, 85–86, 246, 247, 249, 279, 293, 294, 316, 392, 395, 453, 509–10, 533
phase transitions, 444, 491, 535, 576
philosophy: expertise and, 28; natural, 18–19, 35–36, 56, 57, 207,

416; of science, 74–75, 111, 203, 403
phonons, 420, 477, 487
physics: academic, 123, 543; computer simulations in, 200–203; conferences, 114; debates in, 394, 396; electron, 142; experimental, 388, 478; glassware and, 22, 28–29; history and historians of, 8, 29; interdisciplinary interactions, 115, 124, 132, 248–50, 252, 351, 572; laboratories, 451, 454; Nobel Prize in, 8, 342, 362, 413–14, 435, 476, 498, 509, 554, 563; nuclear, 336, 341, 346–47, 381–83, 405, 479, 480, 532, 553, 563–64; of atoms and molecules, 453, 394, 396, 565; of condensed matter, 7–8, 198, 202, 406, 514; of particles and fields, 310, 336, 405–6, 558, 576; of solids 125, 201, 241, 252, 351, 361–62, 383, 415f, 435, 479, 503, 554, 559, 565; of surfaces, 357–58; quantum, 416, 418–19, 422; thought experiments in, 113; wartime, 197, 541, 568
plasmas, 1, 147, 164, 368–74, 369f, 371f, 372f, 374f, 391, 397, 531, 535
plastics, 1, 246, 295, 396, 425, 461
polymers, 1, 7, 113–18, 160, 251, 266, 295, 331f, 396, 468, 501, 535, 576
polythene, 294–96
power sources, 132, 349–54
prestige, 8–10, 405, 506
print culture, 73–75, 80
prisms: dispersion with, 416–17, 417f, 443, 445–46, 459; interferometry and, 443, 445–46, 448–49; Lissajoux, 213; polarization with, 437; thermography and, 538; ultraviolet spectroscopy and, 467; visual effects of, 22
proteins, 4, 65–70, 250–51, 302–3, 321, 396–97, 427–32, 432f, 462, 566, 577
pyrometers, 33, 39–41, 212–13, 213f

quantum theory: applications of, 447, 477, 485–86, 501, 563–64; Nobel Prizes for, 364; of materials, 404–6, 416–20, 497; simulations and, 198–203; superlattices and, 361; tests of, 490–91; thought experiments in, 113
quartz, 370, 465–67, 484, 487

radar, 248, 295, 508, 544, 563–56, 568
radiation: as hazard, 143, 151, 154, 381f, 382f; characterization of materials using, 6–7, 378, 430, 510, 513; damage, 342; discovery of, 404, 411; electromagnetic, 252, 419, 457, 539–40, 546, 561, 564; gamma, 477, 479; medical, 383; nuclear, 380–83, 381f, 382f; scattering and diffraction of, 418–19, 425–26, 437, 439; stimulated emission of, 485; synchrotron, 336, 558, 571–73, 577; thermal, infrared, and heat, 39, 213, 283, 261, 539–41, 543, 545–48, 446–47; visible, 484
railroads, 51, 183–84, 219–20, 227–28, 273, 407
Raman scattering, 435–39, 487
Raman spectroscopy, 435–40, 491
Raman, Chandrasekhara Venkata, 435–37
raw material, *see* materials, raw
recipe, 9, 15, 17–21, 49, 73–80, 114, 372
reference materials, 182–85, 187–89, 191, 534
refrigeration, *see also* cooling, 251, 273–85, 275f, 276f, 277f, 279f, 280f, 282f
reliability: of devices, 23, 40, 59, 330, 354, 451, 500, 556; of knowledge, 30, 112, 422; of materials, 6, 166,

348, 332, 347; of people, 132; of processes, 28, 69, 334; of products, 399, of scales, 34–35
repair, *see also* maintenance 111, 129, 276, 346, 381
research and development (R&D), 147, 259, 263–64, 268–69, 302, 342, 354, 368, 384, 421, 553–54, 558–59
resists, 328–32, 334, 336, 338
Rice University, 172, 176, 243–44, 246
Röntgen, Wilhelm Conrad, 407, 410–14
Royal Society of London, 34, 40, 519–21
ruby, 51, 262, 296, 441, 483–85
Ruska, Ernst, 498, 506, 513
Russia, *see also* Soviet Union, 316–17, 377, 383–84, 437
rust, *see* corrosion

salts, 63, 114–16, 274, 353
sand, 21, 229, 394
scales, 6, 53–61, 182
scattering, 2, 247, 334, 383, 403, 418–19, 426, 435–40, 485–87, 503, 554–56, 574–75
Science, 173, 407, 500, 521
Scotch whisky, 87
secondary school, *see* high school
Second World War, *see* World War II
semiconductors: amorphous, 113; doping, 41, 156, 330, 338, 342–47, 362; experiments with, 243–44, 247, 359–61, 363, 491, 499, 535, 573f; generic, 1, 148, 195, 349, 501, 565, 577; in electronics, 155, 200, 268, 329–30, 334, 338, 343–47, 358, 361–62, 368–70, 509; organic, 203–4; properties of, 201, 444; purity of, 156, 268; surface of, 147, 327, 332, 357, 497
Siemens, 40, 50–52, 235, 259, 261, 496, 506
silicon: charge-couple detector, 439; chemistry of, 115; dioxide, 253, 330; doping, 349, 362; growing, 267; imaging of, 497–499, 498f, 521f; integrated circuits and, 200, 332, 341–343, 345; nitride, 372; purity of, 51, 156; research on properties of, 243, 249; substrate, 330, 367, 370; thin films, 359, 373; transistor, 268
Silicon Valley, 143, 254, 332, 567
silver, 21, 64, 283, 407, 563
solid state physics: basic concepts of, 361–62; computation and, 201, 203; in interdisciplinary relationships, 125, 252, 347, 351, 358; microscopy and, 503; neutron research and, 383, 554, 559; spectroscopy and, 435, 453, 479, 565; X-ray diffraction and, 415f
solvents, 64–65, 83, 278, 300, 305, 338, 353
South Korea, 354, 384
Soviet Union, *see also* Russia, 122, 215, 245, 377, 383–84, 397
space, *see also* aerospace: craft, 15, 248, 322, 335, 343, 383, 384; elevators, 247, 357; environment, 236, 244; programs, 122, 147, 246, 247, 264, 269, 378; science and astronomy, 89, 322, 453, 463, 532, 578
Spain, 245, 305
spallation, 553, 556–58
spectroscopy: Fourier transform infrared, 444–51, 543–44; infrared, 353, 443–55, 463, 565–66; in general, 22, 574; laser, 483, 488–91, 490f; mass, 5, 307, 454; microwave, 125; Mössbauer, 475–80; nuclear magnetic resonance, 4, 307, 310, 454, 561–68; photoacoustic, 454; Raman, 435–40; ultraviolet,

457–70; optical, 310, 417–20, 417f, 461; X-ray, 568, 575
Sputnik, 122, 554
sputtering, 147, 346, 367
Standard Oil of New Jersey, *see* Exxon
standards and standardization: as tools, 6; color, 463; components, 467; electrical, 125; enforcement of, 74; industrial, 106, 231; information, 73; instruments, tools, and equipment, 4, 27–28, 84, 264, 266, 315, 320, 351, 454, 564, 568; international, 112, 181–91; manuals for, 76; materials, 160, 181–82, 184, 186, 189–91, 222, 349; practices, processes, and procedures, 4, 28, 34–36, 64–65, 68–69, 94, 98, 222, 343, 413; professional, 79; control and quality, 220–25, 334, 353, 378; safety, 130f, 132, 135, 137; setting of, 174, 225; temperature, 34–41; texts, 532; units, 190; weighing, 54–61
Stanford University, 345, 489, 491, 498–99, 564, 567
steel: as construction material, 183; corrosion of, 227, 232, 234; for containment vessels, 283, 288, 289, 292; for preventing contamination, 144, 155, 363; production of, 40, 50–51, 86, 212, 231, 234, 261, 263, 265, 509; specialized, 246, 292; testing of, 214, 220–22, 221f, 222f
students: as instrument builders, 508, 510, 513, 517; as laboratory users, 266; as library users, 169, 175–77; as researchers, 288, 319, 427, 470, 475–76, 490, 496; chemistry, 6, 23, 25–26, 78–79; as entrepreneurs, 397, 484; genealogies of, 317, 324, 333–335, 333f, 446, 543; graduate, 8, 92, 122–25, 127, 243–44, 324, 333–35, 333f; mentoring of, 169; safety of, 129, 135f, 137–38
sugar, 16–17, 20, 66, 302, 392
superconductors, 114, 177, 243, 453, 497, 535, 565
superphosphate, 98, 100f, 105
surface science, 147–48, 353–54, 357, 497–501, 576
Sweden, 22, 394, 558, 573
synchrotron radiation, 336–37, 337f, 558, 571–77, 573f
synthesis of materials, *see* materials, synthesis of

tacit knowledge, 4, 117, 129
technicians: as instrument makers, 50, 144, 212, 499; as users and operators, 124, 259, 266, 454, 567; safety of, 129, 133, 135–38, 135f, 136f, 346, 382f; status of, 77–78
telecommunications, 122, 334, 357, 359, 369
telescopes, 8, 22, 212–13, 528
test tubes, 66, 232, 394
Texas Instruments, 247, 268, 369
textiles, 17, 156, 160–61, 258, 163, 392, 427, 510
theory: biological, 18; caloric, 37; chromatography and, 320, 322; density functional, 200; electromagnetic, 540; kinetic, 396; molecular orbital, 199, 201–2; of discharge tubes, 409; of heat, 37; of refrigeration, 278; of the thermometer, 35, 37; oxygen, 56; quantum, 195, 198–99, 203, 405–6, 416–20, 447, 485, 488, 491; relativity, 419, 479; valence bond, 199, 202
thermocouples, 33, 39–41, 372, 446, 540
thermodynamics, 38–39, 91, 282
thermography, 539–51
thermometers, 2–3, 5–6, 15, 17, 19, 22, 33–42, 540

thin films, 147–48, 357, 357, 497, 535
Thomson, J. J. (Joseph John), 407–9, 521
three beams conference, 252–54, 253f
tools, *see also* instruments, measurement: always already, 13–108; for characterizing materials, 401–477; for making materials, 239–400; invisible and infrastructural, 109–238
Topografiner, 496–97, 501
Townes, Charles, 485–88
traceability, 186–89, 191–92
trade: catalogues, 2, 6, 159–67; in antiquity, 53–54; international, 44, 59, 182–83, 187, 189, 190;
training, *see* education
transistors, 247, 255, 268, 327, 330, 331, 341–345, 349, 358, 359, 361, 370, 565
trust, 75, 83, 112, 117, 136, 184, 323, 325
tungsten, 142, 466, 470, 496, 509

ultracentrifuges, *see also* centrifuges, 247, 252, 394–97, 400
United Kingdom, *see also* England, Great Britain
United States: chemistry in, 77; commercial markets in, 90, 295; divisions within, 133; engineering in, 77; fertilizer research in, 101–2, 104–6; government institutions and programs in, 40, 113, 115, 122, 132, 137, 153, 175–76, 264, 266, 322, 342, 362, 421, 463, 478, 496; hegemony of, 269; industry, 223, 265, 268, 359, 363, 447, 497; infrastructure, 227–28, 233; laboratories in, 153, 336, 522, 576; materials production in, 263, 265, 301–2, 305; non-linear optics in, 485, 489; nuclear reactors in, 377–78, 383–84, 386–87, 387f, 553–54, 556–58; patents, 283, 465; policy, 248; professional organizations in, 143–44; Prohibition Era, 91; publication in, 245, 469; research funding in, 94, 255, 268, 310, 354, 357, 519, 563; standards in, 181–83, 186–89; tool development in, 11, 334, 349, 506, 545; universities, 121, 127
United States Air Force, 249, 342, 542–45
United States Army, 79f, 154, 342, 362–63
United States Department of Defense, 122, 126, 181, 186, 193f, 255, 486
United States Department of Energy, *see also* Atomic Energy Commission, 157f, 255, 353, 364f, 522, 556, 558, 560f
United States Office of Naval Research, 115, 486
universities: British, 231; knowledge dissemination into, 114; laboratories in, 1, 130f, 135f, 142, 378; links to government and industry, 46, 122–24, 154, 310, 333, 502, 566, 576; professional societies and, 144; R&D in, 263; research, 50; resources within, 169, 175, 266, 377, 387–88
University of Cambridge, 333–34, 407, 427, 429, 446, 508–11, 517, 520–21
University of Göttingen, 359, 541
University of Illinois, 123, 363, 465, 522
University of Jena, 23, 514
University of Pennsylvania, 113, 115–17, 123–28, 127f, 266
University of Wisconsin-Madison, 386, 572
uranium, 246, 378, 397, 399, 407
users: division of labor among, 4; end, 255; external, 337, 554–55, 558, 572, 576; industrial, 184; interfaces for, 207, 213–13; needs of, 121, 349, 370; of cook books,

78; of imaging tools, 519; of libraries, 169–70, 175–76; of low-profile tools, 15–16, 18; of standards, 220; preparation of, 325; safety of, 121, 134–35, 137–38
USSR, *see* Soviet Union

vacuum: as research tool, 141–48; centrifuges and, 395; chambers, 141–48, 358; distillation, 90–91; flasks, 273, 275–79, 281, 283–84, 284f glove box, 352; mass spectroscopy and, 528; microscopy and, 505, 508, 510, 521; tubes and bulbs, 199, 210, 261, 330, 332, 407, 409, 414, 459, 465, 468; ultra-high, 142–44, 148, 358, 362–62, 499–500; vapor deposition and, 370–71
vacuum flasks, *see* dewars
vapor: as contaminant, 151; as hazard, 152; deposition, 254, 367–74; distillation and, 84–88, 90–92; gas chromatography and, 321; in materials testing, 233, 235; mass spectroscopy and, 530; microscopy and, 510; molecular beam epitaxy and, 358; phase changes and, 37; refrigeration and, 278–81, 283; scattering from, 436
vapor deposition, 254, 367–74
Varian Associates, 143, 334, 363, 567
ventilation, 129–30, 151–54
vitamins, 302, 427, 428, 457, 463, 466, 468
von Hippel, Arthur, 241, 248–49
von Laue, Max, 413–14, 425

waste, human and animal, 49, 98, 307, 392
water: acid and, 328; as corrosive agent, 229, 232–33, 235, 346; cooling, 273–74, 277–79, 281; for fire suppression, 129; for high pressures, 292; for power generation, 98; light scattering by, 437; pollution in, 383; sample preparation and, 65, 69; separation and, 83–85, 90, 392, 394; quality of, 468; thermography and, 544, 548; thermometry and, 34–37; under-, 295, 380, 382; utility, 227–28; vacuum and, 142; vapor, 510
welding, 51, 147, 260, 283
women, 9, 44–45, 77–78
World War I, 16, 78, 154, 231, 290, 293, 301, 346, 349, 541
World War II: anxiety about, 154; as turning point, 2, 143, 264–65, 292, 299, 302, 327, 421, 469; conditions in Germany following, 478; corrosion testing during, 236; electronics research during, 210, 214–15; funding patterns after, 269, 333, 349, 544–45, 563; mass spectroscopy and, 529, 535; materials development during, 255, 294, 468, 534; nuclear research after, 377, 381, 397, 419, 532, 571; nuclear research during, 318, 553; radar research during, 508–9, 568

X-rays: accelerators and, 558, 571, 573–76, 574f; crystallography, 435–31, 566, 568; dental, 383; diffraction and spectroscopy, 4, 125, 265–66, 350–53, 362, 418–21, 425–31, 565–66, 568; discovery of, 404, 406–8, 412–16, 412f; materials manufacture and, 252; microscopy and, 503–4, 509–10, 521f

Yale University, 278, 323–24

zinc, 154

www.ingramcontent.com/pod-product-compliance
Ingram Content Group UK Ltd.
Pitfield, Milton Keynes, MK11 3LW, UK
UKHW022107170926
8624IPUK00024B/177

9 789811 207624